Fundamentals of Electrothermal Atomic Absorption Spectrometry

A Look Inside the Fundamental Processes in ETAAS

Fundamentals of Electrothermal Atomic Absorption Spectrometry

A Look Inside the Fundamental Processes in ETAAS

A-Javier Aller
University of León, Spain

W **World Scientific**

NEW JERSEY · LONDON · SINGAPORE · BEIJING · SHANGHAI · HONG KONG · TAIPEI · CHENNAI · TOKYO

Published by

World Scientific Publishing Co. Pte. Ltd.

5 Toh Tuck Link, Singapore 596224

USA office: 27 Warren Street, Suite 401-402, Hackensack, NJ 07601

UK office: 57 Shelton Street, Covent Garden, London WC2H 9HE

Library of Congress Cataloging-in-Publication Data

Names: Aller, A. J. (A. Javier), author.

Title: Fundamentals of electrothermal atomic absorption spectrometry : a look
    inside the fundamental processes in ETAAS / A-Javier Aller (University of Leon, Spain).

Description: New Jersey : World Scientific, 2018. | Includes bibliographical references.

Identifiers: LCCN 2017044583| ISBN 9789813229761 (hardcover : alk. paper) |
    ISBN 9813229764 (hardcover : alk. paper)

Subjects: LCSH: Atomic absorption spectroscopy. | Atomic spectroscopy. |
    Absorption spectra. | Spectrum analysis.

Classification: LCC QD96.A8 A4225 2018 | DDC 543/.5--dc23

LC record available at https://lccn.loc.gov/2017044583

British Library Cataloguing-in-Publication Data

A catalogue record for this book is available from the British Library.

For any available supplementary material, please visit
http://www.worldscientific.com/worldscibooks/10.1142/10703#t=suppl

To Amelia and Javier

Science is but an image of the truth.

Francis Bacon

Preface

Analytical chemistry and particularly analytical instrumentation, play an important role for solving analytical problems in many different fields, such as biology, biomedicine, chemistry, biochemistry, geochemistry, food, material sciences, among others. The baggage of analytical instrumentation includes a multiple growth package of physico-chemical techniques. Among them, one of the most important technique widely used in different research and industrial laboratories is Electrothermal Atomic Absorption Spectrometry (ETAAS), a versatile and well established analytical technique very useful for the quantitative determination of low levels of metals. The aim of this book is to cover the basic spectroscopic fundamentals together with advanced understanding of the most important instrumentation used and the related chemical processes involved. In this way, a good theoretical knowledge and an efficient practical application is ensured. Consequently, any theoretical or practical problem should be solvable after the book has been read.

The book, which includes recent changes the modern ETAAS has undergone, as well as the usual applications and limitations, is designed for teaching doctoral candidates, masters students, graduates, undergraduates and those with no analytical chemistry background.

Why a new book in ETAAS? Various books have covered the subject of analytical ETAAS. However, this book provides a global vision of the subject, combining theoretical and practical perspectives in a reasonable balance. This comprehensive coverage makes the book valuable for students to acquire a practical perspective, together with a fundamental theoretical knowledge, so students will achieve a high level understanding of all the analytical possibilities of this technique. All major capabalities of the ETAAS as an analytical technique are considered.

Mathematics is kept to a minimum, necessary for an understanding of the key terms and concepts In this way. Some background in calculus, physics, and chemistry is useful, because in several parts of the book, the elemental principles of chemistry and physics are presented.

In each chapter, discussions on the fundamental principles underlying in the subject, as well as detailed descriptions of the instrumentation involved and a large number of theoretical and practical comments are included. In addition, each chapter is completed with a wide bibliography supporting the information provided. My apologies for the possible absence of any paper that would also deserve to be included. However, the huge number of publications related to the subject covered in the book, together with the different accessibility problems, would have fostered such a situation.

Chapter 1 includes a general description of the spectroscopic principles and electrothermal analytical signal characteristics. Chapter 2 deals with the atomizers usually involved in the typical ETAAS and some other associated collateral developments, also outlining the atomization modes and effects of the experimental parameters. In Chapter 3, sample handling and sample introduction methods, with analysis of liquids and solids, are extensively treated. Chapter 4 focuses on the radiation sources, manipulation, isolation and detection of the electromagnetic radiation used for excitation of the gas phase analyte atoms, with some impact on new alternative systems. Chapter 5 treats potential interferences and their overcoming possibilities. Chapter 6 is dedicated to the use and behavior, including advantages and problems, of the chemical modifiers, whose utilization is of imposed fulfillment for a reliable analysis. Chapter 7 tends to disclose a dual vision of the mechanistic pathways for atomization of analytes, including firstly a theoretical look and secondly some practical considerations, both in the absence and in presence of chemical modifiers. At long last, Chapter 8 closse the book with a coverage of the most important analytical characteristics (validation of an analytical method, calibration, sensitivity, atomization efficiency and precision) from an ETAAS perspective. All chapters are discussed in detail, so that students can follow easily every description. Furthermore, the student realizes the overreaching difficulties in the development of the instrumentation and methodologies.

Finally, I hope and wish a fruitful reading for the future readers.

León, March 2018
A.-Javier Aller

Acknowledgements

I would like to express my gratitude to several institutions for their free and friendly permission for reproduction or adaptation of some of the figures and tables included in the book. Acknowledgements are extended to Elsevier (Figures: 2.1, 2.8, 2.10, 2.11, 2.17, 2.19, 2.20, 2.21, 3.3, 3.6, 3.7, 3.8, 3.9, 3.10, 4.4, 5.1, 5.3, 5.6, 5.10, 7.3, and Tables: 5.1, 5.2, 7.2); Royal Society of Chemistry (Figures: 2.4, 2.5, 2.6, 2.15, 2.16, 3.2, 4.3, 5.2, 6.3, 6.4, 6.5, 6.8, 7.4 and Tables: 5.1, 5.2); American Chemical Society (Figures: 2.7, 3.4, 3.5, 4.15, 4.16, 4.22, 4.23 and Table 4.1); Nature-Springer (Figure 2.9 and Table 8.1); Japan Society for Analytical Chemistry (Figures: 6.4 and 6.6); Brazilian Chemical Society (Figure: 4.7); Thermo Jarrell Ash Corporation (Figure 3.1); and Society for Applied Spectroscopy (Figures: 4.17, 4.19, 4.20, 4.21) Special thanks are also gratefully acknowledged to Research Trends (P) Ltd. for allowing me to reuse the two following papers: A.J.Aller (2002). Electrothermal atomization mechanisms: theoretical and practical considerations. *Trends in Applied Spectroscopy* **4**: 101–112, and A.J.Aller (2002). Analytical characteristics of electrothermal atomic absorption spectroscopy. *Current Topics in Analytical Chemistry* **3**: 41–53, as a basic reference for developing Chapters 7 and 8. Last but not least, I'd like also to thank the staff at Scientific World Publisher personalized in Sook Cheng Lim for her patient and assistance.

Contents

Chapter 1

Fundamental Spectroscopic and Analytical Processes in Atomic Absorption Spectrometry

1.1. Introduction

The use of electrothermal (ET) atomization in combination with atomic absorption spectrometry (AAS) has made this technique, electrothermal atomic absorption spectrometry (ETAAS), an indispensable tool for the determination of elements at trace and ultratrace levels, particularly in complex matrices (environmental, biological, chemical, industrial, food, etc.). In comparison with flames and plasmas, ET atomizers allow the sample under analysis to be given thermally controlled handling before and during the vaporization and atomization stages. In this way, better separation between analyte and matrix components (and so a lower likelihood of interference), some economy in sample size, relatively lengthy analyte residence times in the vapor phase, and high absolute atomization efficiencies are possible. The success obtained in the analytical applications of ETAAS has been due to the incorporation of different instrumental and methodological developments [L'vov (1970, 1978); Lundberg and Frech (1981); Welz *et al.* (1992); Aller (1998, 2003)], mainly including: new atomizer designs, rapid devices for data gathering, more efficient background correction methods, use of chemical modifiers, and employment of integrated absorbance.

1.2. Main Processes Occurring in an ET Atomizer: Atomization and Excitation

As a consequence of several physical and chemical transformations of the sample components, during the ET heating, a cloud of free analyte atoms is produced (atomization process). The usually occurring physico-chemical processes behind the analyte atomization are the following:

- dissociation of molecular species on the graphite surface and subsequent vaporization of the atoms formed,
- vaporization of molecular species from the graphite surface and subsequent dissociation,
- vaporization and condensation of molecular species and subsequent re-vaporization as free atoms.

Once the gas phase analyte atoms are formed, they need to be excited by absorbing photons proceeding from a radiation source, normally a hollow cathode lamp (HCL) (Fig. 1.1). In this case, the lamp radiation shows an emission line coincident with the analytical absorption line and in theory only the analyte atoms selectively absorb photons.

$$M + h\nu \leftrightarrow M^* \qquad \{R_{h\nu} = k'_{h\nu}\,[M]\}$$

Hollow Cathode Lamp

Excitation Radiation

P_0

Atomization

ET Atomizer

P

Signal Isolation, Transduction, Detection

Sample Introduction

$$A = \log \frac{P_0}{P} = K\,l\,N_0$$

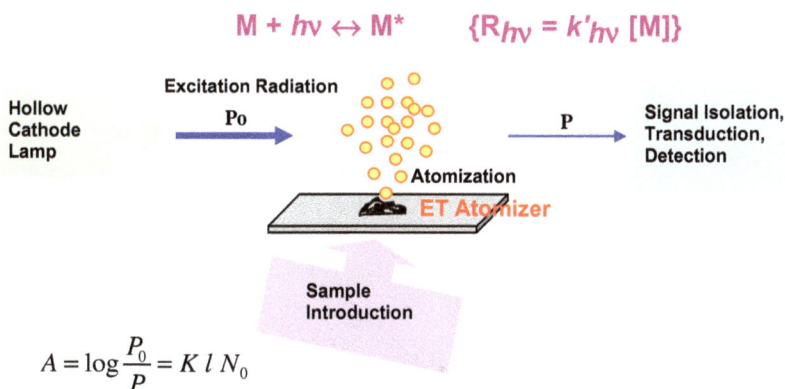

Fig. 1.1.   Schematic representation of the ET atomization and absorption processes.

At the atomization temperatures, a background continuum radiation is emitted by the graphite tube surface, particularly when purged with monoatomic gases (like Ar) [Schwab and Lowett (1990)]. However, absorption of this radiation contributes in much less extension to the excitation of the analyte atoms. Similarly, at the temperatures usually employed in ETAAS ($\leq$3000 K), electron collisions are unlikely to make a significant contribution to the analyte excitation process.

The atomization and excitation mechanisms are regulated by diverse thermodynamic and kinetic aspects (Chapter 7).

All chemical species present in an ET atomizer can only achieve a total balance if atomization is carried out at a constant high temperature [Frech *et al.* (1985)]. However, if the time spent in liberating the energy is too short compared to the transport time and temperature change, we can consider that in the graphite tube a thermal equilibrium (TE) state has been reached in each volume element within the time unit. Each volume element would be in local thermal equilibrium (LTE) if it is characterized by a constant temperature value [Alkemade *et al.* (1982)]. In other words, if LTE conditions would exist in the graphite tube during atomization, the energy distribution associated with: (i) the kinetic energy of electrons and atomic and molecular species; (ii) the rotational and vibrational energy of the molecules; (iii) the excitation and ionization of the atoms; and (iv) the spectral

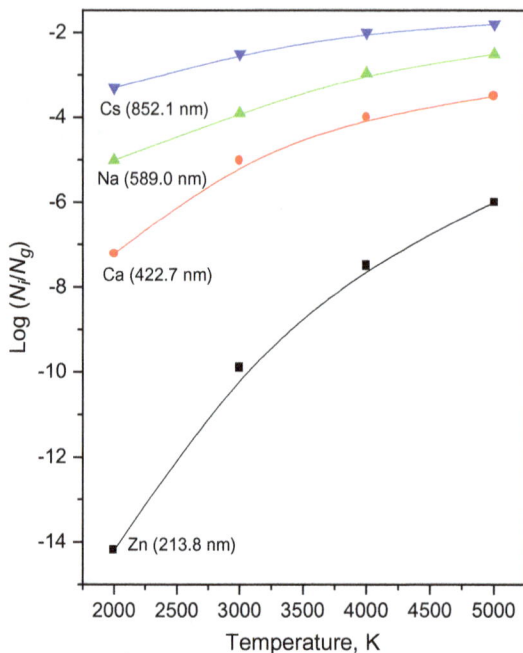

Fig. 1.2. Distribution ratio between two energy levels, N_i (higher energy) and N_g (lower energy) as a function of the temperature for the resonance lines of Cs (852.1 nm; $g_i/g_g = 2$), Na (589.0 nm; $g_i/g_g = 2$), Ca (422.7 nm; $g_i/g_g = 3$), and Zn (213.8 nm; $g_i/g_g = 3$).

distribution of the background radiation from the tube wall, could be described by the same temperature value. This is in accordance with the expressions of Maxwell, Boltzmann, Saha and Planck. Under LTE conditions, chemical species are thermally distributed according to the Boltzmann's distribution law [Boumans (1968)] (Fig. 1.2)

$$N_i^* = \frac{N_t g_i}{Q(T)} \exp(-E_i/kT), \tag{1.1}$$

where N_i^* is the population of the excited state "i" of energy E_i (J), N_t is the total number of analyte atoms in all states at time t, $Q(T)$ is the partition function or state sum at the absolute temperature T (K) of the vapor phase prevailing in the atomizer, g_i is the statistical weight, and k (1.38×10^{-23} J K^{-1}) is the Boltzmann constant. However, the existence of LTE conditions in an ET atomizer still needs to be proved. Deviations from LTE conditions are predominantly due to losses of energy as non-absorbed radiation.

The transmitted radiation at the analytical wavelength is isolated by the monochromator, then reaching the detector. In order to isolate and quantify the transmitted analytical radiation from the continuum radiation arising from the tube, modulation of both radiations is mandatory. Recording the transmitted energy in the absence (P_0) and the presence (P) of the gas phase analyte atoms allows us to deduce the absorbance (A) value which constitutes the analytical signal. The relationship between absorbance and analyte concentration or the analyte atoms (N_0) introduced into the atomizer (Fig. 1.1), is regulated by Beer's law, which is the basis of the practical quantitative analysis.

1.3. Spectroscopic Transitions

If free analyte atoms are excited by light, a change from one energy state to another higher (absorption) results, but then they decay spontaneously to a lower energy state emitting the corresponding radiation (emission). The probability of any atomic transition (absorption and emission), which is commonly called the transition strength, depends on the nature of the initial and final state wave functions, how effectively light can interact with them, and on the intensity of any incident light.

1.3.1. Transition Probability or Transition Strength

The transition probability is expressed by R^2 (J cm^3), where R is the transition moment given by $R = \langle \Psi_i | \mu | \Psi_g \rangle$, as a function of the wave functions of the upper and lower states, Ψ_i and Ψ_g, respectively, and the dipole moment operator, μ. Basically, this equation indicates that the strength of a transition is relative to how strongly the resonant dipole moment between the two energy states can couple to the electric field of the light wave.

1.3.2. Selection Rules

In order to explain transition strengths, the following selection rules determining whether a transition is allowed or disallowed are used.

- The parity of the initial and final wave functions must be different.
- The spin cannot change, $\Delta S = 0$, during the transition.
- The change in orbital angular momentum can be $\Delta L = 0, \pm 1$, but $L = 0$ to $L = 0$ transitions are not allowed.
- The change in the total angular momentum can be $\Delta J = 0, \pm 1$, but $J = 0$ to $J = 0$ transitions are not allowed.

1.3.3. Einstein Coefficients

Practical measurements of transition strengths are usually described in terms of either the Einstein A and B coefficients or the oscillator strength, f. For a two-level system (ground-state level g and upper level i), the rate of an upward stimulated transition (absorption, $-dN_g/dt$ or dN_i/dt) is

$$\frac{-dN_g}{dt} = N_g B_{gi} U_v, \tag{1.2}$$

where N_g is the number density of atoms in the ground state, U_v is the light intensity, and the proportionality factor B_{gi} is the Einstein B coefficient for absorption

$$B_{gi} = \frac{8\pi^3 R^2}{3 h g_g}, \tag{1.3}$$

where h is the Planck constant and g_g is the degeneracy of the ground state.

For stimulated emission, the Einstein coefficient, B_{ig}, becomes

$$B_{ig} = B_{gi} \frac{g_g}{g_i}, \tag{1.4}$$

where g_i is the degeneracy of the excited states. Nonetheless, atoms in the excited state can decay without the presence of an external light field due to "zero-point fluctuations," which are the dynamic variations in the shape of an electronic orbital at any instant in time. The spontaneous decay rate $(-dN_i/dt$ or $dN_g/dt)$ is

$$\frac{-dN_i}{dt} = N_i A_{ig},$$ (1.5)

where N_i is the number density of the atoms in the excited state and A_{ig} is the Einstein coefficient for spontaneous emission

$$A_{ig} = \frac{8\pi h B_{gi} g_g}{g_i \lambda_m^3} = \frac{64\pi^4 R^2}{3 h g_i \lambda_m^3},$$ (1.6)

where λ_m is the wavelength of the radiation emitted. Since atoms in the upper level can decay by both spontaneous and stimulated emission, the total downward rate $(-dN_i/dt$ or $dN_g/dt)$ is ideally given by

$$\frac{-dN_i}{dt} = N_i(A_{ig} + B_{ig} U_v).$$ (1.7)

1.3.4. Oscillator Strength

The oscillator strength, f, of an atomic transition is a very useful dimensionless number for comparing different atomic transitions. The oscillator strength is defined as the ratio of the strength of an atomic transition to the theoretical transition strength of a single electron using a harmonic-oscillator model. For absorption, f_{gi}

$$f_{gi} = \frac{4\varepsilon_0 h c m_e B_{gi}}{e^2 \lambda_m},$$ (1.8)

and for emission, f_{ig}

$$f_{ig} = f_{gi}\frac{g_g}{g_i},$$ (1.9)

where c is the light rate, e and m_e are the electron charge and mass, respectively.

Oscillator strengths can range from 0 to 1. A strong transition will have an f value close to 1. Oscillator strengths greater than 1 result from the degeneracy of real electronic systems. Tabulations in the literature often use gf, where $gf = g_g f_{gi} = g_i f_{ig}$. This parameter has an important analytical meaning, as the analytical signal (and consequently analytical sensitivity) grows with the oscillator strength value.

1.3.5. Excited-State Lifetime

A population of the excited atoms decays exponentially (Fig. 1.3). This means that each excited atom loses its energy excess after a different time interval. In this context, a time interval named lifetime is defined and represents the average time employed by an atom population to be relaxed. So, the excited-state of an atom will have an intrinsic lifetime due to radiative decay given by

$$\frac{-dN_i}{dt} = N_i \sum_{(g<i)} A_{ig}, \tag{1.10}$$

where N_i is the population in the excited state i, and the A_{ig}'s are the Einstein spontaneous emission coefficients for all of the radiative transitions originating from level i. The negative sign arises because the rate decreases with time. Integrating this equation produces

$$N_i(t) = N_i(0)e^{-t/\tau_i}, \tag{1.11}$$

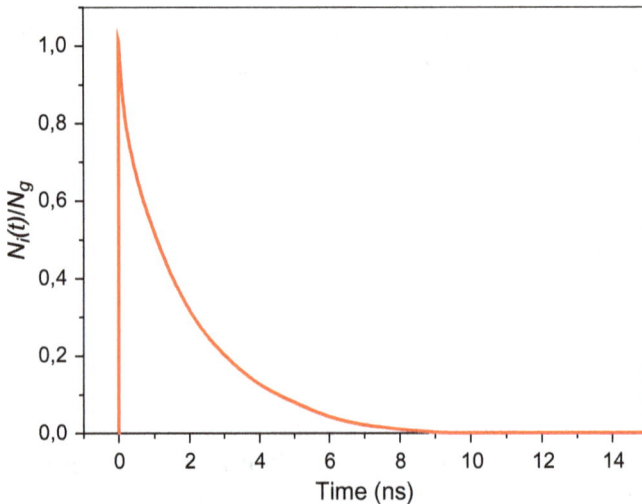

Fig. 1.3. Emission decay curve with a lifetime of 2 ns, after populating the excited state with an excitation pulse at time $t = 0$.

where $N_i(t)$ is the excited-state population at any time t, $N_i(0)$ is the initial excited-state population at $t = 0$, and τ_i is the radiative lifetime defined as

$$\tau_i = \frac{1}{\sum_{(g<i)} A_{ig}}. \tag{1.12}$$

Strong atomic transitions have A_{ig} values of $10^8 \ s^{-1}$ to $10^9 \ s^{-1}$, so lifetimes are 1 ns to 10 ns. The above equation gives only the radiative lifetime, but lifetimes can be shortened by collisions and/or stimulated emission.

1.4. Broadening of the Spectral Lines

From an analytical point of view, the two most important characteristics of the spectral lines are width and intensity (amplitude). The width and amplitude of the spectral lines will affect the ability to extract qualitative and quantitative information from spectra. The spectral line width is related to the analytical selectivity (presence of spectral interferences owing to overlapping), while the intensity of the spectral line is related to the analytical sensitivity.

Spectral lines show profiles with a finite width, $\Delta\lambda$ (Fig. 1.4), at least due to the **natural broadening** (the intrinsic linewidth in the

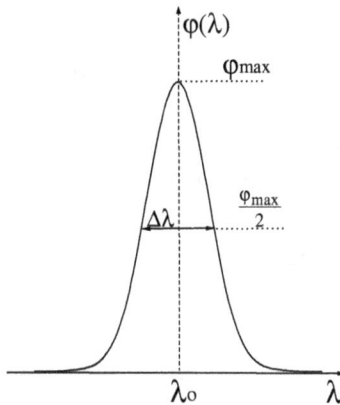

Fig. 1.4. Spectral line profile.

absence of external influences), which is determined by the quantum mechanical uncertainty of the energy levels (Heissenberg's principle).

The natural broadening of an energy level, $\Delta\lambda$ (usually about 10^{-4} Å), is determined by the lifetime due to the Heisenberg uncertainty principle with the following value

$$\Delta E\tau \approx h/2\pi, \tag{1.13}$$

where τ is the half-life time of the excited level ($\tau \approx 10^{-8}$ s). So, the natural broadening of an energy level is

$$\Delta E_i = \frac{h}{2\pi\tau_i}, \tag{1.14}$$

or considering Eq. (1.12)

$$\Delta E_i = \frac{h\sum_{(g<i)} A_{ig}}{2\pi}. \tag{1.15}$$

Since $E = h\nu$ (or $\Delta E = h\Delta\nu$)

$$\Delta\nu = \frac{\sum_{(g<i)} A_{ig}}{2\pi}, \tag{1.16}$$

or alternatively

$$\Delta\lambda_N = \frac{\lambda^2}{2\pi c}A_{ig} = 5,32 \times 10^{-19}\,\lambda^2 A_{ig}, \tag{1.17}$$

where $\Delta\nu$ is the line width in frequency units of a transition from the ground state to an excited state. Since the ground state has an essentially infinite lifetime, the transition line width is governed by the width of the excited state. The natural broadening value is not really the lowest limiting value taken by any spectral line width, because the laser radiation (as constituted by coherent waves) shows shorter natural broadening.

There are other causes giving an additional broadening of the spectral lines. Some of these causes are related to the decrease of τ. When the broadening is the result of several causes, the resultant line profile is the convolution of all the particular profiles. The different broadening mechanisms of the line shapes include: natural broadening, collisional broadening, power broadening, and Doppler broadening.

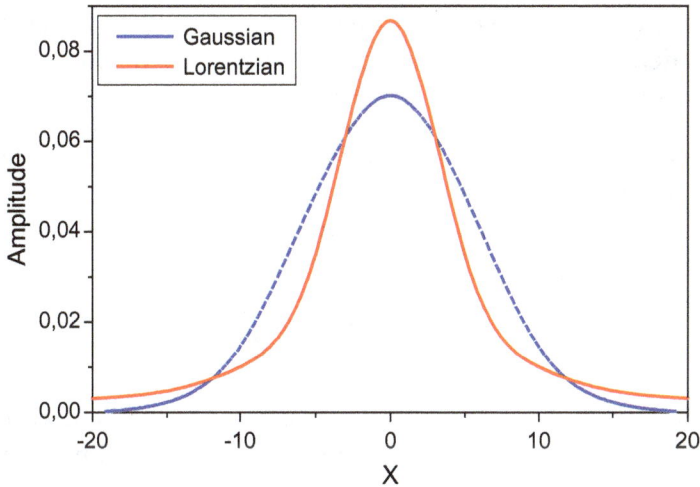

Fig. 1.5.   Gaussian and Lorentzian spectral line shapes.

Natural, collisional, and **power broadenings** are homogeneous mechanisms and produce Lorentzian line shapes, while **Doppler broadening** is a form of inhomogeneous broadening showing a Gaussian line shape (Fig. 1.5). Combinations of Lorentzian and Gaussian line shapes can be approximated by the Voigt profile, which is based on a series expansion for the Voight function [Di Rocco *et al.* (2001)]. The use of the peak intensity and the full width at half maximum (FWHM) is emphasized, because in this region, the signal-to-noise ratio is better than in the wings.

The different causes of broadening of the spectral lines can be described as follows.

- *Power broadening.* Power broadening occurs as a result of the shortening of the fife-time of the excited state due to stimulated emission.
- *Collisional or pressure broadening.* Collisions broaden spectroscopic line widths by shortening the lifetime of the excited states. *Collisions can occur between atoms and charged particles*, provided that the elapsed time between two collisions would be lower or at least similar to the τ value. Interaction between unlike atoms

on the ground state of the same element is known as **resonance broadening** (sometimes called **Holtsmark broadening**), which is symmetric but very small (10^{-8} nm).

- *Resonance broadening* (self-broadening) occurs between identical species involving lines which result from an electric dipole transition (resonance line) to the ground state. The FWHM may be estimated as

$$\Delta\lambda_{1/2}^R \approx 8.6 \times 10^{-30} \left(\frac{g_g}{g_i}\right)^{1/2} \lambda^2 \lambda_r f_r N_g, \tag{1.18}$$

where λ is the wavelength of the observed line, f_r and λ_r are the oscillator strength and wavelength of the resonance line, g_i and g_g are the statistical weights of its upper and lower (ground state) levels, and N_g is the ground state number density. Resonance broadening occurs if the half-width of the absorption line increases with increasing analyte concentration [Rubeska and Svoboda (1965)].

Collision between molecules, atoms and particles of any chemical species with the analyte atoms provides the named foreign-gas broadening (also called Lorentz broadening), which together with the Holtsmark broadening, are usually named pressure broadening. The line half-width due to the Lorentz broadening can be calculated as follows

$$\Delta\lambda_L = \frac{z\lambda_0^2}{\pi c} = \sigma_L^2 N_e \sqrt{\frac{2RT}{\mu}}, \tag{1.19}$$

where, σ_L^2 is the efficient section of collision for the Lorentz broadening; N_e, the atoms or molecules strange density; R, the gas constant; T, the absolute temperature; z, the collision frequency; c, the light speed; and μ, the reduced mass of two species colliding [Galan and Winefordner (1968)].

- *Van der Waals broadening* is the result of the dipolar interaction between excited atoms and the induced dipole over other atoms in the ground state. The van der Waals broadening originates a Lorentzian profile whose FWHM has been deduced recently for atoms other than hydrogen [Yubero *et al.* (2007)].

$$\Delta\lambda_{1/2}^W \approx 8.18 \times 10^{-26}\lambda^2(\alpha\langle R^2\rangle)^{2/5} \left(\frac{T}{\mu}\right)^{3/10} N, \qquad (1.20)$$

where μ is the atom-perturber reduced mass, N the number density of atoms in the ground sate, α the atomic polarizability of the neutral perturber and $\langle R^2\rangle$ the difference of the square radius of the emitting atom in the upper and lower levels (in units of Bohr radius). Van der Waals broadened lines are red shifted by about one-third the size of the FWHM.

- **Self-absorption** is produced when some of the energy emitted by the inside zones of the source is absorbed in the outside zones. The absorption coefficient at the line center is higher than in the wings; consequently, self-absorption increases the line half-width. In heterogenous sources, this process can originate an important decrease of the source luminosity at the line center, which is known as *self-reversal,*

- **Doppler Broadening.** *Doppler broadening* is due to the distribution of atomic speeds and directions, where each has a Doppler shift with respect to an observer. If the pressure is lower than 10 mm Hg and the charged particles concentration, $E_m < 10^{11}$ cm^{-3}, the main cause of the spectral line broadening is Doppler effect. In this case, the broadening is the result of the thermal movement of the emissive atoms, of which the line profile or the distribution of radiation around the center frequency is then given by,

$$i_v = i_0 \exp\left[-\frac{Mc^2}{2RT}\left(\frac{\lambda_0 - \lambda}{\lambda}\right)^2\right] \qquad (1.21)$$

For a Maxwellian velocity distribution, the line shape is Gaussian and the half-width of Doppler contour can be expressed as follows,

$$\Delta\lambda_D = \frac{2\lambda}{c}\sqrt{\frac{2Ln2RT}{M}} = 7.16 \times 10^{-7}\lambda\sqrt{\frac{T}{M}} \qquad (1.22)$$

where R is the gas constant, c, the light speed, T, the absolute temperature of the emissive atoms, and M, the atomic mass of atoms.

The broadening of the lines emitted from a hollow cathode lamp is completely brought by Doppler's effect. In theory, the Doppler broadening is one of the most useful contributions to the broadening for diagnostic purposes, because the shape of the Doppler line can provide directly the rate distribution of those atoms that absorb (or emit). A Gaussian Doppler profile shows that the atoms have a Maxwell–Boltzmann rate distribution. From the width of the Gaussian component, it is possible to deduce a heavy particle temperature. Collisions between particles, where the internal energy from one or both particles colliding is tranformed into translational energy, can modify the rate distribution of both particles and consequently the Doppler broadening of the lines emitted by them. Those collisions with a charge transfer being *exoergics* provide non-Gaussian line profiles with flatter and broadener lines that it would be expected from a Maxwellian distribution. By contrast, the *endoergics* charge transfer processes provide Gaussian profiles with line widths expressed by a temperature only slightly lower than the temperature of the source [Shirts *et al.* (1998)].

- **Stark broadening.** The interaction between two charged particles colliding can be similar to the interaction of their electric fields, by which the spectral line width resulting from the presence of the electric fields can be established as a result of Stark's effect in a heterogeneous and non-stationary electric field (Stark broadening). *Stark broadening* may cause a linear or quadratic shift in energy and usually dominates resonance and van der Waals broadening. The FWHM for hydrogen lines is

$$\Delta\lambda_{1/2}^{S.H} \approx (2.50 \times 10^{-9})\alpha_{1/2}N_e^{2/3}, \tag{1.23}$$

where N_e is the electron density and $\alpha_{1/2}$ is the half-width parameter. Something similar happens if an external magnetic field is applied (Zeeman effect). Under these conditions, not only the width, but also the shift of the spectral lines can be observed. The broadening provided by Stark's effect grows if high energy levels take parts, and increasing the gas pressure and the concentration of charged particles. When the pressure achieves values close to the atmospheric pressure, Stark's effect is the most important contribution to the spectral line broadening. In this case, the spectral linewidth usually takes the value of about 0.1–10 Å. By contrast, at higher pressures ($\approx$10 atm) and for particles concentrations of 10^{18} cm^{-3}, Stark width is usually higher (10–100 Å). The natural broadening and the most general cases of Stark's broadening can be described by the same profile, known as dispersive contour

$$i_v = i_0 \frac{(\Delta v/2)^2}{(v_0 - v)^2 + (\Delta v/2)^2},\tag{1.24}$$

where, Δv is the line half-width, and v_0 is the frequency corresponding to the line center.

- **Autoionization or preionization.** It is known that excited atoms can stay in two different energetic states with the same energy. One of these states is a normal discrete state, while the other is constituted by the ionized atom plus the electron with the corresponding kinetic energy. The proper functions corresponding to both energetic states might be mixed, allowing some non-radiative transition probability and consequently a broadening of the discrete energy transition levels. This effect is important in electronic spectrometry (far UV and X-rays), where it is usually known as Auger effect.

- **Hiperfine or isotopic structure.** Another cause of the spectral line broadening is from the unresolved hiperfine or isotopic structure of the spectral lines. Hiperfine structure is the result of the radiating atoms consisting of a mixture of isotopes or of possessing a non-zero nuclear spin. Hiperfine structure situation is similar to that where more than one transition falls within the spectral bandpass of the monochromator. The line profile resulting from this broadening is assorted and complex, depending on the number and intensity of the components of the hyperfine structure. The broadening due to this cause is usually less than 10 Å.

- **Instrumental broadening — detection devices.** Detector does not respond immediately to the signal changes, but it shows some inertia. This provokes an instrumental distortion of the spectral line profile at the spectrometer output described by a Gaussian distribution

$$I(t) = I_0 \exp(-\gamma v^2 t^2), \tag{1.25}$$

where, I_0 and $I(t)$ are the signals at the monochromator input and output, respectively; $v(= d\lambda/dt)$ is the scanning speed in Å/s; t is time; and $\gamma = 4 \, \mathrm{Ln}(2/a_j^2)$, a_j being the line half-width at the spectrometer output. When the scanning speed is constant, a Gaussian signal profile is obtained at the spectrometer output whenever the true line profile as well as the spectrometer apparatus function can be described by Gaussian curves with half-width a_E and s, respectively. In this case, we can write

$$a_j^2 = a_E^2 + s^2, \tag{1.26}$$

if the inertial properties of the receiver device are brought by the named transitive characteristic, that is

$$h(t) = 1 - \exp\left(-\frac{t}{\tau_r}\right). \tag{1.27}$$

The transitive function, $h(t)$, characterizes the behavior of the system for the unit change in the signal, and the time constant, τ_r, represents the time for which the reply of the recording device achieves the value of 0.632 ($= 1 - \frac{1}{e}$).

1.5. Analytical Signals

The analytical signals obtained in ETAAS are transient signals (Fig. 1.6).

The intensity of a characteristic line is determined by absorbance, A_t, which is dependent on the number of gas-phase analyte atoms, N_t, existing in the graphite furnace at any time. Assuming a homogeneous distribution of the gas-phase atoms over the cross-section of the ET atomizer, S, a constant ratio between the line shape factor, H, and the Doppler width, $\Delta\lambda_D$, it can be expressed as follows [Broek and Galan

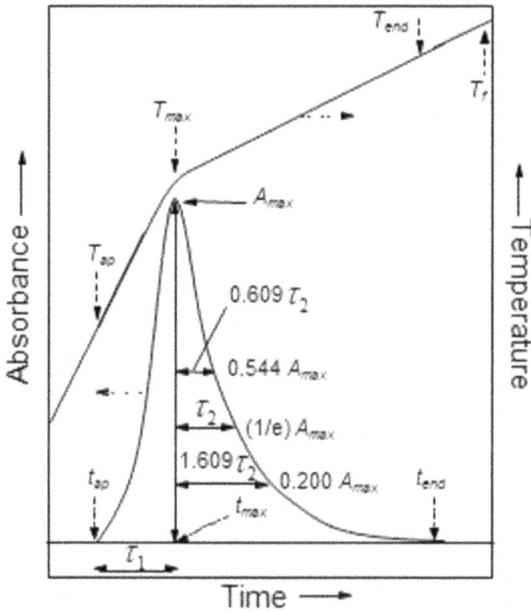

Fig. 1.6. Typical profile of an electrothermal atomic absorption signal.

(1977)]

$$A_t = \frac{0.432(\sqrt{4\pi \, \mathrm{Ln} \, 2})e^2\lambda^2 g_g fH}{mc^2 \Delta\lambda_D Q(T)S} N_t,$$ (1.28)

where g_g and $Q(T)$ are the statistical weight of the ground state and the partition function of the element, m and e are the mass and charge of electron, c is the light speed, while λ and f are the wavelength and oscillator strength of the resonance line. However, the experimental absorbance value is characterized by several terms, which depend on the instrumental and metrological conditions used to analyze the sample [Axner *et al.* (2003, 2004)].

According to Buggert–Lambert–Beer's law, if an incident flux, $\Phi_0(v)$, strikes a homogeneous absorbent layer with the width, l, the emerging flux, $\Phi_l(v)$, will be weakened $e^{K_v l}$ times

$$\Phi_l(v) = \Phi_0(v)e^{-K_v l} = \Phi_0(v)\exp(-K_v l).$$ (1.29)

However, for a heterogeneous absorbent layer, the integral $\int_0^l K_\nu(x)dx$ will be used instead of the $K_\nu l$ product in the exponent. This term is known as the optical width of the absorption layer, while the magnitude K_ν is the absorption coefficient with the $[L^{-1}]$ dimensions. The term

$$K = \int_0^\infty K_\nu d\nu \qquad (1.30)$$

is known as the integral absorption coefficient.

Absorbance is defined as the logarithm of the ratio between the incident, Φ_0, and emerging, Φ_l, fluxes at a particular radiation frequency of the absorption volume

$$A = \log \frac{\Phi_0}{\Phi_l}. \qquad (1.31)$$

However, the detector measures the intensity over the whole spectral range that reaches it, and the absorbance calculation needs to use the functions $\Phi_0(\nu)$ and $\Phi_l(\nu)$ integrated over the whole wavelength range that reaches the detector. In this case, absorbance would be

$$A = \log \frac{\int \Phi_0(\nu)d\nu}{\int \Phi_l(\nu)d\nu} = \log \frac{\int \Phi_0(\nu)d\nu}{\int e^{-K_\nu l}\Phi_0(\nu)d\nu}. \qquad (1.32)$$

The theoretical problem arising with Eq. (1.32) is to calculate the value of the integrals. The most general expression for the absorbance recorded by the detector system would be represented by the following equation [Gilmutdinov *et al.* (1994)]

$$A = \log \left\{ \frac{\int_0^{2R} \int_{-l}^{l/2} \int_{-\lambda^*}^{\lambda^*} J(x, y)J(\lambda)d\lambda dxdy}{\begin{array}{c} \int_0^{2R} \int_{-l}^{l/2} \int_{-\lambda^*}^{\lambda^*} J(x, y)J(\lambda) \\ \times \exp\left(-\int_{-L/2}^{L/2} K(\lambda, x, y, z)dz\right)d\lambda dxdy \end{array}} \right\}, \qquad (1.33)$$

where l is the bandwidth registered by the photomultiplier tube (PMT); R and L are the radius and length of the ET atomizer, respectively; $J(x, y)$ is the transverse distribution of the intensity of incident radiation immediately before the atomizer; x and y are the axes of the perpendicular and vertical coordinates of the diameter of the ET atomizer, respectively; $J(\lambda)$ is a function describing the spectrum profile of the analytical line, and limits $-\lambda^*$ to λ^* represent the spectral bandwidth isolated by the monochromator. The absorption coefficient, $K(\lambda, x, y, z)$ depends on the spatial coordinates and the wavelength, and can be expressed in the following way: $K(\lambda, x, y, z) = c(\lambda, T)n(x, y, z)$, where $n(x, y, z)$ is the local density of the analyte atoms, and $c(\lambda, T)$ is a coefficient depending on the spectral characteristics of the analytical line and on temperature, T.

From this general account, it may be gathered that absorbance depends on three groups of factors

- the spectral characteristics of the analytical line (dependence on the intensity of the incident radiation, J, and the absorption coefficient, K, at λ),
- the spatial distribution of the analyte and the temperature within the atomizer (the dependence of K on x, y, z),
- the transverse distribution of the intensity of the incident beam $(J(x, y))$.

Thus, non-uniform radiation is in use to excite a heterogeneous layer of absorbing atoms in a non-isothermal atomizer [Gilmutdinov et al. (1996a,b,c)]. In these cases, and using conventional detection systems (usually PMT) which detect radiation spatially integrated over their working area, absorbance measurements are subject to photometric errors. This error can be eliminated when using spatially-resolved detection with solid state detectors [Gilmutdinov et al. (2000)]. Of course, these detection systems extend the possibility to carry out multielement ETAAS analysis [True et al. (1999); Lin and Huang (2001)].

In the experimental measurements of absorbance, K_ν and $\Phi_0(\nu)$ can both be considered constants (at least in many cases), because there exists a uniform distribution of the radiation beam, of the analyte atoms and of the temperature in the atomizer, and in addition, it is assumed that the spectral range which reaches the detector is very (infinitely) narrow, and *practically* coincide with the peak profile of the absorption line. Theoretical and experimental values of

K_v at high temperatures are practically constant [Torsi *et al.* (1998)]. As a consequence, for that situation, Eq. (1.32) can be considerably simplified, obtaining the following general expression

$$A = 0.434 K_v l. \tag{1.34}$$

The integral absorption coefficient can also be simplified and related to the population of the levels and other atomic character- istics, such as the oscillator strength for absorption, f_{gi}, wavelength, λ, and the number of atoms in the ground state per volume unit, N_g,

$$\int_0^\infty K_v dv = \frac{\pi e^2}{mc} N_g f_{gi} \left(1 - \frac{g_i}{g_g} \frac{N_g}{N_i} \right), \tag{1.35}$$

where, $N_i y N_g$ are the populations of the upper and lower levels of tran- sition that correspond to the absorption band, and g_i and g_g are the sta- tistical weighs of the levels. When the thermal excitation is produced at relatively low temperature, the term, $\frac{g_i}{g_g} \frac{N_g}{N_i}$, which is responsible for the negative absorption, can be ignored against the unity, by which the abovementioned expression is transformed into the following

$$\int_0^\infty K_v dv = \frac{\pi e^2}{mc} N_g f_{gi} \approx 0.026 \, N_g f_{gi}. \tag{1.36}$$

In ETAAS, N_g can be considered as the total number of atoms per cross- section unit, N; therefore, Eq. (1.34) is transformed into the following using the atomic absorption coefficient related to a single atom, K

$$A = KN = KlC. \tag{1.37}$$

In this way, the concentration of atoms, N, appears independently of the actual shape of the absorption line. The assumption that the absorption coefficient is constant can only be possible assuming that:

- the emission source generates only electromagnetic radiation exactly at the absorption wavelength of analyte, and
- the emission profile is substantially narrower than the absorption profile.

The first condition is fulfilled by the hollow cathode lamps (HCL). For the second condition, it is admitted that the different broadening processes (Doppler, pressure or Lorentz, auto-absorption, fine and hyperfine structures, fundamentally those caused by the Zeeman effect) are under control and they produce minimum effects. The peak absorption coefficient is not dependent on the oscillator strength for pure natural broadened absorption lines [Hannaford (1994)]. Other theoretical considerations about absorption of electromagnetic radiation by the analyte atoms can be further looked into [Aller (1987); Haswell (1991); Hoenig and Kersabiec (1990)].

The heating rate of the atomizer and the atomization efficiency are important instrumental and theoretical parameters which can control not only the free analyte atoms cloud, but also the rate of atom loss due to diffusion. Ideally, an atomizer is required in which the sample would be atomized almost instantaneously into a maximum temperature furnace atmosphere. However, in the usual ETAAS works, the most sensitive wavelength and the slit of monochromator are fixed, so that to increase A_t in Eq. (1.28), N_t and S are the only parameters that can really be modified. Thus, in the ET atomizers, an excessive increase of L (consequently, increased residence time of atoms) is not generally advisable due to the rise in the probability of auto-absorption and the thermal gradients inside the ET atomizer which produce condensation in the cold areas. The maximum temperature of the ET atomizer wall is limited by the half-life of the graphite atomizer that rapidly decreases for temperatures above 3300 K, and the power available in most commercial instruments is restricted to levels that prevent this temperature from becoming greater.

The total free atom concentration at any time, N_t, is the result of a series of complex reactions in the vapor, liquid (melted) and solid phases. These reactions depend on: (i) the physical and chemical properties of analytes and matrices; and (ii) the instrumental parameters, such as the heating rate, final equilibrium temperature and the material of the atomizer surface. According to the L'vov's atomization model, the maximum number of analyte atoms in the furnace, N_{max}, is related to the total number of the analyte atoms put into the furnace, N_0, as follows

$$N_{max} = 2N_0 \left(\frac{\tau_2^2}{\tau_1^2}\right)\left[\frac{\tau_1}{\tau_2} - 1 + e^{-(\tau_1/\tau_2)}\right], \qquad (1.38)$$

where, τ_1 is the atomization time and τ_2 is the residence time (see Fig. 1.6). The loss of the analyte atoms in the graphite atomizer is expressed by τ_2, and it partly originates in the internal gas flow and partly through diffusion. Under gas flow-stop atomization conditions, diffusion and convection caused by gas expansion are the most important factors whose values depend on the internal dimensions of the atomizer. Under diffusion controlling conditions, τ_2 only depends on temperature, so that the diffusion coefficient also depends on temperature

$$\tau_2 \propto T^{-n}, \qquad (1.39)$$

where n usually varies between 1.5 and 2. Hence, the rate of the atom loss increases with temperature, and ETAAS results often show that N_{max}, and consequently the maximum absorption A_{max}, is produced at an optimum temperature considerably smaller than the maximum or equilibrium temperature.

Under normal working conditions, modification of temperature during atomization is less than ideal for ETAAS measurements. For this reason, by increasing the heating rate and also by delaying the vaporization or atomization of the sample, sensitivity can be improved and the matrix interferences eliminated or, at least decreased. In this case, N_{max} is achieved at the equilibrium or maximum temperature, so that those molecules that could be formed as a result of the interfering reactions can be dissociated more easily. However, the use of the highest heating rate and delayed atomization until times in which even the higher temperatures are reached is more advantageous for AES than in AAS, because a simultaneous increase in N_{max} and T exists. In ETAAS, the use of the absolute intensity is advisable [Broeck and Galan (1977); Sturgeon and Berman (1983)]

The general expression for absorbance, A, with continuous sources [Galan *et al.* (1967)] would be the following

$$A = \text{Ln}\left(\frac{I_c^0}{I_c}\right) = -\text{Ln}\left(\frac{1 - A_T}{s}\right), \tag{1.40}$$

where A_T is the total absorption (the product of the path length, l, and the integral of the absorption coefficient, K_λ, over the wavelength interval, $\Delta\lambda$), and s is the spectral bandwidth of the monochromator, which is the decisive parameter in the case of continuum sources.

A comparison between line and continuum sources for measuring the same low atomic concentration would result from the ratio of the absorbances, R_A, (or the ratio of the fractions of the intensity absorbed) [Galan *et al.* (1967)].

$$R_A = \sqrt{\frac{4 \, \text{Ln} \, 2}{\pi} \frac{s\delta(a, \bar{v})}{\Delta\lambda_D}}, \tag{1.41}$$

where the function $\delta(a, \bar{v})$ describes the variation of K_λ over the entire absorption line. The decision factor in this equation is the ratio, $\frac{s}{\Delta\lambda_D}$. A spectral bandwidth, s, of the same order of magnitude can only be obtained with a high resolution monochromator.

1.6. Noise

Measurement errors are usually grouped as systematic and random. The random error is often called noise. Noise is any temporal fluctuation of the signal, which does not provide useful information and, consequently, directly affects both the limit of detection and precision. **Noise** can be **extrinsic** or **intrinsic. Extrinsic noise** can be present in the absence of analyte and is due to a non-specific background signal. It can be produced by dark current, scattered light from the source, background emission and non-selective detection of atoms from a blank. On the other hand, **intrinsic noise**, which cannot be removed, is the result of the statistical condition of the atom detection process itself.

Noise classification is usually based on their spectral noise power distribution as **white, flicker** and **whistle** noises. **White noise** has a flat spectral noise power distribution, while **flicker noise** has a spectral noise power distribution inversely proportional to the electrical frequency of measurement ($\sim 1/f^n$, where n can take values > 0), and **whistle noise** shows discrete noise bands. **White noise** may occur mainly in the radiation source, measurement cell and detector; **flicker noise** may occur in the radiation source, measurement cell, and spectral discrimination device; while **whistle noise** may occur in the measurement cell [Epstein and Winefordner (1984)].

Noise can also be grouped as **chemical** and **instrumental**. **Chemical noise** arises from many uncontrollable variables that affect the chemistry of the system under investigation. The main sources of chemical noise include temperature, pressure, humidity and chemical interactions, being dependent on the method: sample preparation, storage of samples, etc. **Instrumental noise** is contributed by each instrument component (source, input transducer, output transducer, etc.) to the overall noise in an analytical signal. However, it is usual to group instrumental noise into the following types: **thermal noise, shot noise, flicker noise,** and **environmental noise**.

Thermal noise is also referred to as Johnson noise and it is caused by the thermal agitation of electrons or other charge carriers in the electronic components (resistors, capacitors, radiation transducers, etc.) of an instrument. Thermal noise only disappears at absolute zero, and is given by

$$\bar{v}_{rms} = \sqrt{4kTR\Delta v}, \qquad (1.42)$$

where $\bar{v}_{rms}$ is the root-mean-square noise voltage, k the Boltzmann constant (1.38×10^{-23} J/K), T the temperature absolute (K), R the resistance of a resistive element (Ω), and Δv the frequency bandwidth (Hz). According to Eq. (1.42), the magnitude of the thermal noise depends on the frequency bandwidth, but not on the frequency of the signal. The term, Δv, depends on the rise time, τ_r, of the instrument, $\Delta v = \frac{1}{3\tau_r}$. Rise time or response time is the time (in seconds) for the output of an instrument response to change from 10–90% following an abrupt change in input. **Thermal noise** is also referred to as **white noise** since it is not frequency-dependent. **Thermal noise** can be reduced in three ways: (i) narrowing the bandwidth, (ii) reducing the number of resistive elements, and (iii) reducing the temperature of the electronic components.

Shot noise occurs wherever electrons cross a junction. The magnitude of shot noise is given by $i_{rms} = \sqrt{2 I e \Delta \nu}$, where i_{rms} is the root-mean-square current fluctuations, e the electron charge (1.60×10^{-9} C), $\Delta \nu$ the frequency bandwidth (Hz) and I the average direct current (A) given by [Wineland *et al.* (1987)]

$$I = \frac{P(1 - \eta)e\beta}{h\nu}, \tag{1.43}$$

where P is the power of a collimated radiation source, ηP represents the fraction of power absorbed by the atoms ($\eta \ll 1$), and β is the quantum efficiency of conversion of the incident photons into electric current. Other parameters have the usual meaning. As it is not frequency-dependent, it is also considered as white noise.

Flicker noise is inversely proportional to the frequency of the signal observed ($1/f$ noise). Flicker noise is: (i) frequency-dependent, (ii) significant at frequencies <100 Hz, (iii) manifested in the long term shift of the amplifiers, meters, etc., and (iv) reduced with special resistors.

Environmental noise arises from the instrument's surroundings. Most environmental noise results from conductors (circuitry) in the instrument, which acts like antennas to "pick up" signals from power lines, radio transmitters, etc. It is caused by induction (an electric current flowing in a wire will induce a current in a nearly wire).

The major noises in ETAAS are photomultiplier tube (PMT) shot-noise (induced by the radiation source or by the blackbody emission from the atomizer); measurement cell transmission flicker; and radiation source intensity flicker. PMT shot-noise induced by the blackbody emission from the atomizer can be a dominant noise for elements absorbing in the visible region, while in the UV region, scattering and molecular absorption by matrix components can constitute a significant noise. On the other hand, blank absorption flicker noise can become important at absorbances close to the detection limit.

In general, the i_{rms} noise current would be

$$i_{rms} = (i_D^2 + i_T^2 + i_{sh}^2)^{1/2}. \tag{1.44}$$

The contribution from the detector noise, i_D, (Johnson noise) and radiation-source technical noise, i_T, (fluctuations in the radiation) will depend on the experimental configuration, but it is desirable that i_D, $i_T \ll i_{sh}$ (shot noise) for a maximum signal-to-noise ratio.

In ETAAS, increased background radiation reaching the detector is rarely important. Nonetheless, the main sources of residual radiation, with the characteristics of the blackbody radiation and contributing to the background radiation, are usually the following [Frech *et al.* (1992)].

(i) **Blackbody radiation from the atomizer wall**, which is scattered by small molecules and particles present in the hot parts of the atomizer (Rayleigh's scattering). The Rayleigh's scattering predominates at the wavelengths, λ, larger than the particle radius. Even working under optimum conditions, the residual background radiation still exists in the monochromator. The angular distribution of the intensity of scattered light, I_d, in the Rayleigh region ($\lambda \gg$ the particle diameter, d), can be expressed by

$$I_d \approx I_0 \pi^2 (1 + \text{Cos}^2\,\theta)(n^2 - 1)\left(\frac{V^2}{2r^2\lambda^4}\right), \qquad (1.45)$$

where, I_0 is the incident intensity, θ is the angle between the incident and scattered light beams, n is the refraction index of the particle, V is the volume of the particle, λ is wavelength, and r is the distance of the particle from the observation point. So, the intensity of the Rayleigh scattered light is proportional to the incident intensity and to the power sixth of the particle diameter and inversely proportional to the power fourth of the wavelength. The highest intensity occurs for angles close to $0°$ and $180°$. Blackbody radiation emitted or scattered by incandescent carbonaceous particles originating in the decomposition of the graphite atomizer is usually more intense with the graphite atomizer age.

(ii) **Radiation from the atomizer wall scattered by large particles** ($\lambda \ll$ the particle radius), which are formed by condensation in the cold parts of the atomizer (Mie scattering) [Frech *et al.* (1992)]. The intensity of the scattered light produced by the largest particles ($d \gg \lambda$) does not depend on the wavelength, but on the angle,

θ, causing a maximum intensity at small angles and minimum at 180°. The intensity of the Mie scattering is also proportional to the number of scatters and will rise in the presence of a matrix that forms condensed vapors in the cold parts of the atomizer. The Mie scattering will also reduce the intensity, produced by the analyte atoms, that reaches the monochromator and will degrade the signal-to-background and signal-to-noise ratios. To reduce the Mie scattering, modifications in the graphite atomizer or in the gas flow must be made.

(iii) **Radiation from windows** on the atomizer housing and other optical components reflected into the spectrometer.

(iv) **Aberrations and/or dirt particles in the optical system**, focusing lens, furnace windows, etc., that may allow the radiation of the atomizer wall to enter directly into the spectrometer, despite correct alignment.

The intensity of the emission signal given by any of the abovementioned sources is proportional to the radiation intensity that comes from the atomizer wall, i.e. regulated by Planck's law.

1.7. Signal-to-Noise Ratio

The signal-to-noise ratio, S/N, is the most important parameter determining the analytical capabilities of a technique for a particular application. The signal-to-noise ratio represents the ratio of the true signal amplitude to the standard deviation of the noise. The S/N is closely related to the sensitivity, limit of detection and precision, as it is inversely proportional to the relative standard deviation of the signal amplitude. Using an automatic background correction system, the selection of the optimum operating temperature is based on the signal-to-noise ratio, because the detection limit becomes dependent on it. This means that the increase in the analyte signal can be differentiated from the background continuum noise (arising primarily from the sample matrix, stray light and blackbody emission) [Harnly

(1984)]. The S/N can be theoretically predicted by considering the contribution of the instrumental parameters.

For a Gaussian peak, with mean $\cong \sigma$, peak area $\cong A$, and peak height $\cong \frac{A}{\sigma(2\pi)^{0.5}}$, the S/N expression is

$$\frac{S}{N} = \frac{\frac{2A}{\sigma(2\pi)^{0.5}} \tau_i^{-1} \int_0^a e^{-\frac{t^2}{2\sigma^2}} dt}{\left(\frac{\eta 2a}{\tau_i^2}\right)^{0.5}} = \frac{A}{\sigma} \left(\frac{1}{\pi\eta}\right)^{0.5} \frac{\int_0^a e^{-\frac{t^2}{2\sigma^2}} dt}{a^{0.5}}, \qquad (1.46)$$

where $\tau_a \cong 2a$ is the optimum gate window, η is bilateral noise power spectral density of a dominant stationary and white noise, and τ_i is the integration time constant.

The **matched filter**, $S/N_{matched}$, provides the standard against which any peak integration should be compared since it is optimum for peak processing in additive, stationary, white noise. The matched filter $S/N_{matched}$ is

$$\frac{S}{N_{matched}} = A \left(\frac{1}{\sqrt{\pi}2\eta\sigma}\right)^{0.5}. \qquad (1.47)$$

Hence, the ratio would be the following [Voigtman (1991)]

$$\frac{S/N}{S/N_{matched}} = \left(\frac{2}{\sigma\sqrt{\pi}}\right)^{0.5} \frac{\int_0^a e^{-\frac{t^2}{2\sigma^2}} dt}{a^{0.5}}. \qquad (1.48)$$

The signal-to-noise ratio can also be written as the ratio between the signal intensity, I_s, and the noise intensity, I_n [Wineland *et al.* (1987)]

$$\frac{S}{N} = \frac{I_s}{I_n} = \frac{N\sigma_0 Ke\beta P/Ah\nu_0}{\left[i_D^2 + i_T^2 + (2e^2\beta P\Delta\nu/h\nu_0)\right]^{1/2}}. \qquad (1.49)$$

If the minimum detectable atomic absorption, N_{min}, is defined as the concentration that produces a signal equal to twice the noise, the two following expressions were derived [Galan *et al.* (1967)]

$$N_{min,l} = \frac{mc^2 \Delta\lambda_D \chi_{ls} \sqrt{\Delta\nu}}{\sqrt{\pi} \, \text{Ln} \, 2 e^2 \lambda_0^2 fl} \quad \text{(for line sources)} \qquad (1.50)$$

$$N_{min,c} = \frac{2mc^2 s \chi_c \sqrt{\Delta\nu}}{\pi e^2 \lambda_0^2 fl} \quad \text{(for continuum sources)}, \qquad (1.51)$$

where χ_{ls} and χ_c are the source intensity (*ls*, line source and *c*, continuum source), $\Delta\lambda_D$ is the Doppler half-width of the absorption line, *l* is the path length, *s* is the spectral bandwidth of the monochromator, and $\Delta\nu$ is the frequency bandwidth of the readout device.

The ratio, R_L, of the minimum detectable concentrations for a line source and a continuum source is found to be equal to the ratio of the noise ratio, R_N, and the signal ratio, R_A

$$R_L = \frac{N_{min,l}}{N_{min,c}} = \sqrt{\frac{\pi}{4 \, \text{Ln} \, 2}} \frac{\Delta\lambda_D}{s\delta(a,\bar{\nu})} \frac{\chi_l}{\chi_c} = \frac{R_N}{R_A}. \qquad (1.52)$$

From this equation, it is clear that a large value for the signal ratio does not necessarily mean that a lower detection limit is obtained with a line source. The detection limit improves with the square root of the spectral bandwidth and with the square root of the pixel width [Harnly (1993)]. There is an inverse relationship between N_{min} and the ratio S/N.

The above considerations put forward the main theoretical background affecting on an analytical measurement using ETAAS. The negative effect of all these characteristics is displayed by a damage of the S/N ratio. In order to improve the S/N ratio of an analytical measurement it is necessary to increase the analyte signal and to decrease the noise or, alternatively, to better discriminate the analyte signal from the noise, at least. Wavelength modulation-diode laser absorption spectrometry (WM-DLAS) has been used for spectrochemical analysis with low detection limits when coupled to a graphite atomizer. The analytical WM-signal decreases for increased detection order, but the signal-to-background ratio, and thereby also the detectability, are related in the reverse order [Gustafsson and Axner (2003); Gustafsson *et al.* (2003)].

References

Alkemade, C.Th.J.; Hollander, T.; Snelleman, W. and Zeegers, P.J.Th. (1982). *Metal Vapours in Flames*. Pergamon Press, Oxford.

Aller, A.J. (1987). *Espectroscopía de absorción atómica analítica*. Servicio de Publicaciones, Universidad de León, León, Spain.

Aller, A.J. (1998). Looking into the major achievements in the analytical electrothermal atomic spectrometric techniques, A. Hemantaranjan, ed., Scientific Publishers, Jodhpur, India, pp. 479–511.

Aller, A.J. (2003). *Espectroscopía atómica electrotérmica analítica*. Servicio de Publicaciones, Universidad de León, León, Spain.

Axner, O.; Gustafsson, J.; Schmidt, F.M.; Omenetto, N. and Winefordner, J.D. (2003). A discussion about the significance of Absorbance and sample optical thickness in conventional absorption spectrometry and wavelength-modulated laser absorption spectrometry. *Spectrochim. Acta, Part B,* **58**: 1997–2014.

Axner, O.; Gustafsson, J.; Omenetto, N. and Winefordner, J.D. (2004). Line strengths, A-factors and absorption cross-sections for fine structure lines in multiplets and hyperfine structure components in lines in atomic spectrometry — a user's guide. *Spectrochim. Acta, Part B,* **59**: 1–39.

Boumans, P.W.J.M. (1968). Atomic partition functions in spectrochemical analysis. *Spectrochim. Acta, Part B,* **23**: 559–566.

Broek, W.M.G.T. van den and Galan, L. de (1977). Supply and removal of sample vapor in graphite thermal atomizers. *Anal. Chem.,* **49**: 2176–2186.

Di Rocco, H.O.; Iriarte, D.I. and Pomarino, J. (2001). General expression for the Voigt function that is of special interest for applied spectroscopy. *Applied Spectroscopy,* **55**: 822–826.

Epstein, M.S. and Winefordner, J.D. (1984). Summary of the usefulness of signal-to-noise treatment in analytical spectrometry. *Prog. Anal. At. Spectrosc.,* **7**: 67–137.

Frech, W.; Lundberg, E. and Cedergren, A. (1985). Investigations of some methods used to reduce interference effects in graphite furnace atomic absorption spectrometry. *Prog. Anal. At. Spectrosc.,* **8**: 257–370.

Frech, W.; L'vov, B.V. and Romanova, N.P. (1992). Condensation of matrix vapours in the gaseous phase in graphite furnace atomic absorption spectrometry. *Spectrochim. Acta, Part B,* **47**: 1461–1466.

Galan, L. de McGee, W.W. and Winefordner, J.D. (1967). Comparison of line and continuous sources in atomic absorption spectrophotometry. *Anal. Chim. Acta,* **37**: 436–444.

Galan, L. de and Winefordner, J.D. (1968). Slit function effects in atomic spectroscopy. *Spectrochim. Acta, Part B,* **23**: 277–289.

Gilmutdinov, A.Kh.; Nagulin, K.Yu. and Zakharov, Yu.A. (1994). Analytical measurement in electrothermal atomic absorption spectrometry — How correct is it?. *J. Anal. At. Spectrom.,* **9**: 643–650.

Gilmutdinov, A.Kh.; Radziuk, B.; Sperling, M.; Welz, B. and Nagulin, K.Yu. (1996a). Three-dimensional structure of the radiation beam in atomic spectrometry. *Spectrochim. Acta, Part B,* **51**: 931–940.

Gilmutdinov, A.Kh.; Radziuk, B.; Sperling, M.; Welz, B. and Nagulin, K.Yu. (1996b). Spatial distribution of radiant intensity from primary sources for atomic absorption spectrometry. Part II: Electrodeless discharge lamps. *Appl. Spectros.*, **50**: 483–497.

Gilmutdinov, A.Kh.; Radziuk, B.; Sperling, M. and Welz, B. (1996c). Spatially and temporally resolved detection of analytical signals in graphite furnace atomic absorption spectrometry. *Spectrochim. Acta, Part B*, **51**: 1023–1044.

Gilmutdinov, A.Kh.; Nagulin, K.Y. and Sperling, M. (2000). Spatially resolved atomic absorption analysis. *J. Anal. At. Spectrom.*, **15**: 1375–1382.

Gustafsson, J.; Chekalin, N. and Axner, O. (2003). Improved detectability of wavelength modulation diode laser absorption spectrometry applied to window-equipped graphite furnaces by 4th and 6th harmonic detection. *Spectrochim. Acta, Part B*, **58**: 111–122.

Gustafsson, J. and Axner, O. (2003). 'Intelligent' triggering methodology for improved detectability of wavelength modulation diode laser absorption spectrometry applied to window-equipped graphite furnaces. *Spectrochim. Acta, Part B*, **58**: 143–152.

Hannaford, P. (1994). The oscillator strength in atomic absorption spectroscopy. *Spectrochim. Acta, Part B*, **49**: 1581–1593.

Harnly, J.M. (1984). Theoretical comparison of monochromators for wavelength modulated atomic absorption and emission spectrometry. *Anal. Chem.*, **56**: 895–899.

Harnly, J.M. (1993). The effect of spectral bandpass on signal-to-noise ratios for continuum source atomic absorption spectrometry with a linear photodiode array detector. *Spectrochim. Acta, Part B*, **48**: 909–924.

Haswell, S.J. (editor) (1991). *Atomic Absorption Spectrometry; Theory, Design and Applications*. Elsevier, Amsterdam.

Hoenig, M. and Kersabiec, A.-M. de (1990). *L'atomisation électrothermique en spectrométrie d'absorption atomique*. Masson, Paris.

Lin, T.W. and Huang, S.D. (2001). Direct and simultaneous determination of copper, chromium, aluminium, and manganese in urine with a multielement graphite furnace atomic absorption spectrometer. *Anal. Chem.*, **73**: 4319–4325.

Lundberg, E. and Frech, W. (1981). Influence of instrumental response time on interference effects in graphite furnace atomic absorption spectrometry. *Anal. Chem.*, **53**: 1437–1442.

L'vov, B.V. (1970). *Atomic Absorption Spectrochemical Analysis*. Adam Hilger, London.

L'vov, B.V. (1978). Electrothermal atomization — the way toward absolute methods of atomic absorption analysis. *Spectrochim. Acta, Part B*, **33**: 153–193.

Rubeška, I. and Svoboda, V. (1965). Some causes of bending of analytical curves in atomic absorption spectroscopy. *Anal. Chim. Acta*, **32**: 253–261.

Schwab, J.J. and Lovett, R.J. (1990). Graphite furnace atomic emission spectrometry: computer simulations. *Spectrochim. Acta, Part B*, **45**: 281–285.

Shirts, R.B.; Parry, H.P. and Farnsworth, P.B. (1998). Anomalous line shapes caused by charge transfer in low-pressure discharges. *Spectrochim. Acta, Part B*, **53**: 487–498.

Sturgeon, R.E. and Berman, S.S. (1983). Determination of the efficiency of the graphite furnace for atomic absorption spectrometry. *Anal. Chem.,* **55**: 190–200.

Torsi, G.; Fagioli, F.; Landi, S.; Reschiglian, P.; Locatelli, C.; Rossi, F.N.; Melucci, D. and Bernoroli, T. (1998). Theoretical and experimental values of the spectroscopic constant relative to the Hg 253.7 nm line at different temperatures. *Spectrochim. Acta, Part B,* **53**: 1847–1851.

True, J.B.; Williams, R.H. and Denton, M.B. (1999). On the implementation of multielement continuum source graphite furnace atomic absorption spectrometry utilizing an echelle/CID detection system. *Appl. Spectrosc.,* **53**: 1102–1110.

Voigtman, E. (1991). Gated peak integration versus peak detection in white noise. *Appl. Spectrosc.,* **45**: 237–241.

Welz, B.; Schlemmer, G. and Mudakavi, J.R. (1992). Palladium nitrate-Magnesiun nitrate modifier for electrothermal atomic absorption spectrometry. Part 5. Performance for the determination of 21 elements. *J. Anal. At. Spectrom.,* **7**: 1257–1271.

Wineland, D.J.; Itano, W.M. and Bergquist, J.C. (1987). Absorption spectroscopy at the limit: detection of a single atom. *Opt. Lett.,* **12**: 389–391.

Yubero, C.; Dimitrijević, M.S.; Garcia, M.C. and Calzada, M.D. (2007). Using the vand der Waals broadening of the spectral atomic lines to measure the gas temperature of an argon microwave plasma at atmospheric pressure. *Spectrochim. Acta, Part B,* **62**: 169–176.

Chapter 2

Atomizers

2.1. Construction Materials for Atomizers

Electrothermal atomization is usually carried out in graphite atomizers. Nonetheless, atomizers made of metals, quartz and modern ceramic materials (nitrides, carbides, borides, and so forth) have also been tested. The atomizer surface characteristics determine analyte-substrate interactions, which are of great importance in ETAAS. These surface characteristics are fundamentally related to the relative abundance of chemically active centers, the nature of surface defects and dislocations, crystal grain size, and alignment of crystal planes.

2.1.1. Graphite

The graphite type used in making atomizers for ETAAS is a key factor in their behavior. The sorts of graphite utilized for the construction of electrothermal atomizers are normally polycrystalline graphite and particularly pyrolytic graphite.

Polycrystalline graphite shows surfaces with many small dimensions protuberances, 50 nm to 150 nm [Habichi *et al.* (1995)]. This type of graphite is a porous material ($\approx$17%), and hence easily penetrated by liquids and gases, so that the atomic vapor formed easily diffuses through the hot graphite walls (normally 1 mm thick). The high porosity graphite causes a noteworthy delay (tail formation) in the ETAAS signals.

Pyrolytic graphite is obtained from hydrocarbons through pyrolysis at high temperatures (>1400°C) and low pressures ($\approx$0.1 atmosphere). Pyrolytic graphite presents very low porosity (practically zero)

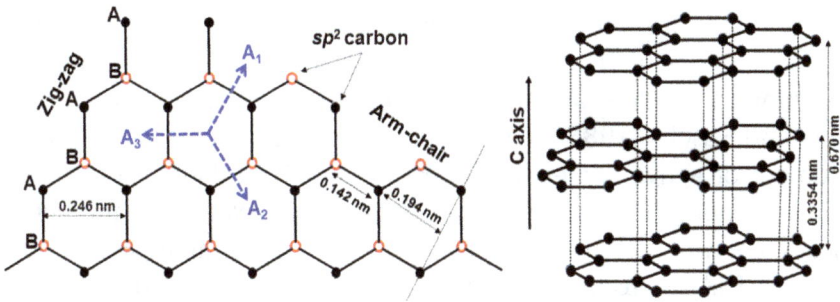

Fig. 2.1. Crystalline structure of hexagonal graphite. Perpendicular (left) and transverse (right) views of a surface. Points indicated with the letter A (solid black circles) represent carbon atoms having neighboring atoms situated directly above and below them in adjacent layers. Points labelled with the letter B (hollow circles) indicate carbon atoms not having immediately neighboring atoms on adjacent layers [Vanderwoort *et al.* (1996)].

and hence low permeability for gases, besides a high sublimation point (3970 K), great resistance to oxidation, and high electrical resistivity [Vanderwoort *et al.* (1996); Slavin *et al.* (1981); Sturgeon and Chakrabarti (1977)]. For these reasons, pyrolytic graphite is almost always used in constructing atomizers. There have also been arrangements involving coating standard graphite tubes with a thin layer ($\approx$50 μm) of pyrolytic graphite so as to reduce the effects of the porosity of the standard graphite.

The perfect hexagonal structure of a graphite crystal is shown in Fig. 2.1. The crystal's structure is made up of layers of carbon atoms joined together by van der Waals forces, while the carbon atoms present on each layer are tightly linked by covalent bonds. The shortest distance between the atoms in a layer is 0.142 nm, while the distance between layers is 0.335 nm. Adjacent layers are staggered with respect of one another, producing an alternating regularly ordered sequence, ABABAB. This atom ordering produces two different centers (two types of carbon atoms) on each layer: the carbon atoms represented by the letter A always have another neighboring carbon atom directly beneath it in the next layer, while the carbon atoms designated B are located over the hexagonal ring of the layer below. The edges of each graphite layer end in a zig-zag pattern (or in crenellation).

These edges, together with the single carbon atoms not fully bound to the lattice, called active sites, are very active centers for chemical reactions, while the carbon atoms within layers are totally inert [Huettner and Busche (1986)]. The use of pyrolytic graphite diminishes the number of active centers but does not eliminate them. Despite the laminar morphology of the pyrolytic graphite, defects at nanometric scale are what determine surface reactivity. Usual analytical conditions in ETAAS produce numerous surface defects as a consequence of interactions with oxygen (always present in the atomizer). Formation of such defects increases the number of active centers, also speeding up the process of intercalation of the analyte and its migration below the surface [Vandervoort *et al.* (1997)]. Not only the analyte, but also other simple species such as water molecules can be retained at the hydrophilic centers of carbon adsorbents [Tarasevich and Aksenenko (2003)].

Atomizers built with pyrolytic graphite offer several advantages: lower net resistance for heating, better confinement of the atomic analyte vapor within the pores of the graphite surface, production of less foam, longer average lifes, shorter tails ETAAS signals, better behavior with the carbides forming elements, greater chemical stability, and a more uniform heating. Consequently, greater sensitivity, lower detection limits, and higher precision can be attained using pyrolytic graphite [Vanderwoort *et al.* (1996)]. Many changes noted in the analytical sensitivity when graphite tubes are used may be related both to differences in the type of graphite employed [Schlemmer and Welz (1986)] and to the way in which the tube is manufactured [Slavin *et al.* (1981)].

Use of graphite with a single crystal structure (more like highly oriented pyrolytic graphite) improves the sensitivity of the ETAAS signals, probably due to one of the following causes. (i) The large grain sizes of single crystals which present a basal plane of the inert surface, and (ii) deriving from a more perfect crystalline structure. One striking disadvantage in the use of highly oriented pyrolytic graphite can be associated with the ease exfoliation of the graphite layers [Vanderwoort *et al.* (1996)].

Although pyrolytic graphite shows very little porosity, analytes migrate rapidly to a depth of at least 3 μm below the surface for sample solution deposition. However, they do not migrate, or at least not so much, if the sample is deposited as a vapor or through laser ablation [Eloi *et al.* (1993, 1995, 1997); Jackson *et al.* (1995)]. When the graphite surface is subjected to thermal treatment in an Ar/H$_2$ atmosphere, the migration phenomenon disappears, probably owing to the formation of C-H bonds as a consequence of the chemisorption of the hydrogen atoms onto the active points located on the edges of the imperfections of the graphite surface. There is also a migration process of oxides, metals or salts from the graphite tube wall to the atomization platform, when present, even though samples would be deposited on the tube wall and at temperatures lower than the atomization temperature [Chen and Jackson (1998)].

The useful half life-time of a graphite tube is determined principally by the maximum working temperature, operation time at maximum temperature, analyte type and matrix composition [Welz *et al.* (1989)]. The physical and chemical properties of the analyte are particularly evident with relatively non-volatile elements, some of which can form refractory carbides, particularly when the graphite tube ages and becomes transformed into a porous material. In such a situation, the graphite tube may present memory effects, unreproducible atomization processes and light scattering phenomena produced by the presence of carbon particles (originating on the incandescent graphite surface) and dependent upon the analytical wavelength [Slavin *et al.* (1981); Sturgeon and Chakrabarti (1977)]. While a graphite tube in this condition is probably of no use for further determination of certain elements, it may still be quite satisfactorily used in testing for more volatile elements. Consequently, it is hard to establish a specific half life-time for a given type of tube. The useful half life-time for graphite tubes can be considerably lengthened by giving the tube a thin coating of pyrolytic graphite or using totally pyrolytic graphite tubes [Littlejohn *et al.* (1985)].

Some matrix components also destroy the pyrolytic graphite coating through mechanisms which have not yet been fully explained. For example, $NaNO_3$ and $HClO_4$ which are particularly destructive, have been widely used to digest pyrolytic graphite in chemical analyses of these materials. The chemical reaction through which perchloric acid acts is the following, $HClO_4 + 4\,C \Leftrightarrow 4\,CO + HCl$, and the hydrochloric acid formed causes further corrosion and intercalation. In comparison with the effects of other acids, such as nitric, hydrofluoric, and hydrochloric, the corrosive action increases in the following order: $HClO_4 > HCl \gg HF > HNO_3$. Morphological changes related to the corrosive attack are different for each acid, so that the analytical behavior of the graphite tubes depends on the acid type used (Table 2.1) [Rohr *et al.* (1999)].

Table 2.1. The Interdependence of Corrosion Phenomena and Analytical Performance in ETAAS [Rohr *et al.* (1999)]

Main Detrimental Parameters	Thermal Load	Chemical Load
		• Low temperature interactions • High temperature corrosion
Primary effects of corrosion	Mass loss	Structural changes
Secondary effects	In gas phase: • C-vapor pressure • CO-content • C-particle emission	In solid phase: • Rising C-reactivity • Thinning of PC-layer • Rising liquid penetration • Rising high temperature diffusion
Effects on analytical performance	Loss of analyte atoms: *Loss of sensitivity*: by reaction and transport	Loss of reproducibility: • by rising interaction with carbon • by rising permeability
Analytical indicator and surface morphology	Lifetime investigations combined with morphological studies by SEM	

Other substances, such as chromic acid, aluminium chloride, sulphuric acid, iron and lanthanum salts, also produce exfoliation of pyrolytic graphite [Ortner *et al.* (2002)].

2.1.2. Metals

Metal atomizers (normally built with tungsten, molybdenum, platinum, tantalum, or zirconium [Xiu-Ping *et al.* (1993); Silva *et al.* (1994); Hou *et al.* (2001a, b); Hou and Jones (2002)]) show certain advantages for some analytes in comparison with graphite atomizers: smaller thermal gradients, samples do not foam, lower background signals, and longer half life-times. When tungsten atomizers are used, no metal carbides are formed with the analyte, so higher rates of heating can be attained with the use of less energy. For many volatile elements, sensitivity and the characteristic mass for the metal tubes (particularly tungsten) are about three times better than using graphite tubes [Ortner *et al.* (1986); Shan *et al.* (1992)].

One advantage of a platinum atomizer is the possibility of using air to purge the tube instead of an inert gas [Ohta *et al.* (1992)]. However, the maximum temperature that can be used is 1837 K, which makes difficult determination of less volatile elements. The use of tantalum has normally been associated with the construction of tubes, rather than with the manufacture of platforms. All the same, metal platforms have also been constructed using other elements and/or compounds: molybdenum, tungsten, carbides (W, Mo, Ta, Ti, Zr, La-Zr, V), and nitrides (boron) [Pantano and Sneddon (1989); Pérez-Parajón and Sanz-Medel (1994); Quan Zhe *et al.* (1994); Pineau *et al.* (1992)].

In order to improve their analytical characteristics, modification of the graphite tube surface by the use of high melting point metal carbides was first proposed in 1975 [Ohta and Suzuki (1975)], and a large number of applications were later developed [Imai *et al.* (1998); Volynsky (1998a, b)]. Metals forming carbides are active catalyzers of graphitization, and so can decrease tube surface reactivity. Metal carbides are formed during the graphite tube pretreatment stage and penetrating via the pores in the tube surface [Almeida and Seitz (1986)] dissolve the unordered (uncrystallized) carbon, precipitating it thereafter as small crystals of non-reactive graphite with scarcely any modification in the composition of the graphite. This process occurs very rapidly above 2000 K and, with continuing thermal treatment the graphite crystals precipitates grow slowly. Hence, the increased sensitivity obtained for elements that are difficult to volatilize when a furnace treated with some metal carbide is used could be the result of a much more ordered graphite surface [Sturgeon and Chakrabarti (1977)].

IUPAC has recommended the general term of "graphite tube coated with metal carbide" to refer to this tube type. However, this terminology is correct only in certain very specific situations, as the treatment of the graphite tube at normal atmospheric pressure does not always create solid carbide layers on the graphite surface, so that recently more detailed terminologies have been suggested [Volynsky (1995, 1998a,b)]. One terminological approach, referring to the structure of the carbide layer, recognizes three principal groups.

— *Graphite tubes **coated** with metal carbides*: This tube type is prepared through application of a chemical and physical process of vapor deposition, which permits the formation of a thin solid film of the given metal carbide on the graphite surface. However, the application of this sort of modification procedure requires special equipment and is rarely used.

— *Graphite tubes **impregnated** with tantalum (wolfram, etc.) carbide*: This modified tube implies treating the graphite surface with a solution (or very rarely a slurry) of compounds containing in their structure elements showing a high melting point (such as: Zr, W, Ta), followed by heat conditioning. Such treatments may be carried out at atmospheric pressure or at low pressure.

— *Graphite tubes **modified** with lanthanum compounds*: This tube category covers all those using lanthanum compounds as modifying agents, independently of the treatment used. Unlike other carbides showing high melting points, lanthanum carbides are easily hydrolysable. Thus, they are transformed into the respective hydroxides (with simultaneous formation of a number of hydrocarbides) during contact with aqueous solutions. However, in the atomization stage, the formation of lanthanum carbides is repeated as a result of the interaction of the lanthanum hydroxide with the atomizer graphite.

2.2. Types of Atomizers

The first graphite tube was used by King around the years 1905–1908 [King (1905, 1908)], although this was to undertake qualitative studies of atomic emission spectra. The earliest quantitative analytical work took place much later, when in 1956, two groups led by Mandelshtam and Zaidel, respectively, first used the technique of electrothermal vaporization to separate and preconcentrate trace elements prior to atomic emission analyses [Mandelshtam *et al.* (1956); Zaidel *et al.* (1956)]. Since then, a whole range of electrothermal atomizers have been developed and applied for analytical purposes.

The original designs have undergone continuous and progressive modifications aimed at improving their technical specification. Many of the modifications brought in have not been really very practical advances for routine analytical work, either owing to the complexity involved or because of the increased cost ensuing for the instrument.

Among the principal criteria taken into account in carrying out developments of the atomizers the following have been prominent:

- easily operation of the equipment,
- variable pyrolysis temperatures,
- atomization under gas stop-flow conditions,
- improved shape designs,
- optimum gas entry routes,
- decreased condensation of the sample vapor on the cold ends of the tube,
- improved sensitivity and precision.

2.2.1. L'vov Furnace and Other Historical Atomizers

The first application of the electrothermal atomizers to perform quantitative atomic absorption analysis was carried out by L'vov [L'vov (1959, 1961)]. L'vov's furnace consisted fundamentally of two parts

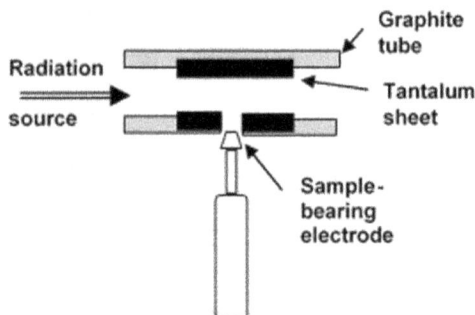

Fig. 2.2.   L'vov furnace [L'vov (1959)].

(Fig. 2.2): an electrode carrying the sample and a graphite tube heated using electrical resistances.

The graphite tube was coated, both inside and outside, with pyrolytic graphite to decrease analyte diffusion. The purge gas used was argon. The sample was introduced into the tube through a hole in the wall of the tube, using the sample-bearing electrode, after the tube had been heated to the atomization temperature during 20 seconds to 30 seconds. The atomizer designed by L'vov in 1959 was based essentially on a principle similar to that described by King in his arc atomizer. However, although this model had extremely high sensitivity, deficiencies in the heating of the sample caused poor reproducibility. For that reason, L'vov developed at a later date (1967) a further version of the graphite tube [L'vov and Lebedev (1967)] in which both the furnace and the sample-bearing electrode were simultaneously heated. In that same year (1967), a copper wire electrothermal atomizer was also designed [Brandenburger and Bader (1967)].

Massmann devised a simplified version of King's furnace for carrying out measurements of atomic absorption and atomic fluorescence [Massmann (1967)]. This atomizer comprised a graphite tube (length, $l = 50$ mm; diameter, $\varnothing = 10$ mm) through which an argon flow circulated (at a rate of 1.5 L/min). The tube was held within a water-cooled metal cylinder. The sample (20 μL) could be introduced into the graphite tube manually (using a jeringe) or automatically. The tube was electrically heated, under a program of temperatures consisting of three stages (drying or elimination of the solvent, char or elimination of the organic material, and atomization or generation of gas-phase analyte atoms). If the temperature rose too high there were losses through volatilization and sample splatter. Nonetheless, Massmann's furnace was rapidly incorporated into commercial instruments.

At almost the same time (in 1968), Woodriff developed a new graphite furnace [Woodriff and Ramelow (1968); Woodriff *et al.* (1968)] to carry out atomic absorption analysis. This was very similar in some of its features to the model devised by L'vov. The Woodriff furnace has been modified and improved on several occasions. One of its versions uses a graphite rod with a recipient, also made of graphite, at one end for introducing and atomizing the sample within the tube. Woodriff's furnace was of some considerable size, making its commercial use difficult.

West and Williams (1969) developed a carbon filament atomizer to undertake atomic absorption and atomic fluorescence analyses. This atomizer consisted essentially of a filament ($l = 20$ mm; $\varnothing = 2$ mm) connected to two stainless steel electrodes, water cooled to prevent them melting during heating. The three key phases in operation include: placing the sample (1–2 μL) on the filament, electrical heating (vaporization/atomization) and absorption measurement. The principal drawback presented by this atomizer was the simultaneous vaporization of both solvent and matrix, causing high background absorption. This problem was partly solved by heating the filament progressively. Nevertheless, there were other minor difficulties relating to the lack of precision and the presence of some chemical interference; an attempt was made to correct this through the use of a protective gas (argon or hydrogen).

The filament atomizer was modified to develop the carbon rod atomizer [Amos *et al.* (1971); Brodie and Matousek (1971)], becoming the Massmann mini-atomizer. This atomizer differs from West's filament atomizer in that the sample injection hole is smaller, and it uses sample volumes of 0.5–1 μL.

The development of filament atomizers went on apace, with graphite being replaced by metals (Pt, W, Ta). The most important model was perhaps the electrically-heated tantalum cup, first described by Donega and Burgess (1970), and then by Matousek and

Stevens (1971). This system is similar in many ways to the carbon filament atomizer. Nevertheless, the temperature is limited to 2500°C to avoid the tantalum melting. The advantage of this system lies in the fact that it avoid carbide formation.

Robinson and Wolcott (1975) constructed an inverted "T" electrothermal carbon furnace, the design of which represents an intermediate stage between the L'vov and Woodriff furnaces. The temperature reached by this system is around 2600°C. The sensitivity attained is similar to that achieved by other furnaces, while precision and reliability are much better.

2.2.2. L'vov Platform Atomizer

The precision obtained with graphite atomizers has been improved upon by the use of L'vov's platform [L'vov *et al.* (1977)] (Fig. 2.3).

In this system, a graphite platform is inserted into a graphite tube. At present, various types of platform exist (Fig. 2.4), although it is still not clear which design is the most adequate. One good selection criterion is whether the platform can achieve a high constant temperature close to that the vapor phase [Frech *et al.* (1989)].

The platform warms up because of radiant heat from inside the furnace which causes its temperature to rise. At high temperatures, the condensed analyte atoms are vaporized and they follow the path of the light beam. At this moment, the background absorption has been at least partly reduced, increasing the method's reliability. The reproducibility improved using the L'vov platform.

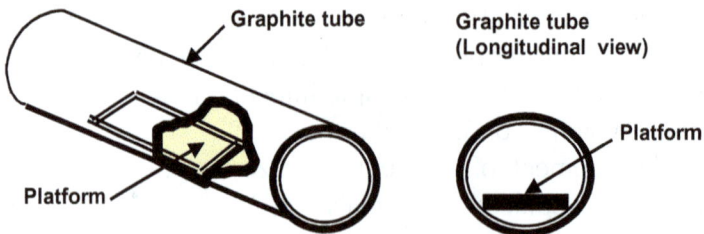

Fig. 2.3.   L'vov platform inside a graphite tube [L'vov *et al.* (1977)].

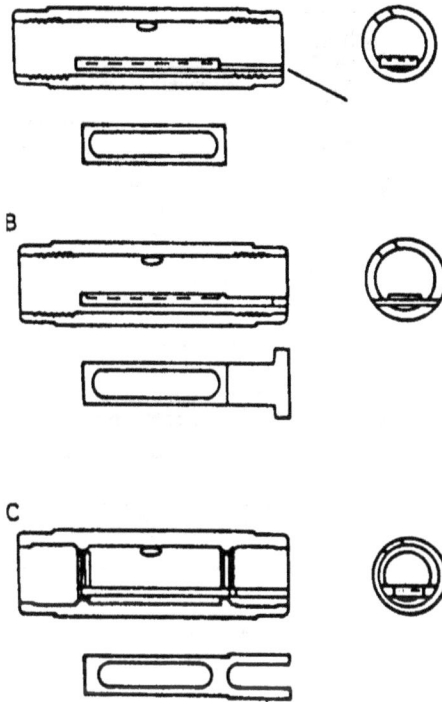

Fig. 2.4. Three types of platform [Frech *et al.* (1989)].

Positioning and alignment of the platform within the graphite tube can be improved through the use of a camera to get images of a transverse section of the tube [Boulo *et al.* (1997)]. Use of these images permits the alignment of: the graphite tube, platform angle relative to the vertical axis, height of the end of an autosampling capillary above the tube wall or the platform, as well as checking of whether all the liquid has been correctly deposited within the tube.

2.2.3. Probe Atomizer

An alternative method of achieving vaporization at higher temperatures in ETAAS is probe atomization [Littlejohn *et al.* (1983, 1984); Carroll *et al.* (1985); Ottaway *et al.* (1986); Gilchrist *et al.* (1992)]. In this procedure, the sample is deposited on a small graphite rod (Fig. 2.5). One end of this rod is introduced through a hole into the tube.

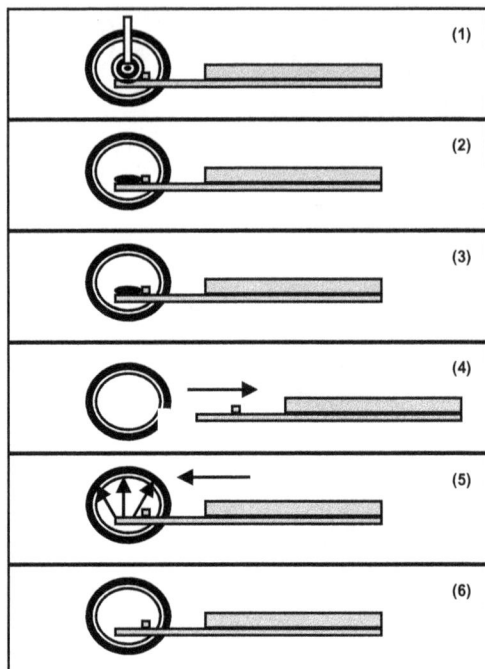

Fig. 2.5. Probe atomizer showing the various stages for analysis. (1) Sample introduction; (2) sample drying and charring; (3) automatic zeroing; (4) removal of the
probe and fixing the atomization temperature; (5) reintroduction of the probe and
analysis; and (6) cleaning of the tube and probe [Littlejohn *et al.* (1983)].

The sequence of the sample drying and charring stages takes place
with both rod and sample inside the tube. Once these stages are complete, the rod and the dried and charred sample are withdrawn from
the tube, which is then reheated until it reaches atomization temperature. When this temperature has stabilized, the probe bearing the dried
sample is reinserted into the atomizer and they are quickly heated up
by radiant heat from the tube walls. In consequence, the sample vaporizes in the hot atmosphere of the tube.

The main advantages of the probe atomization as opposed to the
platform atomization are:

• better control of the temperature met by the analyte atoms as they
  are atomized,

- the atomization surface rapidly heats up to the volatilization temperature of the analyte and volatile matrix, and
- during the heating cycle, there is greater control over the time elapsed from the start of the atomization stage until introduction of the probe (rod), which may aid in obtaining isothermal atomization conditions.

2.2.4. Other Developments

Although there are now a number of varying types of atomizers (Fig. 2.6) developed in different laboratories and inspired by the idea of direct analyses of solids, they are all fundamentally based on the concept of the L'vov tube. Some modifications were signed by the second surface atomizer developed by Holcombe [Rettberg and Holcombe (1984, 1986)], and the two-step atomizer, which can be suitable to carry out double vaporization of the sample. Reduced interferences and better accuracy were derived using this last approach [Grinshtein *et al.* (1999, 2001); Frech *et al.* (2000)]. The ballast furnace constitutes

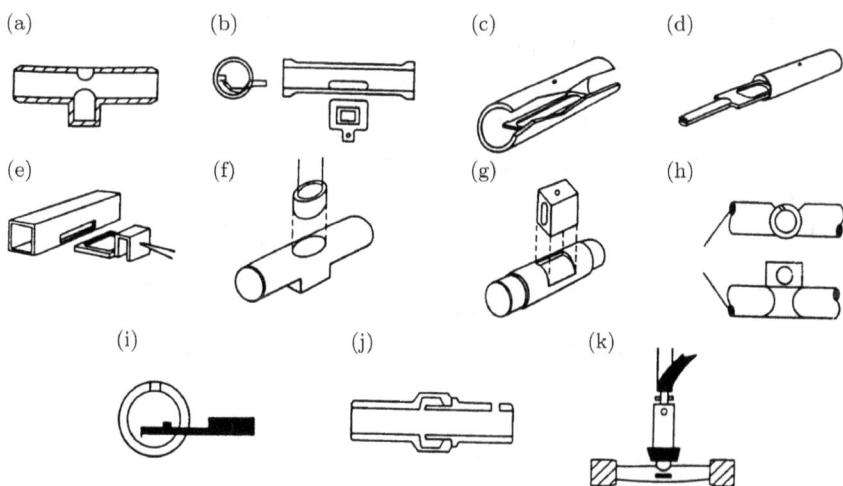

Fig. 2.6. Several atomizers developed: (a) cup atomizer; (b) tube and platform atomizer; (c) L'vov platform atomizer; (d) boat atomizer; (e) microboat atomizer; (f) miniature cup atomizer; (g) cup-in-tube atomizer; (h) carbon rod atomizer; (i) probe atomizer; (j) circular chamber atomizer; and (k) second surface atomizer [Katskov (2005); Katskov *et al.* (2005)].

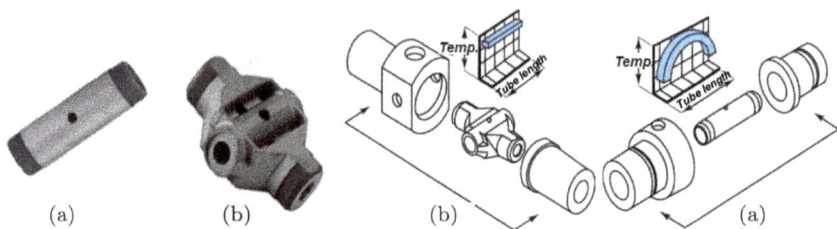

Fig. 2.7. Photograph and representative scheme for the longitudinally (a) and transversally (b) heated graphite tubes [Frech *et al.* (1986)].

another interesting alternative with expected higher and more stable gas temperature than the usual platform [Katskov (2005); Katskov *et al.* (2005)].

Other developments include electrothermal atomizers consisting of a long tube of vitreous carbon packed with activated charcoal. After the sample has been introduced into the atomizer, the temperature is maintained constant (2280°C) and the atomic vapor produced is made to pass through the activated charcoal within the light beam using an argon flow. The advantages of this atomizer are the absence of background absorption and the availability of atomization processes under isothermal conditions [Kitagawa *et al.* (1994)].

Other designs have involved the use of tungsten coil atomizers with multi-channel CCD detectors, and a background correction system. The construction of a portable atomic absorption spectrometer combining this instrumentation and using 12 V batteries, has also been utilized to determine Pb [Sanford *et al.* (1996)] and Cd [Batchelor *et al.* (1998)], attaining absolute detection limits of about 20 pg and 60 pg, respectively. Tungsten coils employed as atomizers have also been combined with diode lasers as radiation sources to develop very powerful, small-size, low-cost atomic absorption spectrometers [Krivan *et al.* (1998)]. Such devices can compete in precision and reliability with conventional ETAAS spectrometers, and can also be used for techniques involving slurry sampling. However, background signal is a limiting factor. This drawback has been reduced by the employment of coils made of tungsten of high purity. Nonetheless, the principal limitation is the lack of availability of diode lasers for a great number of elements.

To increase isothermal atomization conditions, diverse platform shapes have been assayed [Siemer (1982)]: rods, spirals, plates, wire coils, filaments, and the like, with more recent novelties having integrated contact tubes and transverse heating systems (Fig. 2.7) [Frech *et al.* (1986)]. The non-tubular atomizers offer some advantages relative to the tubular furnaces: they are cheap and simple and consume less energy. However, the shortcoming of large thermal gradients in the gaseous phase brings about rapid cooling of the analyte atoms with a consequent likelihood of interference [Ajayi *et al.* (1989)].

Many of the earlier designs can be lumped under the general term of two-component atomizers, which have shown considerable improvements fundamentally in the direct analysis of solid samples [Doçekal (1998)]. Electrothermal atomizers demonstrate great efficiency in the sample vaporization process, but are not so efficient in the atomization process.

Other recently developed atomizers, based on concepts different from those of L'vov's atomizer, include: *filter atomizer*, the graphite *impact furnace*, and the *plasma and glow discharge atomizers*.

2.2.4.1. *Filter atomizer*

The basic idea of the filter atomizer lies in the use of a large surface (Fig. 2.8) to gather fine particles of the sample, with a restricted outlet for the vapor to the analysis zone [Katskov *et al.* (1994, 1995, 1996, 1998)]. The sample is deposited in a cavity located between the tube wall and the filter. When the sample is vaporized, the vapor diffuses through the filter wall into the light path.

Advantages of the graphite filter atomizer: better analysis signals (between 1.6 and 2.8 times stronger), the possibility of introducing larger sample volumes (100 μL), shorter drying time, lower background noise levels, fewer chemical interferences, no need to use chemical modifiers and great potential for direct analysis of organic solutions [Marais *et al.* (2000)]. By contrast, memory effects constitute a considerable problem [Katskov *et al.* (1994)]. This type of diffuser furnace has also been used in atomic fluorescence spectrometry with laser excitation [Gornushkin *et al.* (1996)].

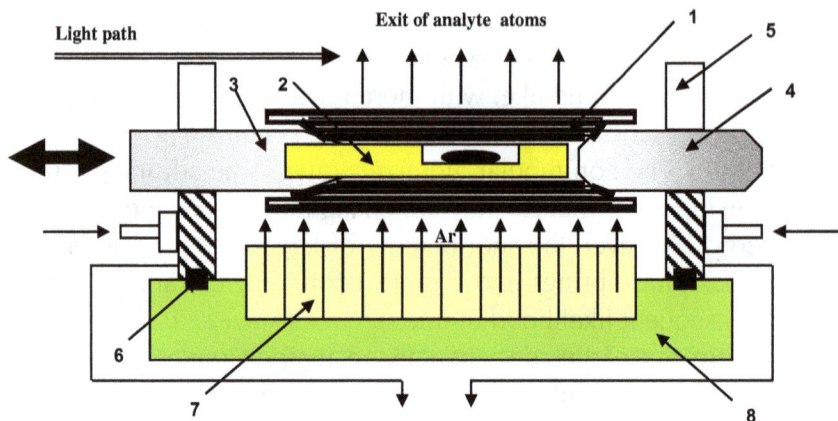

Fig. 2.8. Filter atomizer: (1) porous graphite tube; (2) graphite rod with a sampling delve; (3) and (4) electrodes; (5) steel mounting; (6) insulator; (7) conduit for supplying a laminar flow of argon; (8) stainless steel platform [Katskov *et al.* (1996); Gornushkin *et al.* (1996)].

Another recent atomizer (Fig. 2.9) is the crucible atomizer to carry out fractional evaporation of the sample components and their condensation on a separate cold surface. Then, the condensate is again evaporated/atomized. To reduce interferences, a graphite disk filter is used. Two- and three-chamber electrothermal crucible atomizers with several independently heated zones have also been developed [Oreshkin and Tsizin (2004)].

2.2.4.2. *Impact furnace*

The graphite *impact furnace* (Fig. 2.10) is based on a system for collecting particles from an aerosol.

The movement of particles in an aerosol is governed by Stokes' law

$$S_{tk} = \frac{\rho(d_p)^2\, U\, C_c}{9\,\mu D},\qquad(2.1)$$

where U is the average velocity of the flow leaving the jet (cm/s); μ is the density of the particle (g/cm^3); D is the diameter of the jet or aperture (mm); ρ is the viscosity of air (1.81×10^{-4} g (cm s)$^{-1}$ at 20°C); d_p is the particle size (μm); C_c is Cunningham's correction factor (≈ 1),

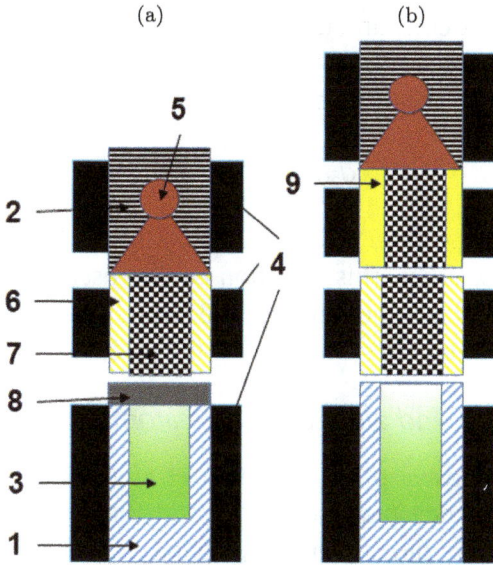

Fig. 2.9. Multifunctional electrothermal graphite crucible atomizers with several independently heated zones; (a) crucible–cylinder–cell with one condensation zone; (b) crucible–cylinder–cell with two condensation zones; (*1*) crucible; (*2*) cell; (*3*) sample concentrate; (*4*) replaceable graphite holders–electrical contacts; (*5*) transmitted hole (analytical filter graphite disk zone), (*6*) first condensation zone; (*7*) microholes; (*8*) graphite disk filter; and (*9*) second condensation zone [Oreshkin and Tsizin (2004)].

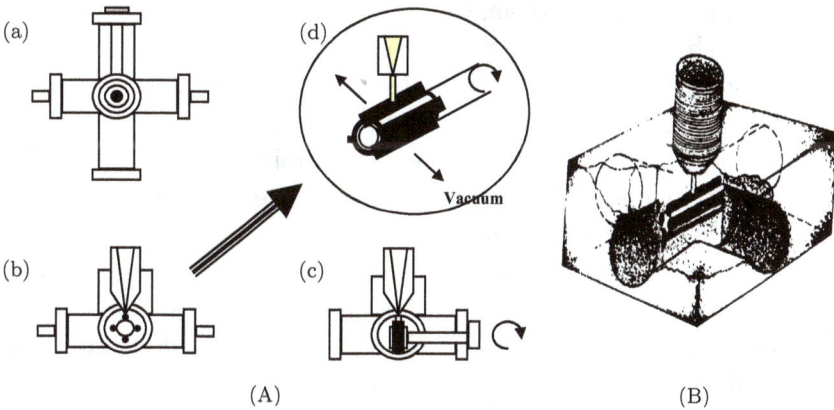

Fig. 2.10. (A) Impact atomizer (a) upper view; (b) right-hand side view; (c) left-hand side view; (d) inside view. (B) Schematic of a graphite impact furnace built in the laboratory [Yong-Ill Lee *et al.* (1996)].

and S_{tk} is Stokes's constant. $(S_{tk})^{1/2}$ is used as a measure of the size of a non-dimensioned particle.

Taking $(S_{tk})^{1/2}$ as equal to 0.475, it is possible to extract

$$d_{50} = \frac{\sqrt{9 S_{tk} \mu D}}{\rho U C_c},$$

(2.2)

where d_{50} is the particle size for which at least 50% of the particles of that size are caught. This equation can be used to obtain theoretical information about the particle size, which can be trapped in the graphite impact furnace system [Yong-III Lee *et al.* (1996)].

2.2.4.3. *Plasma and glow discharge atomizers*

The use of microwave plasmas [Duan *et al.* (1993)], capacitor coupled radio frequency plasmas [Liang and Blades (1988); Chan and Chan (1998)] and direct current plasmas [Oikari *et al.* (2001)] as atomization systems have also been evaluated for ETAAS. Glow discharge atomizers (GDAs) have recently aroused considerable interest for atomic absorption spectrometry (GDAAS). In one of the designs, a thin-walled hollow metal cathode aligned with the optical axis is used (Fig. 2.11a). With this geometry, the atoms discharged, (once they are in the gaseous phase) immediately generate the analytical signal. Thus, thanks to the elimination of transport losses typical of a GDA with a flat geometry, the sensitivity of analysis is increased. Atomization with this system can be carried out in pulse or continuous mode [Ganeyev and Sholupov (1998)]. Zeeman atomic absorption spectrometry using high frequency modulated light polarization is the most suitable technique for work in combination with this type of atomizer, as it provides relatively high frequencies (30–100 kHz). The relative distribution of atoms in the graphite tube for different temperatures is shown in Fig. 2.11b.

It has been demonstrated that the pressure and velocity of the gas flow of discharge, as also the radio frequency power applied and the extent of negative luminescence, have an important effect on the atomization and distribution of the atoms sparked within the atomizer [Parker and Marcus (1994)]. Other parameters, related to the geometry of the sampling hole, also influence absorbance and thus sensitivity. In this way, the larger the sampling hole, the greater the

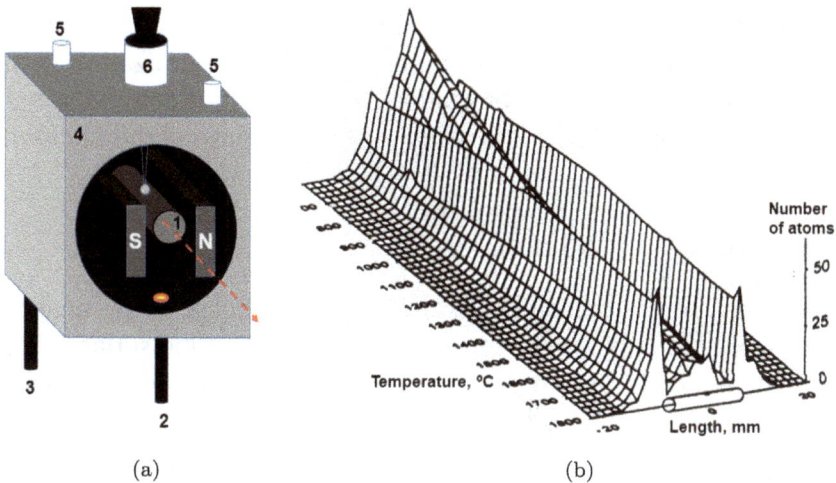

Fig. 2.11. Glow discharge atomizer. (a): (1) TMHC (thin metal hollow cathode); (2) electrode with negative potential; (3) electrode with positive potential; (4) discharge tube; (5) gas flow; (6) microsample; (7) sample introduction port. (b) Spatial atom distribution along the optical axis calculated for different temperatures of the TMHC [Ganeyev and Sholupov (1998)].

absorbance, while on the other hand, the disc thickness of the sampling hole presents a negative effect due to the fact that the distance from the atomizer region investigated to the cathode surface grows with increasing thickness of the disc sampling hole [Absalan *et al.* (1998)].

2.3. Vaporization/Atomization Modes

The atomization process implies a previous and, in many cases, a simultaneous vaporisation stage which is linked to the mechanism for production of the gas phase analyte atoms. For this reason, it would be more appropriate to use the joint term vaporization/atomization in referring to the obtaining of the analyte atoms. The vaporization/atomization process takes place from the surface of the graphite tube in those atomizers consisting of a single tube. In contrast, in atomizers incorporating a platform, filament, rod, or similar component, the vaporization/atomization process is from the platform surface. Consequently, the vaporization/atomization processes can be assigned to one of the two categories or modes of atomization: wall atomization and platform atomization. The atomization process occurring with a probe atomizer is really a particular case of the platform atomization.

2.3.1. Wall Atomization

The name "wall atomization" is given to the process of the analyte transformation into atomic vapor directly from the tube surface (any platform is absent). The vaporization/atomization processes take place on and from the tube wall.

The principal problem found with this type of atomization is the presence of a thermal gradient during the atomization process. In this case, the atomic absorption signal appears during the heating, making it more difficult to interpret, or to give a theoretical description of the process. Interest in developing atomization systems working at a controlled or stabilized temperature has led to the design and use of platforms.

2.3.2. Platform Atomization

Attempts have been made to improve the atomization of volatile elements through the sample vaporization into an environment with a temperature higher than what could be hoped for under wall atomization. The approach used for this purpose is the platform atomization method. In this system, a pyrolytic graphite platform (L'vov's platform) is introduced into the atomization tube, and on this is placed the sample. During the heating of the atomizer, the platform warms up at a much lower rate than the rest of the tube, so that when vaporization/atomization takes place from this platform, the analyte and the matrix components volatilize into a gas whose temperature is greater than could have been achieved with wall atomization.

The relative volatilization from the tube wall and from the platform is shown in Fig. 2.12. The signal produced with the L'vov platform is bigger, while the delay in its appearance contributes to cutting down the background signals found in wall atomization.

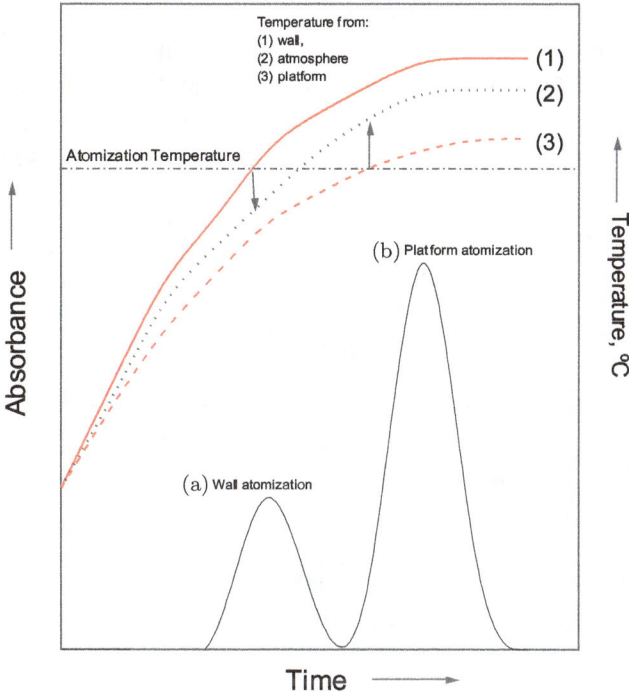

Fig. 2.12. Atomic absorption signal produced after the atomization process from the tube wall (a) and from a platform (b).

Heat is transferred to the platform almost exclusively by radiation, and the change in the platform temperature over time is shown by the following equation [Chakrabarti *et al.* (1984)]

$$\frac{dT_p}{dt} = \frac{\xi k S (T_w^4 - T_p^4)}{CQV}, \qquad (2.3)$$

where the parameters ξ, C, Q, S and V refer to the platform, with ξ being its emissivity, C its specific heat, Q its density, S its surface area, and V its volume, while k is Boltzmann's constant, and T_w and T_p are the temperatures of the tube wall and of the platform, respectively.

Variations of the parameters in Eq. (2.3) allows the regulation of the time at which the platform is subjected to the atomization temperature. However, only some of the parameters in the equation are of real relevance. Made-material selection is normally restricted to the various forms of graphite and a few refractory metals (tantalum, tungsten, and so on); other alternative materials are subjected to chemical and/or thermal decomposition. For any given material, the only way of lengthening the platform temperature delay is to increase the thickness of the platform.

The use of longer platforms of the same thickness is not of much use, as the overall surface area grows almost proportionately to the mass or volume, and the temperature delay does not increase greatly. Other disadvantages of longer platforms are:

- greater spread of the sample within the tube,
- variable heating rates due to the temperature gradient present in all graphite tubes, but particularly in these longer forms, and
- lower angular aperture.

A similar delay can be obtained by utilizing platforms with reduced edge contact points, but this increases the time needed for platform cooling between successive sample injections. On the other hand, some advantages may be gained by using two platforms set one above the other. The temperature delay achieved in this case is almost the same as when a single platform of double the mass is used. The two platforms utilized touch the tube only at their edges, so that the internal wall of the tube heats both sides of the two platforms equally; in consequence, there is no heat transference between them, since the temperature on each platform is the same.

2.3.2.1. *Probe atomization*

The temperature distribution (heat transfer) on the surface of the graphite probe as a function of time is affected by two main factors [Chakrabarti *et al.* (1996)]: (i) radiation, and (ii) thermal conduction. The graphite probe is heated firstly by radiation proceeding from the hot wall of the graphite tube. The heat transfer rate ($Q_{rad,w-p}$) by radiation from the tube wall to the surface of the graphite probe depends fundamentally upon the temperature difference between the tube wall and the probe surface, according to the following equation

$$Q_{rad,w-p} = \sigma A_p F_{w-p}(T_w^4 - T_p^4), \qquad (2.4)$$

where σ is the Stefan–Boltzmann constant, A_p, the total surface of the probe (of the inner part of the graphite tube), F_{w-p}, the total radiation interchange factor between the probe and the graphite tube, and T_w and T_p, the temperatures of the tube wall and the probe surface, respectively.

The heat transfer rate through thermal conduction along the probe to the part outside the graphite tube lowers the temperature on the probe's surface. Heat transfer rate by thermal conduction along the length of the probe (in direction x), $Q_{cond,p-p}$ is given by the following equation

$$Q_{cond,p-p} = -\frac{\partial}{\partial x}\left[w(x)b(x)k_p(x,t)\frac{\partial T_p(x,t)}{\partial x}\right]dx, \qquad (2.5)$$

where $w(x)$, $b(x)$, k_p and T_p represent the width, height, thermal conductivity and temperature of the graphite probe, respectively, in direction x. The heat by thermal conduction turns out to be related to the temperature gradient along the probe from its inner to its outer end. However, the temperature gradient of the part of the graphite probe inside the graphite tube is not large, especially in the sample deposition zone. The rate at which the probe heats up is determined principally by the final temperature of the tube wall. A rise in this temperature causes an increase both in the heating rate of the probe surface in the initial stage and in the temperature of the steady state of the probe surface in the final stage, as well as a decrease in the time necessary to attain the final temperature. Unlike a graphite platform, the final steady state temperature for a graphite probe is not equal to the ultimate temperature of the tube wall, because thermal conduction relating to the part of the probe which always remains outside the graphite tube eliminates heat through a conductive mechanism [Chakrabarti *et al.* (1996)].

2.4. Experimental Parameters

A number of experimental parameters from the atomizer (atomization system, purge gases, temperature program, type of sample) can show some influence over the vaporization/atomization process.

2.4.1. Gases

Owing to considerations such as cost, diffusion loss rate, heat conductivity and reactivity, the gas normally used in controlling the furnace atmosphere is argon. Nitrogen and helium decrease the maximum temperature that can be reached in the furnace as a consequence of their diffusivity and thermal conductivity. Air and pure oxygen are used during the char stage (and during the cleaning phase). On some occasions, mostly when some interferences need to be controlled, other gases such as H_2, CO_2, CH_4 and so on may be added.

During the atomization stage, the gas flow passing through the graphite tube usually reduces to zero (zero flow or stopped gas flow conditions). In this way, elimination of the atomic vapor takes place principally through a diffusion mechanism, and consequently, the time the atomic vapor remains in the absorption cell increases. In contrast, if the gas flow during the atomization stage is not set at zero, elimination of the atomic vapor also occurs through a convection mechanism, decreasing the analytical sensitivity.

Atomization at low pressure reduces the atom residence time in the atomizer, being possible in this way to separate the processes of generation and dissipation of the analyte atoms. This operation mode yields very sharp peaks and, in consequence, allows events to be picked out even if they change rapidly or are very close together in time [Hassell *et al.* (1988); Wang and Holcombe (1992, 1994); Holcombe and Wang (1993)]. Atomization at reduced pressure allows us to work with isotope resolution, because the broadening of analyte lines is mainly owing to the Doppler's effect. However, further reductions of the line width can be attained by using Doppler-free techniques [Wizemann and Haas (2003); Maniaci and Tong (2004)].

On the other hand, use of high pressure — where dissipation is controlled by diffusion, the analyte's chemical interactions are not altered and convection is not an important analyte loss mechanism — gives lower diffusion coefficients, longer analyte residence times and higher integrated absorbance. Nevertheless, at times, reduced sensitivity for some elements is found at high pressures. This is caused by the broadening and shift of the absorption profile due to an increase in the line width through collision. The peak profile center position is shifted towards longer wavelengths in argon and nitrogen atmospheres, and towards shorter wavelengths in hydrogen and helium. As a result, it would appear that the only way of measuring absorbance effectively at high pressures is by using a continuous radiation source. This is because with a broad band spectrum, the deconvoluted absorption profile on the focal plane is hardly affected at all by changes in pressure [Smith and Harnly (1995)].

2.4.1.1. *The atmosphere temperature*

Knowledge about temperature of the atmosphere inside the graphite tube is essential in order to predict the characteristics of the atomic absorption signal. In particular, it is necessary to know the effect produced by temperature on:

- the atomization mechanism,
- interferences from the matrix components, and
- transport processes in the gaseous phase.

The temperature-dependent chemical reactions include dissociation of the analyte molecules, ionization of atoms and reactions between the analyte species and species from the matrix components, either into the gaseous phase or in the condensed phase.

The thermal gradient existing in the gaseous phase within the atomization cell depends on the temperature distribution along the graphite tube wall, the purge gas characteristics, the presence/absence of any convection flow of the purge gas, matrix composition, heating time, and whether a platform is used [Koirtyohann *et al.* (1984); Falk *et al.* (1985); Welz *et al.* (1988)]. The temperature gradient along the graphite tube's wall is dependent on the geometry and design of the tube, the heating program applied, and the material used to make the tube [Chakrabarti *et al.* (1984, 1985)]. The transversely heated graphite tubes yield atomization systems with narrower temperature gradients, being virtually isothermal atomizers. The platform is the main cause of thermal gradients into the tube [Sperling *et al.* (1996)]. Emitted radiation is directly related to the atomizer temperature and radiation from the tube wall and the platform can provide useful information about the dynamics of the graphite heating [Somov and Gilmutdinov (1997)]. The geometry of the graphite tube is an important factor in determining the magnitude of the temperature distribution.

For the graphite tubes that operate under stop-flow gas conditions, the vapor temperature, T, is closely related to the tube wall temperature, T_w. It is admitted that $T_w \approx T + 73$ K. Differences in the vapor temperature between the middle and ends of the tube can reach values as high as 1200 K with tube heating about 2700 K. Graphite tubes can be considered as blackbody radiators when the wall temperature is constant and the tube length is more than three times the inner diameter. In this case, the wall temperature can be measured using an optical pyrometer, whose error is less than 1% [Falk (1984)].

The vapor temperature (T) (also termed as any of the following ways: excitation temperature, gaseous phase temperature, furnace atmosphere temperature, effective gaseous phase temperature, gas temperature and effective vapor temperature), may be determined using the method of the two wavelengths (λ_1, λ_2), as described by Siemer (1982), taking the following equation

$$T = \frac{E_i - E_g}{\mathrm{Ln}\left(\frac{BA_{\lambda_1}}{A_{\lambda_2}}\right)} \tag{2.6}$$

where A_{λ_1} and A_{λ_2} represent atomic absorbance at two wavelengths (λ_1, λ_2), E_g represents the ground level energy; E_i is the energy of the excited state; and the constant B is a correction factor (Table 2.2) linked to the statistical weight (g), the oscillator strength (f) and the analytical wavelength.

Table 2.2. Wavelength Couples and Correction Factor (*B*) for Some Elements (See Eq. (2.6))

Element	Wavelengths, nm	*B*
In	410.5/451.1	6.84
Ni	323.3/356.7	4.29
Pb	368.3/280.2	1.89
Pd	276.3/340.5	18.10
Se	196.0/204.0	0.30
Sn	224.6/284.0	2.50

A consequence of the fact that all the analyte atoms, distributed throughout the tube, contribute to the total absorbance signal measured, is that the vapor temperature deduced by this method does not correspond exactly with the spatial distribution of temperature within the tube [Terui *et al.* (1991)].

2.4.2. Temperature Program

Although the atomic absorption instrumentation offers the possibility to operate using different heating stages, normally the experimental temperature program usually consists of four stages (Fig. 2.13): drying, pyrolysis, atomization and cleaning (Table 2.3). The temperature, heating rate and duration of each stage can be adjusted at will over a wide range of values, so that before each analysis, it is necessary to optimize these instrumental parameters [Halls (1995)]. Use of fast heating programs allows short analytical cycles (up to 40%) [Hoenig and Cilissen (1993); Granadillo *et al.* (1994)].

2.4.2.1. *Drying*

The drying stage is introduced so as to eliminate the solvent from the sample deposited. The temperature at this stage is usually some 100°C to 150°C when the solvent is water. As a general rule, the drying temperature is normally set a little higher than the boiling point of the solvent in use.

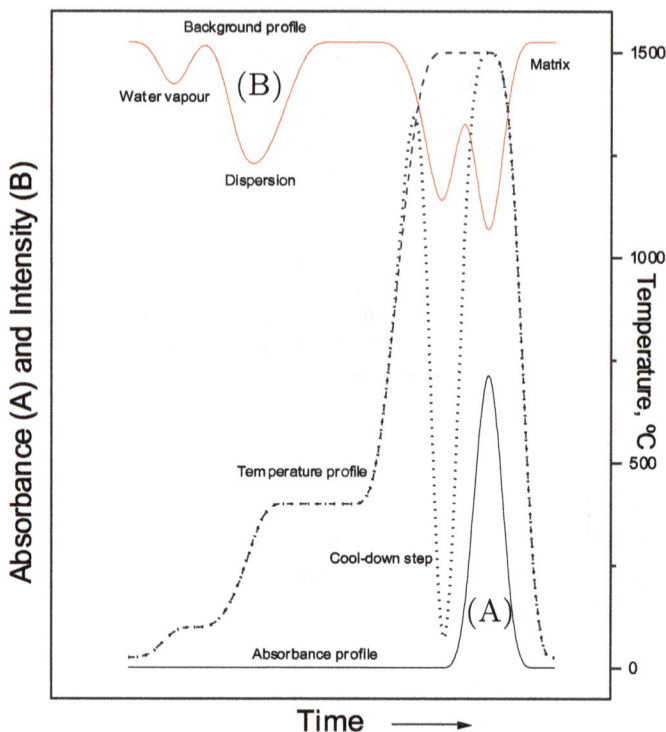

Fig. 2.13. Temperature program for an electrothermal atomizer (background and analyte absorption signals are also included).

2.4.2.2. *Pyrolysis*

The aim of the pyrolysis stage is to eliminate most of the more volatile matrix components (organic matter and volatile elements), while attempting to avoid analyte losses. The pyrolysis temperature usually varies between 350°C and 1200°C. The pyrolysis time and temperature used depend on the matrix and analyte type, with different heating rates being possible. The incorporation of a low temperature stage (*cool-down step*) (Fig. 2.13), after rapid heating in the pyrolysis stage and prior to atomization, creates a wide isothermal zone within the graphite tube.

Pyrolysis with air or oxygen is sometimes necessary for the analysis of certain biological materials that leave carbon residues after atomization. Pyrolysis with air or oxygen extends in time the heating

Table 2.3. Temperature Program (T, °C) and the Ramp (t_{Tc}, s) and Hold (t_{Tm}, s) Times at each Temperature for some Analytes

Analyte	Dry			Pirolysis I			Pirolysis II			Atomization			Cleaning	
	T	t_{Tc}	t_{Tm}	T	t_{Tc}	t_{Tm}	T	t_{Tc}	t_{Tm}	T	t_{Tc}	t_{Tm}	T	t_{Tm}
Ag	150	2	0	300	10	0	400	10	0	1400	0	4	2000	0
Al	150	2	0	900	20	0	1100	10	25	2400	0	4	2500	0
As	150	2	0	375	15	0	500	15	0	2300	0	4	2500	0
Au	150	2	0	450	20	0	600	20	0	2200	0	4	2400	0
Ba	150	2	0	900	25	0	1100	10	15	2500	0	1	2500	3
Be	150	2	0	900	20	0	1000	10	10	2400	0	4	2500	0
Bi	150	2	0	300	20	0	400	10	15	2000	0	4	2200	0
Ca	150	2	0	350	15	0	525	15	0	2400	0	4	2500	0
Cd	150	2	0	200	10	0	250	10	0	1400	0	4	2200	3
Co	150	2	0	500	20	0	900	20	0	2300	0	4	2500	0
Cr	150	2	0	900	15	0	1000	15	0	2300	0	4	2500	2
Cu	150	2	0	550	15	0	750	15	0	2200	0	4	2400	0
Fe	150	2	0	650	15	0	900	15	0	2300	0	4	2500	0
Ga	150	2	0	650	15	0	900	15	0	2400	0	4	2500	0
Ge	150	2	0	650	15	0	900	15	0	2300	0	4	2500	0
Hg	150	2	0	175	10	0	200	10	0	750	1	3	1600	2
In	150	2	0	275	10	0	600	20	0	2300	0	4	2500	0
La	150	2	0	950	10	0	1000	10	10	2550	0	1	2550	3
Mg	150	2	0	750	20	0	1000	20	0	2200	0	4	2400	0
Mn	150	2	0	400	20	0	600	20	0	2200	0	4	2400	0
Mo	150	2	0	1000	15	0	1200	20	0	2550	0	1	2550	3
Nb	150	2	0	750	20	0	800	10	10	2200	0	4	2400	0
Ni	150	2	0	475	15	0	600	15	0	2200	0	4	2400	0
Os	150	2	0	450	10	0	550	10	0	2550	0	1	2550	3
Pd	150	2	0	750	20	0	1000	20	0	2500	0	4	2550	0
Pt	150	2	0	750	20	0	1000	20	0	2500	0	4	2550	0
Rb	150	2	0	750	20	0	1000	20	0	2500	0	4	2550	0
Sb	150	2	0	325	15	0	400	15	0	2050	0	4	2250	0
Se	150	2	0	400	20	0	500	20	0	2300	0	4	2500	0
Si	150	2	0	800	15	0	1100	20	0	2550	0	1	2550	3
Sn	150	2	0	475	15	0	600	15	0	2300	0	4	2400	0
Sr	150	2	0	750	15	0	1000	15	0	2450	0	4	2500	0
Ti	150	2	0	1000	15	0	1200	10	10	2550	0	1	2550	3
Tl	150	2	0	275	10	0	325	15	0	1600	0	4	2200	3
V	150	2	0	500	15	0	750	15	0	2550	0	1	2550	3
Zn	150	2	0	325	15	0	425	15	0	1400	0	4	2200	3

program and may seriously shorten the tube life-time where a very high temperature is used.

An alternative approach, which skips the pyrolysis step and injects the slurry into a pre-heated furnace, is also possible, decreasing the analysis time and improving the sample throughput [Halls (1984); Hoenig and Van Hoeyweghen (1986); Hoenig and Cilissen (1993)]. For both situations, accuracy and precision for various analytes are comparable whenever peak area measurements are used. Fast furnace programs, in which the drying and pyrolysis steps are joined or even skipped, have proved to be very useful in many cases [Halls (1995); López-García *et al.* (1996); Feo *et al.* (2003); Filgueiras *et al.* (2004).]. Occasionally, however, high backgrounds could remain, but only if the D_2 lamp correction system is used. The optimum pyrolysis temperature can be found from the pyrolysis graph (Fig. 2.14a).

2.4.2.3. *Atomization*

In the atomization stage, the analyte compounds formed during the pyrolysis stage are decomposed, yielding the free analyte atoms. The temperatures used during the atomization stage usually vary over a

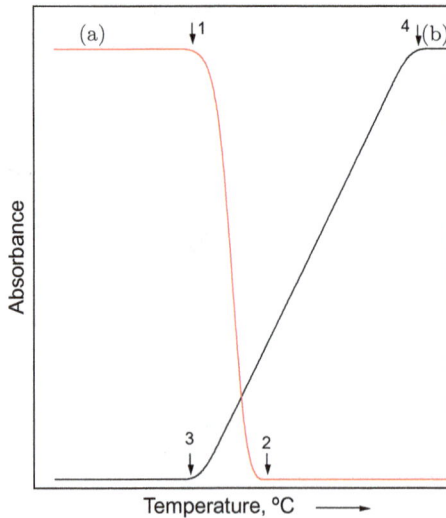

Fig. 2.14. (a) Pyrolysis, and (b) atomization curves. Better (1, 4) and poor (2, 3) pyrolysis (1, 2) and atomization (3, 4) temperatures.

wide range (800°C to 3000°C), depending on the analyte characteristics (volatile or non-volatile), matrix composition, and whether chemical modifiers are employed. The furnace temperature always shows a changing profile over the whole heating cycle, except in virtually isothermal atomizers such as those organized to use integrated contact atomizers. The optimum atomization temperature can be found from the relevant atomization graph (Fig. 2.14b).

2.4.2.4. *Cleaning*

This stage is included to eliminate any sample residue that has not been volatilized at the previous stages from the graphite tube. The temperatures used generally vary between some 200°C and 300°C higher than the atomization temperature. Finally, the temperature is reduced to room temperature or at least to the drying temperature, also activating water and gas flows.

2.4.3. **Parameters Related to the Type of Atomizer and Sample**

The design of the graphite tube may show a decisive influence on the temperature distribution of both the graphite tube atmosphere and the graphite tube wall. However, very often analytical chemists are not able to do much with regard to this parameter, since tube design is fixed by the manufacturer. Moreover, the analytical signal logically depends on the sort of atomization carried out (wall, platform and probe), as the different atomization modes alter the thermal conditions of the analyte vaporization/atomization. Other parameters that may influence the analytical signals include:

(i) history and age of the graphite tube (number of firing cycles), because the graphite crystallinity changes with use principally as a consequence of the appearance of crystal defects, cracks, and so forth;

(ii) material used in the construction of the atomizer, whether it is a graphite tube (non-pyrolytic graphite, pyrolytic graphite) or a metal tube; and of course

(iii) composition of the sample matrix, since its components change the atomization mechanisms of the analyte, owing to different chemical processes.

2.5. Associated ET Atomizers

The high efficiency provided by ET vaporization for generating aerosols has been successfully used for many years in combination with various spectroscopic techniques, especially ETAAS, as has been previously noted. There are several couplings between ET vaporization and molecular spectrometries, such as infrared emission spectrometry [Busch and Busch (1991); Tilotta *et al.* (1991)]. However, atomic emission spectrometry (AES) has benefitted less from this coupling, although there are several symbiotic attempts giving rise to ET atomic emission spectrometry (ETAES).

Thus, in order to improve the ET atomization efficiency, these vaporizer systems have been used in combination with other sources of atomization/excitation, such as plasmas, primarily inductively-coupled plasma (ICP), employed as a sample introduction system, leading to the technique ET-ICP-AES. Nevertheless, combined use of ET vaporization and plasmas has also taken other alternative forms, in which plasmas are generated inside the ET atomizer. Among these couplings the following may be fundamentally noted: FANES [**F**urnace **A**tomization **N**on-thermal **E**xcitation **S**pectrometry] and FAPES [**F**urnace **A**tomization **P**lasma **E**mission **S**pectrometry]. Moreover, replacement of the photoelectric detector (or detectors), photomultiplier tubes, with other alternative detectors such as charge transfer device (CTD) and mass spectrometry (MS), has allowed the development of other techniques such as ET-AES-CTD, ET-MS, and ET-ICP-MS.

2.5.1. Electrothermal Atomic Emission Spectrometry (ETAES): *Thermal Excitation*

Although the use of electrically heated graphite tube atomizers began as far back as the 19th century [Liveing and Dewar (1882)], it was

Table 2.4. Overview of the Historical Development of ET Instrumentation

Development	Year
Introduction of commercial instrumentation into ETAAS	1970
First publication on ETAES	1975
First analytical application of ETAES	1975
Wavelength modulation for background correction in ETAES	1976
Platform atomization for ETAAS	1977
Probe atomization for ETAAS	1978
Application of platform atomization in ETAES	1980
Application of probe atomization in ETAES	1982

primarily King (1905, 1908) with his studies on emission lines, even if they were not intended for analytical purposes, which encouraged the use of such atomizers. Further, the work done by Ottaway and Shaw (1975) was the first to incorporate conventional ETAAS instrumentation to make analytical measurements by atomic emission spectrometry (Table 2.4).

In the ETAES technique, atoms in the vapor phase existing in the graphite tube are excited (thermally or otherwise), and then the excess of energy is emitted. The energy emitted reaches the monochromator (or polychromator), where the photons are separated as a function of their associated wavelength, and the number of photons (energy) corresponding to each wavelength determined and recorded. The relationship between the energy emitted and the quantity of analyte deposited on the atomizer surface permits quantitative analytical measurements [Hui-Ming and Yao-Han (1984)].

Conventional atomic absorption spectrometers were occasionally replaced with spectrometers modified to be able to modulate the wavelength by insertion of a vibrating refractory quartz sheet or a rotating quartz chopper placed immediately after the entry slit or before the exit slit. Nonetheless, some limitations were the following:

- the peak corresponding to the maximum emitted radiation would appear immediately after the maximum atom concentration in the graphite tube (Fig. 2.15) [Bezur *et al.* (1983)],
- the graphite tube presented a pronounced temperature gradient,

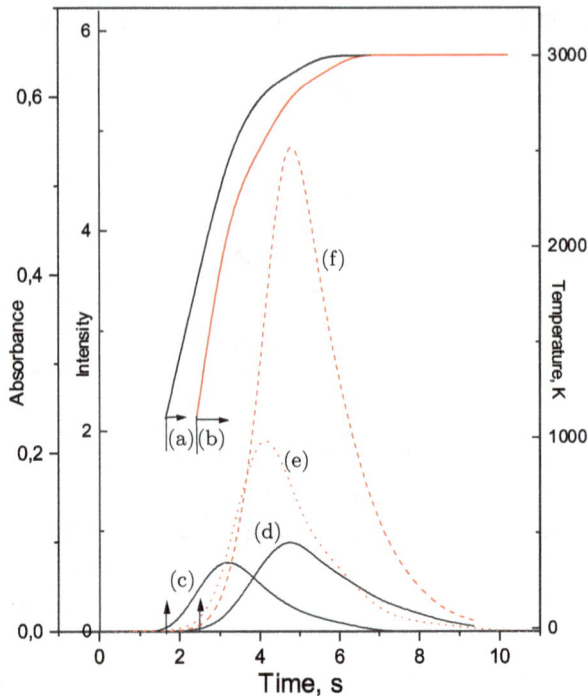

Fig. 2.15. Atomic absorption and atomic emission profiles originated from 50 μL gallium solutions at concentrations of 50 μg/L. (a) Temperature profile of the graphite tube. (b) Temperature profile of the platform. The other lines represent atomic absorption (c) and (e) and atomic emission (d) and (f) signals using wall (c) and (d) and platform (e) and (f) atomization [Bezur *et al.* (1983)].

- at high temperatures, the graphite tube itself contributed to the signal observed in the detector (Fig. 2.16).

The first two difficulties noted above were directly related to the use of Massmann type graphite tubes. To solve these problems and to optimize the excitation conditions and to achieve volatilization of the analyte within a higher temperature environment, the geometry of the graphite tubes routinely used in ETAAS was altered (Table 2.5) [Littlejohn (1988); Littlejohn and Ottaway (1979)].

With these modifications, it was possible to improve the heating characteristics, providing better detection limits for ETAES (and also for ETAAS). However, this gain is not really a generalized advance

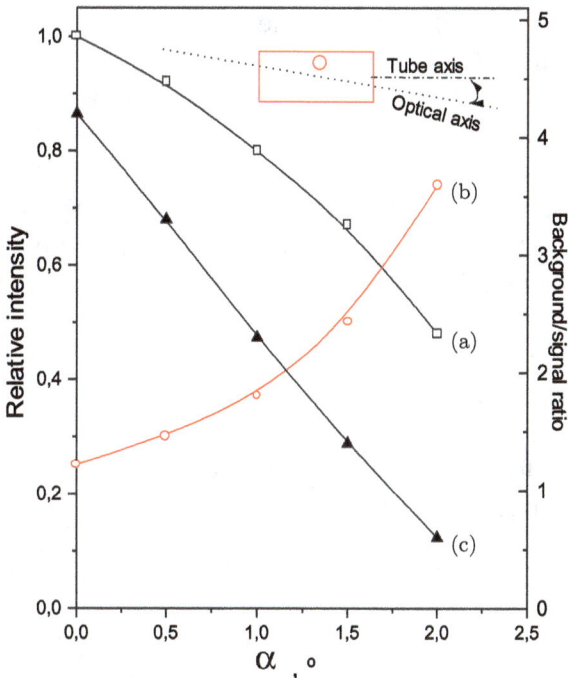

Fig. 2.16. Effect of the lateral alignment of the graphite tube on the atomic emission signal for manganese (a), the background signal measured at 403.08 nm (b), and the signal-to-background ratio (c) [Bezur *et al.* (1983)].

Table 2.5. Developments in ET Atomizers Intended to Improve Excitation Conditions [Littlejohn (1988)]

1. New designs of tubes for wall atomization
2. Use of platform atomization
3. Use of probe atomization
4. Two-stage constant-temperature atomizer
5. FANES atomizer
6. FAPES atomizer

for ETAES, because any particular design of the graphite tube will not necessarily offer optimum conditions for all elements (or even for a large number of them), and so will degrade the best analytical feature of ETAES (multi-element detection). As consequence, it was necessary to separate the volatilization from the atomization and excitation processes of the analyte.

The earliest heating systems applied a constant voltage through the graphite tube with the aim of controlling both the final temperature and the heating rate of the graphite tube. Later on, an infrared photodiode was used to regulate the radiation emitted by the hot tube, with a strong current being passed through the tube until the ideal temperature was reached. Hence, the current was regulated by means of a feedback circuit with temperature coupling through the photodiode. The utilization of controlled temperature heating enhanced the detection limits for ETAES even using measurements without background correction.

Platform atomization also offers advantages for ETAES: delay is noted in the appearance of the signal because:

- during the evolution of the signal pulse the temperature is almost constant,
- when the analyte is vaporized it reaches a gaseous phase with a higher temperature thermodynamically favoring formation of free atoms, and
- more effective excitation and higher emission intensities (better detection limits) are obtained for ETAES with higher temperatures for the vapor phase, in accordance with Boltzmann's distribution law.

Despite these advantages, the use of platform atomization in ETAES suffers the limitation that the delay in getting the analyte into the gaseous phase is strictly related to the heating up rate and cannot be controlled independently of the heating of the tube.

An increase in sensitivity is also attained for ETAES probe atomization as a result of the exposure of the analyte atoms to a higher excitation temperature, when the heating rate of the probe is greater than that of the platform or of the tube wall in the same atomizer [Carroll *et al.* (1992)]. It should be kept in mind that when the sample introduction is under isothermal conditions, the emission signal becomes independent of the furnace heating rate, depending exclusively on the final temperature reached [Marshall *et al.* (1983)].

Even though the platform and probe atomizers might be considered as quasi-constant temperature atomizers, the constant temperature atomizer most widely used in ETAES is the two-stage atomizer. This atomizer consists of a graphite cup in which the sample is placed, adjusted within a hole in the lower part of the graphite tube. The tube and cup are heated separately. The graphite tube is kept at a previously selected equilibrium temperature and the cup is heated up later at an appropriate rate. In this way, the sample is vaporized in the cup and the analyte vapor passes into the graphite tube, where it is atomized and excited. The various developments in the design of two-phase atomizers have given rise to integral contact tubes (laterally heated) similar to those developed for ETAAS [Frech *et al.* (1986); Lundberg *et al.* (1988)].

2.5.2. Coupling ET Vaporization and Plasma Atomic Emission Spectrometry

Electrothermal vaporization/atomization is normally used in ETAAS and ETAES for the production of aerosols, but it can also be used in combination with inductively coupled plasma atomic emission spectrometry (ICP-AES). The combination is achieved by using the ET atomizer as a substitute for, or combined with, pneumatic nebulization, usually employed to produce and transport the atomic vapors from the sample in routine analysis by ICP-AES. Volatilization of the sample is carried out by heating the support bearing the sample within the graphite tube. These supports come in a variety of shapes: filaments, capsules, wires, rods, and tubes made of graphite or some refractory element. Heating may be achieved by following a thermal program similar to those used in ETAAS. The aerosol (liquid or solid particles) produced during the vaporization stage is carried out to the plasma by a transporter gas (normally argon) (Fig. 2.17A) [Fonseca *et al.* (1997)].

Many applications have been made by combining different ET systems with commercially available ICP-AES instrumentation [Luge *et al.* (1995); Katschthaler *et al.* (1995); Moens *et al.* (1995); Záray and Kántor (1995); Golloch *et al.* (1995); Kickel and Zadgorska (1995); Schäffer and Krivan (1998)]. Among these ET systems, tungsten (wolfram) filaments (Fig. 2.17B) and graphite rods have been the

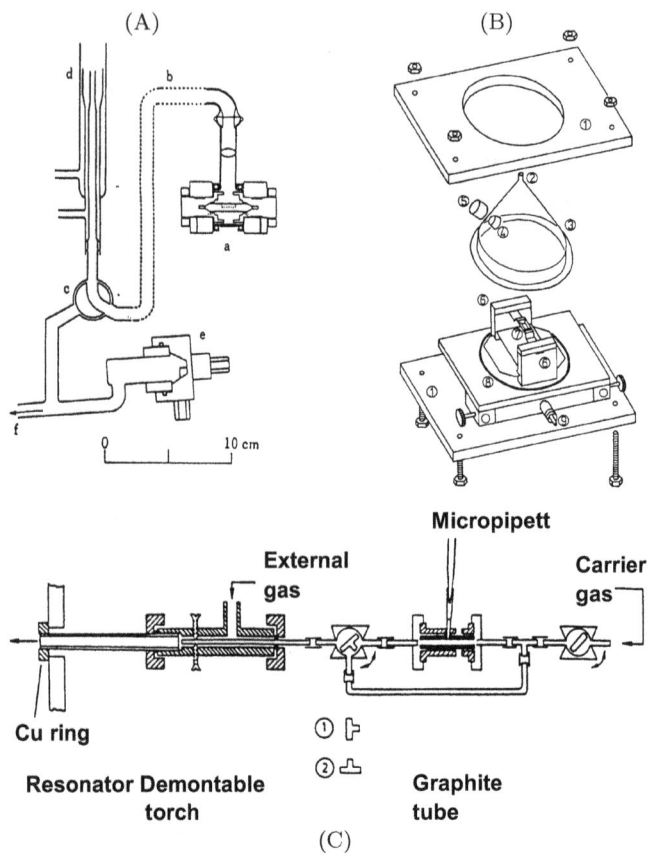

Fig. 2.17. **(A)** Scheme for ET-ICP-AES coupling {(a) graphite atomizer; (b) carrier tube; (c) three-way valve; (d) torch; (e) pneumatic nebulizer and chamber; (f) waste}. (B) Tungsten filament vaporization system {(1) bakelite sheets; (2) port for argon entry and introduction of sample into the plasma; (3) quartz body; (4) sample introduction port; (5) quartz stopper; (6) furnace electrode; (7) tungsten cup; (8) ring; (9) argon entry port}. (C) ET-MIP coupling {(1) position during drying and decomposition; (2) position during the vaporization stage}. [Fonseca *et al.* (1997); Okamoto *et al.* (1990); Heltai *et al.* (1990)].

most widely used, even for the vaporization of microsamples. However, besides inductively coupled plasmas (ICP) [Okamoto *et al.* (1990, 1993, 1994); Isoyama *et al.* (1990); Ng and Caruso (1982); Huang *et al.* (1991)], some interest has been aroused among researchers in the use of microwave plasmas (MIP) (Fig. 2.17C) [Heltai *et al.* (1990)].

Mass spectrometry (MS) is the only multi-element technique allowing the simultaneous or sequential determination of all elements together with their isotopes. In combination with ICP, the ICP-MS technique [Hulmston and Hutton (1991); Lamoureux *et al.* (1994a, 1994b); Grégoire and Sturgeon (1993)] achieves detection limits lying between 0.1 and 1 ng/mL for the majority of elements. This is approximately two to three orders of magnitude lower than those found with ICP-AES. Gray and Date (1983) were the first to couple up an ET system and an ICP-MS. Since then many applications and experimental studies have been performed with this type of coupling [Hoffmann *et al.* (1994)]. The kind of graphite utilized has a considerable effect on the production of the ion type reaching the detector [Majidi and Miller-Ihli (1998)]. Thus, the peak intensity drops considerably when oxygenated pyrolytic-graphite coated surfaces are used.

Couplings between the different atom or ion sources and atomic spectroscopic techniques have been designated by the general term of "combined sources" or "tandem sources." A first source produces free atoms and/or fragments of molecules available to enter a second source without changing the state. The second source serves to excite or ionize these species for emission analysis or for mass spectrometry, respectively. Any tandem source would have to include the following four components: a sampling system, an atomization cell, an ionization device, and an excitation source. In this sense, systems with pre-vaporization sample introduction should be included within the term "tandem source." Nevertheless, to avoid confusion, further descriptions should be added to indicate the function of each source in the tandem [Kantor and Hieftje (1995)].

2.5.3. FANES

In *furnace atomization non-thermal excitation spectrometry* (FANES), a Massmann-type graphite furnace was used for conventional electrothermal vaporization. However, the tube also acts as an electrode (cathode) in such a way that a gas discharge can be started within it (Fig. 2.18) [Banks *et al.* (1992)].

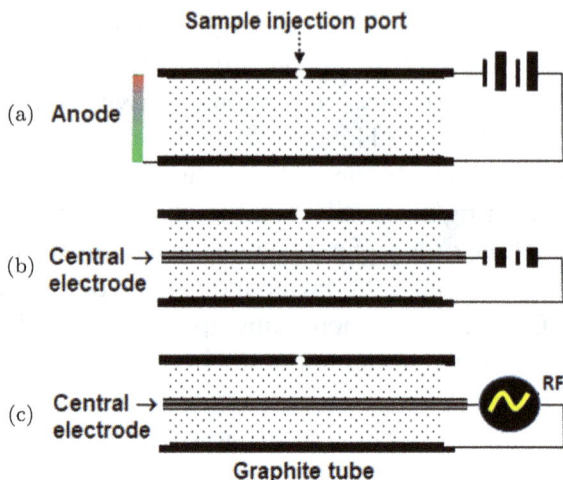

Fig. 2.18. Schematic comparison of FANES and FAPES systems: (a) HC-FANES, (b) HA-FANES and (c) FAPES [Banks *et al.* (1992)].

If the atmosphere inside the furnace is made up of an inert gas at low pressure (1–20 torr), a luminescent discharge (hollow cathode discharge) can be set going there as long as the anode is appropriately placed. Hence, by heating the furnace during the atomization phase while the discharge is occurring, the atoms from the sample are thermally vaporized within the negative luminosity of the luminescent discharge, where efficient excitation takes place. In this way, the volatilization and atomization processes associated with the graphite tube are not coupled to the excitation processes in the discharge. Excitation in these sources depends principally upon the discharge and is not a direct result of the electrothermal heating of the furnace.

The use of glow discharges within a graphite tube employed as a source of excitation/atomization in atomic emission spectrometry has been widespread and extensively evaluated from an analytical point of view [Falk (1977, 1986); Falk *et al.* (1979, 1981, 1983, 1984, 1986); Naumann *et al.* (1988); Harnly *et al.* (1990); Falk and Tilsch (1987); Ballou *et al.* (1988); Liang and Blades (1989); Smith *et al.* (1990); Sturgeon *et al.* (1989, 1990); Dittrich *et al.* (1994)]. In the first applications of FANES [Falk (1977); Falk *et al.* (1979, 1983, 1984, 1986)], a low-pressure luminescent discharge in which the furnace acted as a hollow cathode was used. For this reason, the technique was called hollow-cathode FANES (HC-FANES) (Fig. 2.18a).

2.5.3.1. *HA-FANES*

Later, a low-pressure direct current discharge similar to the former but with inverse polarity was described [Littlejohn and Ottaway (1978); Harnly *et al.* (1990)]. In this case, the furnace acts as a hollow anode while the cathode is formed by a coaxial graphite rod running the length of the tube (Figs. 2.18b, 2.19). This technique has been termed hollow-anode FANES (HA-FANES).

The HA-FANES system differs significantly in several ways from the HC-FANES system. The hollow-anode geometry of the HA-FANES system excludes the intense emission that comes from the classic hollow-cathode design of HC-FANES. In contrast, the discharge in HA-FANES forms a corona surrounding the length of the graphite rod that serves as the cathode (Fig. 2.19). The intense emission region of interest for HA-FANES is found near the cathode, unlike HC-FANES, where the discharge is to be seen in the center of the furnace. The surface area of the cathode for HA-FANES is significantly smaller than for HC-FANES. In consequence, for the same current, the current density on the surface of the cathode is greater than that for HA-FANES.

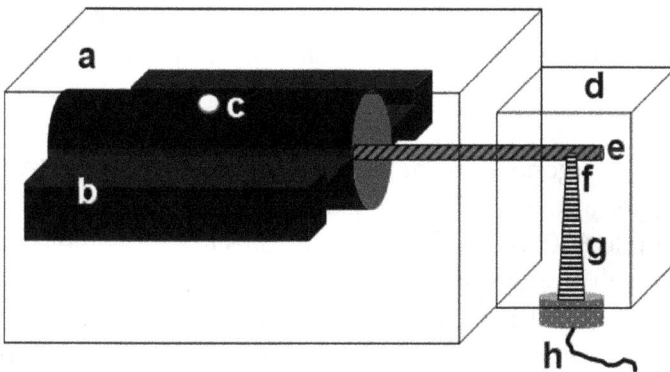

Fig. 2.19. Scheme of an HA-FANES system {(a) integrated contact cuvette; (b) connectors for power supply to the atomizer; (c) sample injection hole; (d) block holding cathode; (e) cathode; (f) adjuster screw; (g) contact pin; (h) power supply connections} [Riby *et al.* (1991)].

The discharges in HA-FANES differ from conventional glow discharges in several features.

(i) Parameters affecting HA-FANES characteristics must be optimized for the analyte excitation, but not for the sputtering process. The mass introduction rate for thermal atomization, from the cathode, exceeds that for sputtering by approximately three orders of magnitude [Falk *et al.* (1988)]. In HA-FANES, sputtering of the analyte is negligible as the sample is deposited on the anode. Nevertheless, sputtering cannot be ignored when the carbon sputtered from the surface of the cathode contributes significantly to the background signal.

(ii) In HA-FANES, a "hot" cathode is used and reaches temperatures varying from 800°C to 2000°C, depending on the element to be determined. Above 1500°C, the considerable production of thermionic electrons drastically reduces the potential through the discharge. Under these conditions, the discharge has been related to low-voltage arcing.

(iii) HA-FANES operates at pressures considerably higher than those used in conventional glow discharges. For many elements, the emission intensity (peak area) increases linearly with pressure [Harnly *et al.* (1990)]. Consequently, the best detection limits with HA-FANES have been achieved at pressures of 160–200 mm Hg, in contrast with the 1–25 mm Hg for HC-FANES [Riby *et al.* (1991)].

Although argon (Ar) is the filler gas normally used in HA-FANES, helium (He) also provides adequate analytical discharges at pressures of up to 600 mm Hg. Higher pressures may nonetheless also be obtained with a power supply capable of providing stronger currents [Riby and Harnly (1993)]. The drop-off in discharge potential and fluctuation in excitation temperature with the appearance of thermionic electrons are noticeably reduced in operating at a pressure ranging between 400 mm Hg and 600 mm Hg and currents between 140 mA and 200 mA. Excitation temperatures decline somewhat as a function of increases in pressure and rise as a function of stronger current.

These high working pressures and currents are of importance in analysis, since the overall analytical signal grows linearly with pressure.

The determination of non-volatile elements by HA-FANES involves two problems.

- There can be a large background emission because of the black body emission from the carbon surface, and because the discharge is close to the cathode. Black body emission increases with atomization temperature. However, HA-FANES allows atomization temperatures lower than in conventional furnaces [Frech *et al.* (1986)].
- Precision may drop as a consequence of the discharge potential change for a high-temperature atomization cycle which results from the development of thermionic electrons.

A new hollow cathode has been designed in which a graphite cup (of small size: 6 × 3 mm) is employed in place of the usual graphite tube (with relatively large dimensions, size: 60 × 6 mm) [Papp and Bánhidi (1998)]. In this new atomizer, the diameter (*d*) of the cavity of the emission zone is smaller, so that the intensity (*I*) of the emitted radiation will be greater, as the product of these two items should remain constant (*Id* = constant).

2.5.4. FAPES

This technique has been termed furnace atomization plasma excitation spectrometry [FAPES] and graphite furnace capacitively-coupled plasma (GF-CCP) (Figs. 2.18c and 2.20). The FAPES technique, similar to HA-FANES, uses a radio-frequency (*r.f.*) discharge in helium at atmospheric pressure. A voltage sufficient to decompose the gas is applied with a match in dynamic impedance between the *r.f.* circuit and the discharge (Fig. 2.20) [Liang and Blades (1989); Smith *et al.* (1990); Sturgeon *et al.* (1989, 1990)]. In the FAPES technique, there is also an electrode set lengthwise within the graphite tube.

Radio-frequency (*r.f.*) discharges offer no difficulties in maintaining uniform capacitively coupled plasmas at atmospheric pressure within the graphite tube, without the necessity for the large current densities inherent in arc discharges. This sort of plasma is more prone

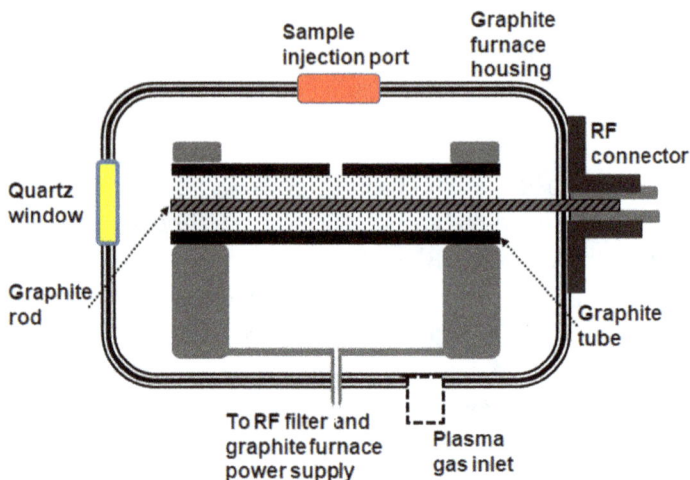

Fig. 2.20.   Scheme of a FAPES system [Smith *et al.* (1990)].

to use the advantages inherent in a graphite tube and is easier to use than low-pressure luminescent discharge sources [Sturgeon *et al.* (1989)].

Differences between ETAAS and ETAES lie exclusively with the stage after vaporization, i.e. is, the excitation phase, since at the *r.f.* potentials normally used the plasma does not significantly affect the atomization process from the tube wall. These discharges offer characteristics similar to MIP plasma (and sometimes ICP), and in many ways they are much more versatile in their operational procedures when compared with other plasma discharges [Blades (1994)].

The atomization and excitation mechanisms in FAPES are kept separated [Falk (1991)]. With such a separation, it is possible (at least in theory) to optimize these two processes individually and gain in operational flexibility. However, a complete separation is very rarely in place. The FAPES technique combines high efficiency at the atomization stage (as in ETAAS and ETAES), with great efficiency at the excitation stage, as occurs with plasmas at atmospheric pressure. This (atomic emission) analytical technique also combines the sensitivity of ETAAS with the capability of ICP-AES to perform multi-element analyses (simultaneously) and hence can come close to its high speed of analysis [Gilchrist *et al.* (1993)].

Helium is the gas normally used, but it is sometimes replaced by argon. However, there are differences in the way these two gases are used.

- Initiation of helium plasma is easier than with argon [Hettipathirana and Blades (1992a, b); Sturgeon *et al.* (1991)]. Argon has a first ionization energy (15.75 eV) lower than that of helium (24.59 eV) and so it should be much easier to initiate *r.f.* plasma for Ar than for He. Nonetheless, the electron ionization probability, defined as the number of collisions of an electron leading to ionization per cm and Torr is about 120 times greater for He than for Ar, which is the reason that He plasmas are easier to initiate in practice. The ionization rate of helium plasma is two orders of magnitude greater than for argon, at the voltages and pressures normally utilized for FAPES. However, operation at higher working frequencies would make it easier to initiate an argon plasma due to the reduction in the loss of electrons in the furnace walls [Hettipathirana and Blades (1992a, b)], as has been experimentally verified [Sturgeon *et al.* (1991)]. In short, the problems of coupling the impedances are reduced by operating at 27 MHz, which allows easy initiation and maintenance of argon plasma at atmospheric pressure.
- Using argon, greater sensitivity can be achieved for some metals (Ag, Cu, Mn, Pb, Ni and Fe) [Sturgeon *et al.* (1991)].
- Argon has lower price compared to helium.
- Emission proceeding from some varieties of molecule (Cl, CN, and C_2) has been observed when an argon (Ar) atmosphere is used, but not with helium (He). Nonetheless, the species CO^+ (19.39 eV and 19.66 eV) is not excited in Ar but is in He. If the central electrode is not heated by the plasma, it may act as a second surface for condensation and re-vaporization.

Furnace heating rates of the order of 2500°C/s can be attained, what is a definite advantage [Sturgeon and Berman (1983)]. Transversally heating of the graphite tube provides greater isothermal conditions [Frech *et al.* (1986)], reduces interference in the gaseous phase, and consequently facilitates the use of shorter cycles.

Very long tubes with smaller internal diameters produce optically broad plasmas which undergoes the process of self-absorption, being not really suitable for AES and formation of argon *r.f.* plasmas. In contrast, short tubes, and hence plasmas that are optically thinner, would reduce the absorption path, cutting back on the probability of self-absorption, and generating longer linear dynamic intervals. However, short tubes also decrease the analyte residence time in the plasma and would thus reduce sensitivity. Hence, the most appropriate tube is of an intermediate length, selected experimentally for each case.

Tubes with small internal diameters and using argon at high potentials form filaments and arcs between the central electrode and the tube surface. Such localized discharges cause a drop in sensitivity and precision. The use of graphite tubes with larger internal diameters may reduce or even eliminate these filaments and arcs when argon is employed, while sensitivity and precision are maintained.

Study of the He 667.82 nm line reveals the presence of two different plasmas: (i) an intense luminosity around the central electrode; and (ii) a diffuse plasma near the tube walls. In a graphite tube, the plasma shows a major dependence on the *d.c.* bias of the central electrode for a constant *r.f.* potential of 50W [Hettipathirana and Blades (1992a, b)]. The background poly-atomic species show distributions which overall are similar and respond to *d.c.* bias of the central electrode [Sturgeon *et al.* (1996)]. The condensation and re-volatilization on the central electrode diminishes when the analyte is atomized in high power plasma [Pavski *et al.* (1994, 1997)].

An increase in *r.f.* potential raises the molecular rotation temperature and the atomic excitation temperature, also favoring molecular dissociation, improved sensitivity and elimination of some molecular interferences [Le Blanc and Blades (1995)]. The emission intensity of the He I line grows significantly for higher *r.f.* potentials, particularly in the zone adjacent to the central electrode. However, less intense emission is observed in the zone next to the graphite tube wall, what suggests that FAPES sources operate like radio-frequency luminescent discharges [Le Blanc and Blades (1995)].

Separation of vaporization, excitation and/or ionization processes can be achieved by using the two-stage atomizer. The cup and tube are heated independently, permitting vaporization of the analyte from the cup into the tube, which is kept at a previously selected temperature and where excitation and/or ionization occurs. Operation at high atomization temperatures (2600 K) allows the determination of less volatile elements [Ohlsson *et al.* (1993)]. FAPES plasmas have also been utilized as sources of ions for inorganic mass spectrometry [Sturgeon and Guevremont (1997)].

2.5.5. Coherent Forward Scattering (CFS): Magneto-Optical Phenomena

The theoretical basis of atomic detection by forward scattering is related to the alteration of the polarization state of a linearly polarized radiation when transmitted through a resonance atomic absorption cell, located within a suitably aligned external magnetic field. A typical instrumental set up is shown in Fig. 2.22 [Grimm *et al.* (1988)].

The external magnetic field may be oriented either perpendicular or parallel to the optical path. Materials exhibit optical activity when subjected to an external magnetic field, which is known as the magneto-optic effect. Optical activity in a transverse magnetic field is called the Voigt's effect, while optical activity in a longitudinal field is called the Faraday's effect.

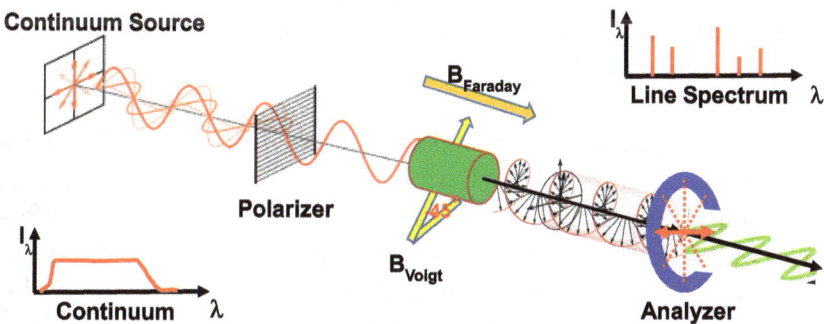

Fig. 2.22. Principle of CFS spectroscopy [Grimm *et al.* (1988)].

Owing to the magneto-optic effect, the polarization state of incident light is altered when scattered by atomic vapor in a magnetic field. Changes in the polarization state of the scattered light are analyzed with a birefringent prism placed behind the scattering vapor, which is followed by a photomultiplier [Yamamoto *et al.* (1980)]. The transmitted coherent forward scattering (CFS) line intensity, I_T, is a measure of the related analyte concentration [Hermann (1992)] ($I_T \propto I_0 N^2 L^2 f^2$), where I_0 is the spectral radiation intensity of the exciting light, N is the number of atoms in the unit volume of atomic vapor, L is the length of the magneto-optic interaction region, and f is the oscillator strength of the observed spectral transition.

Under the effect of an external magnetic field, the absorption lines split following the rules of the Zeeman's effect. In the so-called normal Zeeman's effect, splitting in the transverse field occurs into three polarized components, one parallel and the other two perpendicular to the field, while in the longitudinal field, into two reverse circular components. The mathematical treatment of the more general case of the anomalous Zeeman's effect follows analogous lines.

For a transverse magnetic field configuration, the electric vector of the incident radiation is solved into two linear components, one parallel (π component) and one perpendicular (σ component) to the field axis. It is apparent that the phase and amplitude of the former will be influenced by the refractive index and absorptivity exclusively from the π components, while the phase and amplitude of the later will be similarly influenced by the σ multiplet components. Since the two sets of interactions are not equivalent, it is evident that the optical transmission properties of the vapor are anisotropic. Hence magnetically induced linear birefringence and dichroism will be observed (the Voigt's effect).

If a longitudinal magnetic field configuration is considered, exactly analogous arguments apply. That is, left and right circularly polarized components of the electric vector of the source are now influenced by the left and right circularly polarized σ^+, σ^- components of a longitudinal Zeeman absorption multiplet. Thus, magnetically induced optical rotation and circular dichroism occurs (the Faraday's effect).

The interactions responsible for the Voigt's effect are clearly linear, and the corresponding Jones matrix is given for a birefringent and dichroic medium (different refractive indices and absorption coefficients for x- and y-directions) by the following equation [Kankare and Stephens (1980)],

$$V = \begin{pmatrix} \exp\left[i(\widehat{n}_x - 1)\dfrac{\omega b}{c}\right] & 0 \\ 0 & \exp\left[i(\widehat{n}_y - 1)\dfrac{\omega b}{c}\right] \end{pmatrix}, \tag{2.7}$$

where c is the light velocity, b is the position of the electric vector, ω is the angular frequency of the wave ($\omega = 2\pi\nu$), i is the imaginary unit, and $\widehat{n}_x$ and $\widehat{n}_y$ are the refractive indices for x- and y-directions.

Similarly, the interactions responsible for the Faraday's effect must be described by circular vectors, where the matrix G (Eq. (2.8)) is suited to this purpose

$$G = U^{-1} \begin{pmatrix} \exp\left[i(\widehat{n}_+ - 1)\dfrac{\omega b}{c}\right] & 0 \\ 0 & \exp\left[i(\widehat{n}_- - 1)\dfrac{\omega b}{c}\right] \end{pmatrix} U$$

$$= \exp\left[i\dfrac{\omega b}{2c}(\widehat{n}_+ + \widehat{n}_- - 2)\right] \begin{pmatrix} \text{Cos } \chi & -\text{Sin } \chi \\ \text{Sin } \chi & \text{Cos } \chi \end{pmatrix} \tag{2.8}$$

where, $\chi = \chi' + i\chi'' = \frac{\omega b}{2c}(\widehat{n}_- - \widehat{n}_+)$ and $\widehat{n}_+$ and $\widehat{n}_-$ are analogous to $\widehat{n}_x$ and $\widehat{n}_y$.

In the simplest configuration, the polarization analyzer is placed perpendicular to the incident light polarization and no light is transmitted unless a sample is atomized (i.e. scattering signals appear on the dark background).

Multi-element analysis with a continuum radiation source can be made in either the Faraday or the Voigt configuration [Yamamoto *et al.* (1980)].

(i) In the Faraday configuration, two eigenwaves propagating along the magnetic field, i.e. left- and right-circularly polarized waves have refractive indices n^+ and n^-, respectively. The mean, $\bar{n}$, and differential, Δn, refractive indices are defined by $\bar{n} = \frac{(n^+ + n^-)}{2}$ and $\Delta n = \frac{(n^+ - n^-)}{2}$, respectively, while the transmitted light intensity, $I_F(k)$, is obtained as follows

$$I_F(k) = I_0(k)[Sin(kL\Delta n)]^2 \exp[-2kL\bar{n}^{(i)}], \qquad (2.9)$$

where $I_0(k)$ is the incident light intensity, which depends in general on the wavenumber k; L is the magneto-optic interaction length; and index i represents the imaginary unit.

(ii) In the Voigt configuration, two eigenwaves polarize either parallel (ordinary wave) or perpendicularly (extraordinary wave) to the magnetic field. The mean, $\bar{n}$, and differential, Δn, refractive indices are defined by $\bar{n} = \frac{(n_o + n_e)}{2}$ and $\Delta n = \frac{(n_o - n_e)}{2}$, where n_o and n_e are ordinary and extraordinary refractive indices, respectively. The transmitted light intensity, $I_V(k)$, has the same form as Eq. (2.9)

$$I_V(k) = I_0(k)[Sin(kL\Delta n)]^2 \exp[-2kL\bar{n}^{(i)}]. \qquad (2.10)$$

In Eqs. (2.9) and (2.10), the sinusoidal function represents polarization alteration caused by magneto-optic effect. The exponential function represents attenuation by atomic absorption. Analysis of forward scattering spectra, $I_F(k)$ and $I_V(k)$, requires calculation of $\bar{n}$ and Δn as functions of k and the number of scattering atoms. Several studies have been made using graphite tubes as atomizers [Ito *et al.* (1977); Kitagawa *et al.* (1981a,b); Hirokawa and Namiki (1982); Hermann *et al.* (1990)] even for simultaneous multi-element determinations [Hermann (1992); Bernhardt *et al.* (1999)].

References

Absalan, G.; Chakrabarti, C.L.; Headrick, K.L.; Parker, M. and Marcus, R.K. (1998). Effect of anode dimensions and location of the discharge gas inlet port on the spatial distribution of copper atoms in a radio frequency glow discharge atomizer for atomic absorption spectrometry. *Anal. Chem.*, **70**: 3434–3443.

Ajayi, O.O.; Littlejohn, D. and Boss, C.B. (1989). Influence of graphite furnace tube design on vapour temperatures and chemical interferences in ETA-AAS. *Talanta*, **36**: 805–810.

Almeida, M.C. and Seitz, W.R. (1986). Carbide-treated graphite cuvettes for electrothermal atomization prepared by impregnation with metal chlorides. *Appl. Spectrosc.*, **40**: 4–8.

Amos, M.D.; Bennett, P.A.; Brodie, K.G.; Lung, P.W.Y. and Matoušek, J.P. (1971). Carbon rod atomizer in atomic absorption and fluorescence spectrometry and its clinical application. *Anal. Chem.*, **43**: 211–215.

Ballou, N.E.; Styris, D.L. and Harnly, J.M. (1988): Hollow-anode plasma excitation source for atomic emission spectrometry, *J. Anal. At. Spectrom.*, **3**: 1141–1143.

Banks, P.R.; Liang, D.C. and Blades, M.W. (1992). Graphite-furnace capacitively coupled plasma spectrometry. A new, fast furnace technique, *Spectroscopy*, **7**: 36

Batchelor, J.D.; Thomas, S.E. and Jones, B.T. (1998). Determination of cadmium with a portable, battery-powered tungsten coil atomic absorption spectrometer. *Appl. Spectrosc.*, **52**: 1086–1091.

Bernhardt, J.; Hermann, G. and Lasnitschka, G. (1999). Simultaneous multi-element determination with coherent forward scattering spectrometry employing chromatically corrected polarizers and a fast scanning spectrometer. *Spectrochim. Acta, Part B*, **54**: 645–656.

Bezur, L.; Marshall, J.; Ottaway, J.M. and Fakhrul-Aldeen, R. (1983). Platform atomisation in carbon furnace atomic-emission spectrometry. *The Analyst*, **108**: 553–572.

Blades, M.W. (1994): Atmospheric-pressure, radio frequency, capacitively coupled helium plasmas, *Spectrochim. Acta, Part B*, **49**: 47–57.

Boulo, P.R.; Soraghan, J.J.; Sadler, D.A.; Littlejohn, D. and Creeke, A. (1997). Use of Image Processing to Aid Furnace set-up in Electrothermal Atomic Absorption Spectrometry. *J. Anal. At. Spectrom.*, **12**: 293–300.

Brandenberger, H. and Bader, H. (1967). The determination of nanogram levels of mercury in solution by a flameless atomic absorption technique. *At. Absorpt. Newslet.*, **6**: 101–103.

Brodie, K.G. and Matousek, J.P. (1971). Application of the carbon rod atomizer to atomic absorption spectrometry of petroleum products. *Anal. Chem*, **43**: 1557–1560.

Busch, M.A. and Busch, K.W. (1991): Analytical applications of flame/furnace infrared emission spectrometry. *Spectrochim. Acta Rev.*, **14**: 303–336.

Carroll, J.; Marshall, J.; Littlejohn, D. and Ottaway, J.M. (1985). Automatic end-entry sample atomisation in electrothermal atomic absorption spectrometry. *Fresenius Z. Anal. Chem.*, **322**: 145–150.

Carroll, J.; Miller-Ihli, M.J.; Harnly, J.M.; O'Haver, T.R. and Littlejohn, D. (1992). Comparison of sodium chloride and magnesium chloride interferences in continuum source atomic absorption spectrometry with wall, platform and probe electrothermal atomization. *J. Anal. At. Spectrom.*, **7**: 533–538.

Chakrabarti, C.L.; Chen, J. and Grenier, M. (1996). Determination of the temperature of the graphite probe surface in graphite probe furnace atomic absorption spectrometry. *Spectrochim. Acta, Part B*, **51**: 1335–1343.

Chakrabarti, C.L.; Wu, S.; Karwowska, R.; Rogers, J.T. and Dick, R. (1985). The gas temperature in and the gas expulsion from a graphite furnace used for atomic absorption spectrometry. *Spectrochim. Acta, Part B*, **40**: 1663–1676.

Chakrabarti, C.L.; Wu, S.; Karwowska, R.; Rogers, J.T.; Haley, L.; Bertels, P.C. and Dick, R. (1984). Temperature of platform, furnace wall and vapour in a pulse-heated electrothermal graphite furnace in atomic absorption spectrometry. *Spectrochim. Acta, Part B*, **39**: 415–448.

Chan, G.C.Y. and Chan, W.T. (1998). Determination of lead in a chloride matrix by atomic absorption spectrometry using electrothermal vaporization and capacitively coupled plasma atomization. *J. Anal. At. Spectrom.*, **13**: 209–214.

Chen, G. and Jackson, K.W. (1998). Wall-to-platform migration in electrothermal atomic absorption spectrometry. Part 2. Low-temperature migration characteristics of thallium, lead, cadmium, and manganese during thermal pretreatment. *Spectrochim. Acta, Part B*, **53**: 981–991.

Dittrich, K.; Franz, T. and Wennrich, R. (1994): Determination of mercury by furnace atomic nonthermal-excitation spectrometry, *Spectrochim. Acta, Part B*, **49**: 1695–1705.

Doçekal, B. (1998). A new design of the two-component atomizer for the direct determination of medium and volatile elements in high-purity solid refractory metals by electrothermal atomic absorption spectrometry. *Spectrochim. Acta, Part B*, **53**: 427–435.

Donega, H.M. and Burgess, T.E. (1970). Atomic absorption analysis by flameless atomization in a controlled atmosphere. *Anal. Chem.*, **42**: 1521–1524.

Duan, Y.; Hou, M.; Du, Z. and Jin, Q. (1993). Evaluation of the performance of microwave-induced plasma atomic absorption spectrometry (MIP-AAS). *Appl. Spectrosc.*, **47**: 1871–1879.

Eloi, C.C.; Robertson, J.D. and Majidi, V. (1993). Investigation of high temperature reactions on graphite with Rutherford backscattering spectrometry: interaction of cadmium, lead and silver with a phosphate modifier. *J. Anal. At. Spectrom*, **8**: 217–222.

Eloi, C.C.; Robertson, J.D. and Majidi, V. (1995). Rutherford backscattering spectrometry investigation of the effects of oxygen and hydrogen pretreatment of pyrolytically coated graphite on Pb atomization. *Anal. Chem.*, **67**: 335–340.

Eloi, C.C.; Robertson, J.D. and Majidi, V. (1997). Rutherford Backscattering Spectrometry Investigation of the permeability of pyrolytically coated graphite substrates. *Appl. Spectrosc.*, **51**: 236–239.

Falk, H. (1977). Einige theoretische Überlegungen zum vergleich der physikalischen Grenzen thermischer und nicht-thermischer spektroskopischer Strahlungsquellen. *Spectrochim. Acta, Part B*, **32**: 437–443.

Falk, H. (1984). On the pyrometric temperature measurements in graphite furnaces-theoretical approach. *Spectrochim. Acta, Part B*, **39**: 387–396.

Falk, H. (1986): Hollow-cathode discharge within a graphite furnace: furnace atomic nonthermal excitation spectrometry (FANES), in *Improved Hollow Cathode Lamps for Atomic Spectroscopy*, Ed. S. Caroli, p. 74. Ellis Harwood, Halstead Press, New York.

Falk, H. (1991): Tandem sources using electrothermal atomizers: Analytical capabilities and limitations, *J. Anal. At. Spectrom.*, **6**: 631–635.

Falk, H. and Tilsch, J. (1987). Atomization efficiency and over-all performance of electrothermal atomisers in atomic-absorption furnace atomization non-thermal excitation and laser-excited atomic fluorescence spectrometry. *J. Anal. At. Spectrom.*, **2**: 527–531.

Falk, H.; Glismann, A.; Bergann, L.; Minkwitz, G.; Schubert, M. and Skole, J. (1985). Time-dependent temperature distribution of graphite-tube atomizers. *Spectrochim. Acta, Part B*, **40**: 533–542.

Falk, H.; Hoffmann, E. and Lüdke, Ch. (1981). FANES (Furnace atomic nonthermal excitation spectrometry) — A new emission technique with high detection power. *Spectrochim. Acta, Part B*, **36**: 767–771.

Falk, H.; Hoffmann, E. and Lüdke, Ch. (1988): Experimental and theoretical investigations relating to FANES, *Prog. Analyt. Spectrosc.*, **11**: 417–480.

Falk, H.; Hoffmann, E.; Jaeckel, I. and Lüdke, Ch. (1979). Atomic emission trace analysis by non-thermal excitation. *Spectrochim. Acta, Part B*, **34**: 333–339.

Falk, H.; Hoffmann, E. and Lüdke, Ch. (1984). A comparison of furnace atomic nonthermal excitation spectrometry (FANES) with other atomic spectroscopic techniques. *Spectrochim. Acta, Part B*, **39**: 283–294.

Falk, H.; Hoffmann, E.; Lüdke, Ch.; Ottaway, J.M. and Giri, S.K. (1983). Furnace atomisation with non-thermal excitation — Experimental evaluation of detection based on a high-resolution echelle monochromator incorporating automatic background correction. *Analyst*, **108**: 1459–1465.

Falk, H.; Hoffmann, E.; Lüdke, Ch.; Ottaway, J.M. and Littlejohn, D. (1986). Studies on the determination of cadmium in blood by furnace atomic non-thermal excitation spectrometry. *Analyst*, **111**: 285–290.

Feo, J.C.; Castro, M.A.; Lumbreras, J.M.; de Celis, B. and Aller, A.J. (2003). Nickel as a Chemical Modifier for Sensitivity Enhancement and Fast Atomization Processes in Electrothermal Atomic Absorption Spectrometric Determination of Cadmium in Biological and Environmental Samples. *Anal. Sci.*, **19**: 1631–1636.

Filgueiras, A.V.; Lavilla, I. and Bendicho, C. (2004). Development of fast thermal programs in electrothermal atomic absorption spectrometry using hot injection and removal of the ashing stage for determination of heavy metals in sequential extracts from sediments. *Anal. Chim. Acta*, **508**: 217–223.

Fonseca, R.W.; Miller-Ihli, N.J.; Sparks, C.; Holcombe, J.A. and Shaver, B. (1997): Effect of oxygen ashing on analyte transport efficiency using ETV-ICP-MS. *Appl. Spectros.*, **51**: 1800–1806.

Frech, W.; Arshadi, M.; Baxter, D.C. and Hütsch, B. (1989). Vapour-phase temperature measurements in the evaluation of platform designs for graphite furnace atomic absorption spectrometry. *J. Anal. At. Spectrom.*, **4**: 625–629.

Frech, W.; Baxter, D.C. and Hütsch, B. (1986). Spatially isothermal graphite furnace for atomic absorption spectrometry using side-heated cuvettes with integrated contacts. *Anal. Chem.*, **58**: 1973–1977.

Frech, W.; Hadgu, N.; Henriksson, D.; Radziuk, B.; Rödel, G. and Tamm, R. (2000). Characterization of a pressurizable two-step atomizer for atomic absorption spectrometry. *Spectrochim. Acta, Part B*, **55**: 461–472.

Ganeyev, A.A. and Sholupov, S.E. (1998). A thin-walled metallic hollow cathode as an atomizer for Zeeman atomic absorption spectrometry. *Spectrochim. Acta, Part B*, **53**: 471–486.

Gilchrist, G.F.R.; Celliers, P.M.; Yang, H.; Yu, C. and Liang, D.C. (1993): Simultaneous multielement determination using helium or argon plasma for graphite furnace capacitively coupled plasma atomic emission spectrometry. *J. Anal. At. Spectrom.*, **8**: 809–814.

Gilchrist, G.F.R.; Chakrabarti, C.L.; Ashley, J.T.F. and Hughes, D.M. (1992). Vaporization and atomization of lead and tin from a pyrolytic graphite probe in graphite furnace atomic absorption spectrometry, *Anal. Chem.*, **64**: 1144–1153.

Golloch, A.; Haveresch-Kock, M. and Plantikow-Voßgätter, F. (1995): Optimization of a novel ETV system for solid sample introduction into an ICP and its application to the determination of trace impurities in SiC. *Spectrochim. Acta, Part B*, **50**: 501–516.

Gornushkin, I.B.; Smith, B.W. and Winefordner, J.D. (1996). Use of laser-excited atomic fluorescence spectrometry with a novel diffusive graphite tube electrothermal atomizer for the direct determination of silver in sea water and in solid reference materials. *Spectrochim. Acta, Part B*, **51**: 1355–1370.

Granadillo, V.A.; Parra de Machado, L. and Romero, R.A. (1994). Determination of total chromium in whole blood, blood components, bone, and urine by fast furnace program electrothermal atomization AAS and using neither analyte isoformation nor background correction. *Anal. Chem.*, **66**: 3624–3631.

Gray, A.L. and Date, A.R. (1983). Inductively coupled plasma source — mass spectrometry using continuum flow ion extraction. *Analyst*, **108**: 1033–1050.

Grégoire, D.C. and Sturgeon, R.E. (1993): Background spectral features in electrothermal vaporization inductively coupled plasma mass spectrometry: molecular ions resulting from the use of chemical modifiers. *Spectrochim. Acta, Part B*, **48**: 1347–1364.

Grimm, W.; Hermann, G.; Jung, M.; Krüger, R.; Lasnitschka, G.; Scharmann, A. and Seib, M. (1988). Coherent forward scattering spectrometry by means of a commercial furnace electromagnet unit. *Spectrochim. Acta, Part B*, **43**: 1269–1272.

Grinshtein, I.L.; Vilpan, Y.A.; Vasilieva, L.A. and Kopeikin, V.A. (1999). Reduction of matrix interference during the atomic absorption determination of lead and cadmium in strongly interfering matrix samples using a two-step atomizer with vaporizer purging. *Spectrochim. Acta, Part B*, **54**: 745–752.

Grinshtein, I.L.; Vilpan, Y.A.; Saraev, A.V. and Vasilieva, L.A. (2001). Direct atomic absorption determination of cadmium and lead in strongly interfering matrices by double vaporization with a two-step electrothermal atomizer. *Spectrochim. Acta, Part B*, **56**: 261–274.

Habichi, J.; Prohaska, Th.; Friedbacher, G.; Grasserbaner, M. and Ortner, H.M. (1995). Nanotopographical changes on graphite tube surfaces in electrothermal atomic absorption spectrometry experiments as studied by atomic force microscopy. *Spectrochim. Acta, Part B*, **50**: 713–723.

Halls, D.J. (1984). Speeding up determinations by electrothermal atomic-absorption spectrometry. *Analyst*, **109**: 1081–1084.

Halls, D.J. (1995). Analytical minimalism applied to the determination of trace elements by atomic spectrometry. *J. Anal. At. Spectrom.*, **10**: 169–175.

Harnly, J.M.; Styris, D.L. and Ballou, N.E. (1990): Furnace atomic non-thermal excitation spectrometry with the furnace as a hollow anode. *J. Anal. At. Spectrom.,* **5**: 139–144.

Hassell, D.C.; Rettberg, T.M.; Fort, F.A. and Holcombe, J.A. (1988). Low-pressure vaporization for graphite furnace atomic absorption spectrometry. *Anal. Chem.,* **60**: 2680–2683.

Heltai, Gy.; Broekaert, J.A.C.; Burba, P.; Leis, F.; Tschöel, P. and Tölg, G. (1990): Study of a toroidal argon MIP and a cylindrical helium MIP for atomic emission spectrometry-II. Combination with graphite furnace vaporization and use for analysis of biological samples. *Spectrochim. Acta, Part B,* **45**: 857–866.

Hermann, G. (1992). Coherent forward scattering atomic spectrometry. *Anal. Chem.,* **46**: 571A–579A.

Hermann, G.; Jung, M.; Lasnitschka, G.; Moder, R.; Scharmann, A. and Zhou, X. (1990). Multielement determination on strong and weak lines by coherent forward scattering spectroscopy with continuum sources. *Spectrochim. Acta, Part B,* **45**: 763–768.

Hettipathirana, T.D. and Blades, M.W. (1992a). Furnace atomization plasma excitation spectroscopy — spectral, spatial, and temporal characteristics. *Spectrochim. Acta, Part B,* **47**: 493–503.

Hettipathirana. T.D. and Blades, M.W. (1992b). Temporal emission and absorption characteristics of silver, lead and manganese in furnace atomization plasma excitation spectrometry. *J. Anal. At. Spectrom.,* **7**: 1039–1046.

Hirokawa, K. and Namiki, M. (1982). Coherent forward scattering spectrometry of manganese and chromium using a xenon lamp and a tellurous oxide acousto-optic tunable filter. *Spectrochim. Acta, Part B,* **37**: 165–170.

Hoenig, M. and Cilissen, A. (1993). Electrothermal atomic absorption spectrometry: fast or conventional programs?. *Spectrochim. Acta, Part B,* **48**: 1003–1012.

Hoenig, M. and Van Hoeyweghen, P. (1986). Alternative to solid sampling for trace metal determination by platform electrothermal atomic absorption spectrometry: direct dispensing of powdered samples suspended in liquid medium. *Anal. Chem.,* **58**: 2614–2617.

Hoffmann, E.; Lüdke, Ch. and Scholze, H. (1994). Electrothermal vaporization for simultaneous multielement determination. *J. Anal. At. Spectrom.,* **9**: 1237–1241.

Holcombe, J.A. and Wang, P. (1993). Direct solid sample analysis using pressure regulated electrothermal atomization with atomic absorption spectrometry. *Fresenius' J. Anal. Chem.,* **343**: 1047–1053.

Hou, X.; Levine, K.E.; Salido, A.; Jones, B. T.; Ezer, M.; Elwood, S. and Simeonsson, J.B. (2001b). Tungsten coil devices in atomic spectrometry: Absorption, fluorescence and emission. *Anal. Sci.,* **17**: 175–180.

Hou, X.; Yang, Z. and Jones, B.T. (2001a). Determination of selenium by tungsten coil atomic absorption spectrometry using iridium as a permanent chemical modifier. *Spectrochim. Acta, Part B,* **56**: 203–214.

Hou, X. and Jones, B.T. (2002). Tungsten devices in analytical atomic spectrometry. *Spectrochim. Acta, Part B,* **57**: 659–688.

Huang, M.; Jiang, Z. and Zeng, Y. (1991): A new method for determination of rare earth elements vaporized in graphite furnace with a polytetrafluoroethylene slurry fluorinating reagent by inductively coupled plasma atomic emission spectrometry. *Anal. Sci.*, **7**: 773–778.

Huettner, W. and Busche, C. (1986). Structure and reactivity of carbon materials used in atomization furnaces. *Fresenius Z. Anal. Chem.*, **323**: 674–680.

Hui-Ming, H. and Yao-Han, L. (1984). Determination of trace Na, K, Ba and Li by graphite furnace atomic emisión spectrometry. *Spectrochim. Acta, Part B*, **39**: 493–499.

Hulmston, P. and Hutton, R.C. (1991): Analytical capabilities of electrothermal vaporization-inductively coupled plasma-mass spectrometry. *Spectroscopy International*, **3**: 35–38.

Imai, S.; Kubo, Y.; Yonetani, A.; Ogawa, N. and Kikuchi, Y. (1998). Effect of refractory element carbide coating of a pyrolytically coated graphite furnace on injectable sample volume in the electrothermal atomic absorption spectrometric determination of lead. *J. Anal. At. Spectrom.*, **13**: 1199–1202.

Isoyama, H.; Okuyama, S.; Uchida, T.; Takeuchi, M.; Iida, C. and Nakagawa, G. (1990): In-furnace standard addition method for the determination of trace metals in biological samples by electrothermal vaporization-inductively coupled plasma atomic emission spectrometry. *Anal. Sci.*, **6**: 555–560.

Ito, M.; Murayama, S.; Kayama, K. and Yamamoto, M. (1977). The detection of cadmium by employing the forward scattering of resonance radiation. *Spectrochim. Acta, Part B*, **32**: 347–355.

Jackson, J.G.; Fonseca, R.W. and Holcombe, J.A. (1995). Migration of Ag, Cd and Cu into highly oriented pyrolytic graphite and pyrolytic coated graphite. *Spectrochim. Acta, Part B*, **50**: 1837–1846.

Kankare, J.J.; Stephens, R. (1980). A unified theory of magneto-optic phenomena in analytical atomic spectroscopy. *Spectrochim. Acta, Part B*, **35**: 849–864.

Kantor, T. and Hieftje, G.M. (1995): "Tandem" versus "combined" sources in atomic spectrometry. *Spectrochim. Acta, Part B*, **50**: 961–962.

Katschthaler, C.; Quan, X.; Krizová, H.; Gross, R. and Knapp, G. (1995): Evaluation of an electrothermal vaporization sample introduction system into a stabilized capacitively coupled He-plasma (SCP) for the determination of chlorine, *Spectrochim. Acta, Part B*, **50**: 453–462.

Katskov, D.A. (2005). Fast heated ballast furnace atomizer for atomic absorption spectrometry. Part 1. Theoretical evaluation of atomization efficiency. *J. Anal. At. Spectrom.*, **20**: 220–226.

Katskov, D.A.; Marais, P.J.J. and Tittarelli, P. (1996). Design, operation and analytical characteristics of the filter furnace, a new atomizer for electrothermal atomic absorption spectrometry. *Spectrochim. Acta, Part B*, **51**: 1169–1189.

Katskov, D.A.; Marais, P.J.J.G. and Ngobeni, P. (1998). Transverse heated filter atomizer for electrothermal atomic absorption spectrometry. *Spectrochim. Acta, Part B*, **53**: 671–682.

Katskov, D.A.; McCrindle, R.I.; Schwarzer, R. and Marais, P.J.J.G. (1995). The graphite filter furnace: a new atomization concept for atomic spectroscopy, *Spectrochim. Acta, Part B*, **50**: 1543–1553.

Katskov, D.A.; Sadagov, Y.M. and Banda, M. (2005). Fast heated ballast furnace atomizer for atomic absorption spectrometry. Part 2. Experimental assessment of performances. *J. Anal. At. Spectrom.*, **20**: 227–232.

Katskov, D.A.; Schwarzer, R.; Marais, P.J.J.G. and McCrindle, R.I. (1994). Use of a furnace with a graphite filter for electrothermal atomic absorption spectrometry. *J. Anal. At. Spectrom.*, **9**: 431–436.

Kickel, H. and Zadgorska, Z. (1995): A new electrothermal vaporization device for direct sampling of ceramic powders for inductively coupled plasma optical emission spectrometry, *Spectrochim. Acta, Part B*, **50**: 527–535.

King, A.S. (1905). Some emission spectra of metals as given by an electric oven. *Astrophys. J.*, **21**: 236–257.

King, A.S. (1908). The production of spectra by an electrical resistance furnace in hydrogen atmosphere. *Astrophys. J.*, **27**: 353–362.

Kitagawa, K.; Nanya, T. and Tsuge, S. (1981a). Application of the atomic Faraday effect to the trace determination of lead. *Spectrochim. Acta, Part B*, **36**: 9–20.

Kitagawa, K.; Ohta, M.; Kaneko, T. and Tsuge, S. (1994). Packed glassy carbon tube atomizer for direct determinations by atomic absorption spectrometry, free from background absorption. *J. Anal. At. Spectrom.*, **9**: 1273–1277.

Kitagawa, K.; Suzuki, M.; Aoi, N. and Tsuge, S. (1981b). Analytical and spectral features of atomic magneto-optical rotation spectroscopy (the atomic Faraday effect) of Sb, Bi, Ag and Cu with a hollow cathode lamp operated in a pulse mode. *Spectrochim. Acta, Part B*, **36**: 21–34.

Koirtyohann, S.R.; Giddings, R.C. and Taylor, H.E. (1984). Heating rates in furnace atomic absorption using the L'vov platform. *Spectrochim. Acta, Part B*, **39**: 407–413.

Krivan, V.; Barth, P. and Schnürer-Patschan, Ch. (1998). An electrothermal atomic absorption spectrometer using semiconductor diode lasers and a tungsten coil atomizer: Design and first applications. *Anal. Chem.*, **70**: 3525–3532.

Lamoureux, M.M.; Grégoire, D.C.; Chakrabarti, C.L. and Goltz, D.M. (1994a): Modification of a commercial electrothermal vaporizer for sample introduction into an inductively coupled plasma mass spectrometer. 1. Characterization. *Anal. Chem.*, **66**: 3208–3216.

Lamoureux, M.M.; Grégoire, D.C.; Chakrabarti, C.L. and Goltz, D.M. (1994b): Modification of a commercial electrothermal vaporizer for sample introduction into an inductively coupled plasma mass spectrometer. 2. Performance evaluation. *Anal. Chem.*, **66**: 3217–3222.

Le Blanc, C.W. and Blades, M.W. (1995): Spatially resolved temperature measurements in a furnace atomization plasma excitation spectrometry source, *Spectrochim. Acta, Part B*, **50**: 1395–1408.

Liang, D.C. and Blades, M.W. (1988). Atmospheric pressure capacitively coupled plasma atomizer for atomic absorption spectrometry. *Anal. Chem.*, **60**: 27–31.

Liang, D.C. and Blades, M.W. (1989). An atmospheric-pressure capacitively coupled plasma formed inside a graphite-furnace as a source for atomic emission-spectrometry. *Spectrochim. Acta, Part B*, **44**: 1059–1063.

Littlejohn, D. (1988): Graphite furnace atomic emission spectrometry — the rediscovery of a technique. *Anal. Proceed.*, **25**: 217–220.

Littlejohn, D. and Ottaway, J.M. (1978). Improved detection of volatile elements by carbon furnace atomic-emission spectrophotometry using a carbon tube fitted with a sample cup. *Analyst,* **103**: 662–665.

Littlejohn, D. and Ottaway, J.M. (1979): Investigation of atomiser tube design for carbon furnace atomic-emission spectrometry. *Analyst,* **104**: 1138–1150.

Littlejohn, D.; Cook, S.; Durie, D. and Ottaway, J.M. (1984). Investigation of working-conditions for graphite probe atomization in electrothermal atomic-absorption spectrometry. *Spectrochim. Acta, Part B,* **39**: 295–304.

Littlejohn, D.; Duncan, I.S.; Hendry, J.B.M.; Marshall, J. and Ottaway, J.M. (1985). Comparison of uncoated, pyro-coated and totally pyrolytic graphite tubes for the HGA-500 electrothermal atomiser. *Spectrochim. Acta, Part B,* **40**: 1677–1687.

Littlejohn, D.; Marshall, J.; Carroll, J.; Cornack, W. and Ottaway, J.M. (1983). Automatic graphite probe sample introduction for electrothermal atomic-absorption spectrometry. *Analyst,* **108**: 893–896.

Liveing, G.D. and Dewar, J. (1882). On an arrangement of the electric arc for the study of the radiation of vapours, together with preliminary results. *Proc. R. Soc. Lond.,* **34**: 119–122.

López-García, I.; Sánchez-Merlos, M. and Hernández-Córdoba, M. (1996). Slurry sampling for the determination of lead, cadmium and thallium in soils and sediments by electrothermal atomic absorption spectrometry with fast-heating programs. *Anal. Chim. Acta,* **328**: 19–25.

Luge, S.; Broekaert, J.A.C.; Schalk, A. and Zach, H. (1995): The use of different sample introduction techniques in combination with the low power stabilized capacitive plasma (SCP) as a radiation source for atomic emission spectrometry, *Spectrochim. Acta, Part B,* **50**: 441–452.

Lundberg, E.; Frech, W.; Baxter, D.C. and Cedergren, A. (1988). Spatially and temporally constant-temperature graphite furnace for atomic absorption/emission spectrometry. *Spectrochim. Acta, Part B,* **43**: 451–457.

L'vov, B.V. (1959). Исследование атомных спектров поглощения путем полного испарения вещества в графитовой кювете. *Inzh. Fiz. Zh.,* **2(2)**: 44–52.

L'vov, B.V. (1961). The analytical use of atomic absorption spectra. *Spectrochim. Acta, Part B,* **17**: 761–770.

L'vov, B.V. and Lebedev, G.G. (1967). Pulsed electroheating electrode in the graphite coated atomic absorption spectral analysis. *Zhur. Priklad. Spektroskopii,* **7**: 264–265.

L'vov, B.V.; Pelieva, A.L. and Sharmopolsky, A.I. (1977). Уменьшение влияния основы при атомно-аъсоръционном анализе растворов в труъчатых печах путем испарения проъ с графитовой подложки. *Zh. Prikl. Spektrosk.,* **27**: 395–399.

Majidi, V. and Miller-Ihli, N.J. (1998). Influence of graphite substrate on analytical signals in electrothermal vaporization-inductively coupled plasma mass spectrometry. *Spectrochim. Acta, Part B,* **53**: 965–980.

Mandelshtam, S.L.; Semenov, N.N. and Turovtseva, Z.M. (1956). Evaporation method and its use to determine boron and other uranium mixtures. *Zhur. Analit. Khim.,* **11**: 9–20.

Maniaci, M.J. and Tong, W.G. (2004). Multiphoton laser wave-mixing absorption spectroscopy for samarium using a graphite furnace atomizer. *Spectrochim. Acta, Part B*, **59**: 967–973.

Marais, P.J.J.G.; Panichev, N.A. and Katskov, D.A. (2000). Performance of the transverse heated filter atomizar for the atomic absorption determination of mercury. *J. Anal. At. Spectrom.*, **15**: 1595–1598.

Marshall, J.; Giri, S.K.; Littlejohn, D. and Ottaway, J.M. (1983): An investigation of graphite-probe atomisation for carbon-furnace atomic emission spectrometry. *Anal. Chim. Acta*, **147**: 173–184.

Massmann, H. (1967). Bestimmung von Arsen mittles Atomabsorption. Fresenius' *Z. Anal. Chem.*, **225**: 203–213.

Matoušek, J.P. and Stevens, B.J. (1971). Biological applications of the carbon rod atomizer in atomic absorption spectroscopy. *Clin. Chem.*, **17**: 363–368.

Moens, L.; Verrept, P.; Boonen, S.; Vanhaecke, F. and Dams, R. (1995): Solid sampling electrothermal vaporization for sample introduction in inductively coupled plasma atomic emission spectrometry and inductively coupled plasma mass spectrometry, *Spectrochim. Acta, Part B*, **50**: 463–475.

Naumann, B.; Knull, B.; Kerstan, F. and Opfermann, J. (1988). Multivariate optimisation of simultaneous multi-element analysis by furnace atomic non-thermal excitation spectrometry (FANES). *J. Anal. At. Spectrom.*, **3**: 1121–1126.

Ng, K.C. and Caruso, J.A. (1982). Microliter sample introduction into an inductively coupled plasma by electrothermal carbon cup vaporization. *Anal. Chim. Acta*, **143**: 209–222.

Ohlsson, K.E.A.; Sturgeon, R.E.; Willie, S.N. and Luong, V.T. (1993): Figures of merit for two-step furnace atomization plasma emission spectrometry. *J. Anal. At. Spectrom.*, **8**: 41–43.

Ohta, K. and Suzuki, M. (1975). Determination of selenium in metallurgical samples by flameless atomic absorption spectrometry. *Anal Chim Acta.*, **77**: 288–292.

Ohta, K.; Itoh, S. and Mizuno, T. (1992). Electrothermal atomization atomic absorption spectrometry with platinum tube atomizer in air. *Anal. Lett.*, **25**: 745–752.

Oikari, R.; Häyrinen, V.; Parviainen, T. and Hernberg, R. (2001). Continuous monitoring of toxic metals in gas flows using direct-current plasma excited atomic absorption spectroscopy. *Appl. Spectrosc.*, **55**: 1469–1477.

Okamoto, Y.; Kakigi, H. and Kumamaru, T. (1993): Determination of cadmium and lead by inductively coupled plasma atomic emission spectrometry with tungsten boat furnace vaporizer. *Anal. Sci.*, **9**: 105–109.

Okamoto, Y.; Murata, H.; Yamamoto, M. and Kumamaru, T. (1990): Determination of vanadium and titanium in steel by inductively coupled plasma atomic emission spectrometry with modified use of a tungsten boat furnace atomizer for atomic absorption spectrometry. *Anal. Chim. Acta*, **239**: 139–143.

Okamoto, Y.; Sugawa, K. and Kumamaru, T. (1994): Chemical modification for inductively coupled plasma atomic emission spectrometric determination of boron with tungsten boat furnace vaporizer. *J. Anal. At. Spectrom.*, **9**: 89–92.

Oreshkin, V.N. and Tsizin, G.I. (2004). Determination of the total trace elements in natural waters by the sorption–atomic absorption method with the fractional evaporation of concentrates in a crucible atomizer. *J. Anal. Chem.*, **59**: 890–894.

Ortner, H.M.; Birzer, W.; Welz, B.; Schlemmer, G.; Curtius, J.A.; Wegscheider, W. and Sychra. V. (1986). Surfaces and materials of electrothermal atomic absorption spectrometry — more than a merely morphological study. *Fresenius' Z. Anal. Chem.*, **323**: 681–688.

Ortner, H.M.; Rohr, U.; Schlemmer, G.; Weinbruch, S. and Welz, B. (2002). Corrosion of transversely heated graphite tubes by iron and lanthanum matrices. *Spectrochim. Acta, Part B,* **57**: 243–260.

Ottaway, J.M. and Shaw, F. (1975). Carbon furnace atomic-emission spectrometry — preliminary appraisal. *Analyst,* **100**: 438–445.

Ottaway, J.M.; Carroll, J.; Cook, S.; Corr, S.P.; Littlejohn, D. and Marshall. J. (1986). Developments in probe design for electrothermal atomic-absorption spectrometry. *Fresenius Z. Anal. Chem.*, **323**: 742–747.

Pantano, P. and Sneddon, J. (1989). Effect of atomization surface on the quantitation of vanadium by electrothermal atomization atomic absorption spectrometry. *Appl. Spectrosc.*, **43**: 505–511.

Papp, L. and Bánhidi, O. (1998). Computer-controlled electrothermal-hollow cathode emission spectrometric source for the simultaneous multi-element determination of trace elements in environmental liquid samples in microlitre amounts. *J. Anal. At. Spectrom.*, **13**: 653–657.

Parker, M. and Marcus, R.K. (1994). Role of discharge parameters and limiting orifice diameter in radio-frequency glow discharge-atomic absorption spectrophotometry (rf-GD-AAS). *Appl. Spectrosc.*, **48**: 623–629.

Pavski, V.; Chakrabarti, C.L. and Sturgeon, R.E. (1994): Spatial imaging of the furnace atomization plasma emission spectrometry source. *J. Anal. At. Spectrom.*, **9**: 1399–1409.

Pavski, V.; Sturgeon, R.E. and Chakrabarti, C.L. (1997). Effect of bias voltage and easily-ionized elements on the spatial distribution of analytes in furnace atomization plasma emission spectrometry. *J. Anal. At. Spectrom.*, **12**: 709–723.

Perez-Parajón, J.M. and Sanz-Medel, A. (1994). Determination of silicon in biological fluids using metal carbide-coated graphite tubes. *J. Anal. At. Spectrom.*, **9**: 111–116.

Pineau, A.; Chappuis, P.; Arnaud, J.; Baruthio, F.; Zawislak, R. and Jaudon, M.C. (1992). Interlaboratory tests: development of a method for measuring aluminium in serum by electrothermal atomic absorption spectrometry. *Ann. Biol. Clin.*, **50**: 577–585.

Quan Zhe; Ni Zhe-ming and Yan Xiu (1994). Influence of atomizer surface on the kinetics of tin atomization in electrothermal atomic absorption spectrometry. *Can. J. Appl. Spectrosc.*, **39**: 54–59.

Rettberg, T.M. and Holcombe, J.A. (1984). A temperature controlled, tantalum second surface for graphite furnace atomization. *Spectrochim. Acta, Part B,* **39**: 249–260.

Rettberg, T.M. and Holcombe, J.A. (1986). Direct analysis of solids by graphite furnace atomic absorption spectrometrry using a second surface atomizer. *Anal. Chem.*, **58**: 1462–1467.

Riby, P.G. and Harnly, J.M. (1993): Characterization of a Helium discharge for hollow anode furnace atomization non-thermal excitation spectrometry, *J. Anal. At. Spectrom.*, **8**: 945–953.

Riby, P.G.; Harnly, J.M.; Styris, D.L. and Ballou, N.E. (1991): Emission characteristics of chromium in hollow anode-furnace atomization non-thermal excitation spectrometry, *Spectrochim. Acta, Part B*, **46**: 203–215.

Robinson, J.W. and Wolcott, W. (1975). A hollow-T carbon atomizer for atomic absorption spectrometry. *Anal. Chim. Acta*, **74**: 43–52.

Rohr, U.; Ortner, H.M.; Schlemmer, G.; Weinbruch, S. and Welz, B. (1999). Corrosion of transversely heated graphite tubes by mineral acids. *Spectrochim. Acta, Part B*, **54**: 699–718.

Sanford, C.L.; Thomas, S.E. and Jones, B.T. (1996). Portable, battery-powered, tungsten coil atomic absorption spectrometer for lead determinations. *Appl. Spectrosc.*, **50**: 174–181.

Schäffer, U. and Krivan, V. (1998): A graphite furnace electrothermal vaporization system for inductively coupled plasma atomic emission spectrometry. *Anal. Chem.*, **70**: 462–490.

Schlemmer, G. and Welz, B. (1986). Influence of the tube surface on atomization behaviour in a graphite tube furnace. *Fresenius' Z. Anal. Chem.*, **323**: 703–709.

Shan, X.-q.; Radziuk, B.; Welz, B. and Sychra, V. (1992). Application pf palladium as a chemical modifier in electrothermal absorption spectrometry with a tungsten tube atomizer. *J. Anal. At. Spectrom.*, **7**: 389–396.

Siemer, D.D. (1982). Four rod carbon rod atomizer for AAS. *Anal. Chem.*, **54**: 1659–1663.

Silva, M.M.; Silva, R.B.; Krug, F.J.; Nobrega, J.A. and Berndt, H. (1994). Determination of barium in waters by tungsten coil electrothermal atomic absorption spectrometry. *J. Anal. At. Spectrom.*, **9**: 861–865.

Slavin, W.; Manning, D.C. and Carnrick, G.R. (1981). Effect of graphite furnace substrate materials an analysis by furnace atomic absorption spectrometry. *Anal. Chem.*, **53**: 1504–1509.

Smith, C.M.M. and Harnly, J.M. (1995). Effect of elevated gas pressure on atomization in graphite furnace continuum source atomic absorption spectrometry with linear photodiode array detection. *J. Anal. At. Spectrom.*, **10**: 187–195.

Smith, D.L.; Liang, D.C.; Steele, D. and Blades, M.W. (1990). Analytical characteristics of furnace atomization plasma excitation spectrometry (FAPES). *Spectrochim. Acta, Part B*, **45**: 493–498.

Somov, A.R. and Gilmutdinov, A.Kh. (1997). Irradiance distribution in the image of a tube electrothermal atomer. *Spectrochim. Acta, Part B*, **52**: 1413–1420.

Sperling, M.; Welz, B.; Hertzberg, J.; Rieck, C. and Marowsky, G. (1996). Temporal and spatial temperature distribution in transversely heated graphite tube atomizers and their analytical characteristics for atomic absorption spectrometry. *Spectrochim. Acta, Part B*, **51**: 897–930.

Sturgeon, R.E. and Berman, S.S. (1983). Determination of the efficiency of the graphite-furnace for atomic-absorption spectrometry. *Anal. Chem.*, **55**: 190–200.

Sturgeon, R.E. and Chakrabarti, C.L. (1977). Evaluation of pyrolytic-graphite-coated tubes for graphite furnace atomic absorption spectrometry. *Anal. Chem.,* **49**: 90–97.

Sturgeon, R.E. and Guevremont, R. (1997). Furnace atomization plasma ionization mass spectrometry. *Anal. Chem.,* **69**: 2129–2135.

Sturgeon, R.E.; Pavski, V. and Chakrabarti, C.L. (1996). Two-dimensional imaging of excited analyte species in electrothermal vaporizers. *Spectrochim. Acta, Part B,* **51**: 999–1006.

Sturgeon, R.E.; Willie, S.N.; Luong, V. and Berman, S.S. (1990): Determination of cadmium and lead in sediment and biota by furnace atomisation plasma emission spectrometry. *J. Anal. At. Spectrom.,* **5**: 635–638.

Sturgeon, R.E.; Willie, S.N.; Luong, V. and Dunn, J.G. (1991). Influence of the generator frequency on the analytical characteristics of Fapes. *Appl. Spectrosc.,* **45**: 1413–1418.

Sturgeon, R.E.; Willie, S.N.; Luong, V.; Berman, S.S. and Dunn, J.G. (1989): Furnace atomisation plasma emission spectrometry (FAPES). *J. Anal. At. Spectrom.,* **4**: 669–672.

Tarasevich, Y.I. and Aksenenko, E.V. (2003). Interaction of water, methanol and benzene molecules with hydrophilic centres at a partially oxidised model graphite surface. *Colloids and Surfaces A: Physicochem. Eng. Aspects,* **215**: 285–291.

Terui, Y.; Yasuda, K. and Mirokawa, K. (1991). Measurement of effective vapor temperature in graphite furnace of atomic absorption spectrometry. *Anal. Sci.,* **7**: 599–604.

Tilotta, D.C.; Busch, M.A. and Busch, K.W. (1991): A miniature electrical furnace as an excitation source for low-temperature, gas phase, infrared emission spectrometry. *Appl. Spectrosc.,* **45**: 178–185.

Vanderwoort, K.G.; Butcher, D.J.; Brittain, C.T. and Lewis, B.B. (1996). Scanning Tunneling Microscope Images of Graphite Substrates used in Graphite Furnace Atomic Absorption Spectrometry. *Appl. Spectrosc.,* **50**: 928–938.

Vanderwoort, K.G.; McLain, K.N. and Butcher, D.J. (1997). Scanning Tunneling Microscope Study of Etch Pits on Highly Oriented Pyrolytic Graphite Heated in an Atomic Absorption Electrothermal Analyzer. *Appl. Spectrosc.,* **51**: 1896–1904.

Volynsky, A.B. (1995). Terminology for the modification of graphite tubes with high-melting carbides used in electrothermal atomic absorption spectrometry. *Spectrochim. Acta, Part B,* **50**: 1417–1419.

Volynsky, A.B. (1998a). Application of graphite tubes modified with high-melting carbides in electrothermal atomic absorption spectrometry. I. General approach. *Spectrochim. Acta, Part B,* **53**: 509–535.

Volynsky, A.B. (1998b). Graphite atomizers modified with high-melting carbides for electrothermal atomic absorption spectrometry. II. Practical aspects. *Spectrochim. Acta, Part B,* **53**: 1607–1645.

Wang, P. and Holcombe, J.A. (1992). Pressure-regulated electrothermal atomizer for atomic absorption spectrometry. *Spectrochim. Acta, Part B,* **47**: 1277–1286.

Wang, P. and Holcombe, J.A. (1994). Electrothermal atomization with atomic absorption at reduced pressures for studies of analyte distribution in solids. *Appl. Spectrosc.,* **48**: 713–719.

Welz, B.; Schlemmer, G.; Ortner, H.M. and Birzer, W. (1989). Scanning electron microscopy studies on surfaces from electrothermal atomic absorption spectrometry — IV. Total pyrolytic graphite tubes. *Spectrochim. Acta, Part B*, **44**: 1125–1161.

Welz, B.; Sperling, M. and Schlemmer, G. (1988). Spatially and temporally resolved gas phase temperature measurements in a massmann-type graphite tube furnace using coherent anti-Stokes Raman scattering. *Spectrochim. Acta, Part B*, **43**: 1187–1207.

West, T.S. and Williams, X.K. (1969). Atomic absorption and fluorescence spectroscopy with a carbon filament atom reservoir. Part I. Construction and operation of atom reservoir. *Anal. Chim. Acta*, **45**: 27–41.

Wizemann, H.D. and Haas, U. (2003). The rare earth elements as candidates for isotope selective graphite furnace applications. *Spectrochim. Acta, Part B*, **58**: 931–947.

Woodriff, R. and Ramelow, G. (1968). Atomic absorption spectroscopy with a high-temperature furnace. *Spectrochim. Acta, Part B*, **23**: 665–671.

Woodriff, R.; Stone, R.W. and Held, A.M. (1968). Electrothermal atomization for atomic absorption analysis. *Appl. Spectrosc.*, **22**: 408–411.

Xiu-Ping, Y.; Zhe-Ming, N.; Xiao-Tao, Y. and Guo-Quiang, H. (1993). Kinetics of indium atomization from different atomizer surfaces in electrothermal atomic absorption spectrometry (ETAAS). *Talanta*, **40**: 1839–1846.

Yamamoto, M.; Murayama, S.; Ito, M. and Yasuda, M. (1980). Theoretical basis for multielement analysis by coherent forward scattering atomic spectroscopy. *Spectrochim. Acta, Part B*, **35**: 43–50.

Yong-III Lee; Smith, M.V.; Indurthy, S.; Deval, A. and Sneddon, J. (1996). An improved impaction-graphite furnace system for the direct and near real-time determination of cadmium, chromium, lead and manganese in aerosols and cigarette smoke by simultaneous multielement atomic absorption spectrometry. *Spectrochim. Acta, Part B*, **51**: 109–116.

Zaidel, A.N.; Kaliteevskii, N.I.; Lipis, L.V.; Chaika, M.P. and Belyaev, Y.I. (1956). Спектралъный анализ по методу испарения 1. Принцип метода испарения примесей в вакууме И некоторые ео приложения.. *Zhur. Analit. Khim.*, **11**: 21–29.

Záray, G. and Kántor, T. (1995): Direct determination of arsenic, cadmium, lead and zinc in soils and sediments by electrothermal vaporization and inductively coupled plasma excitation spectrometry. *Spectrochim. Acta, Part B*, **50**: 489–500.

Chapter 3

Sample Introduction

3.1. Introduction

Before generating the analytical signal in ETAAS, the sample must be introduced into the tube and then transformed into an atomic vapor cloud. These processes are critical, and consequently adequate procedures for introducing the sample into the atomizer are required.

The sample introduction into the graphite tube (or onto the platform within the tube) is the first stage in ETAAS measurements. The usual way of introducing a sample for ETAAS is as a solution (normally an aqueous solution), although it may also be introduced in the solid or gaseous state.

3.2. Solutions: Liquids

The commonest way of introducing samples into an electrothermal atomizer is in the liquid state. However, there are various operational options.

3.2.1. Introduction of Solutions and Aerosols

There are two widely used ways for introducing liquid samples:

- as a small volume (1–50 μL) of liquid (liquid drop approach), either manually using a micropipette or automatically utilizing an autosampler, and
- as aerosol (aerosol approach), where the graphite tube is maintained at a temperature above 100°C, so that the aerosol dries rapidly in contact with the graphite wall [Alary and Salin (1995)] (Fig. 3.1).

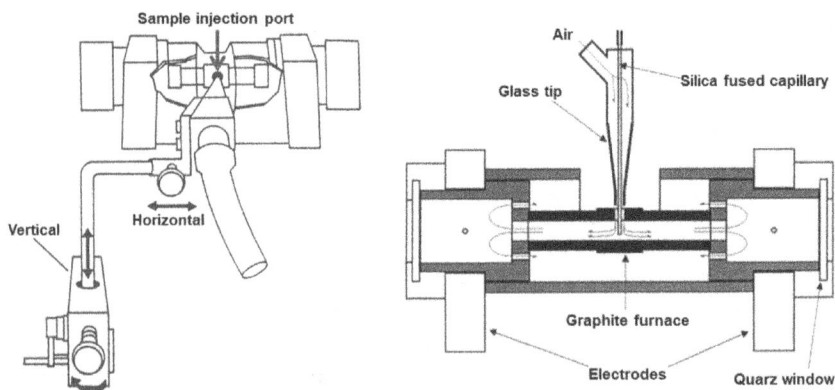

Fig. 3.1. (Left) Schematic for an aerosol introduction system (Fastac system, used in Thermo Jarrel Ash instruments). (Right) Positioning of the silica fused capillary in the graphite furnace (arrows show the air path). [Reproduced from Thermo Jarrel Ash.]

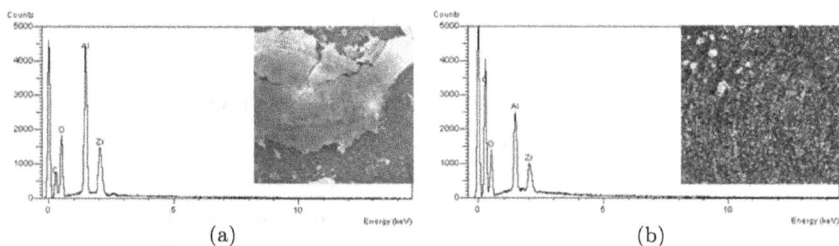

Fig. 3.2. Secondary electron pictures (X27) with the corresponding ED X-ray spectra, which show general morphological and chemical changes of the solid residue constituted of aluminium and zirconium salts (100 μg each) on a (a) non-pyrolytic, and (b) pyrolytic graphite platform, after heating to 1000°C [Castro *et al.* (2005)].

Any ideal introduction technique of liquid samples ought to deposit them into the graphite tube in such a way that once the solvent (normally water) evaporates there would be reproducible distribution of the analyte and the salts constituting the matrix, forming a homogeneous thin layer over the graphite surface. This would allow near-simultaneous volatilization of all the analyte atoms, so that during the analysis stage an intense narrow peak would be obtained. However, practically the dry residue from a sample volume of 10 μL forms relatively large-sized crystals during the drying stage, even after heating at higher temperatures (Fig. 3.2), but this also depends on the platform substrate [Castro *et al.* (2005)].

The sample introduction as an aerosol has been described as a method presenting certain improvements with respect to the liquid drop introduction approach. This is because, if the aerosol is deposited onto the hot graphite surface, each aerosol drop should dry during contact with that surface. Owing to the small drop size, and the small quantity of solid within each aerosol particle, this procedure should be able to leave residues with smaller crystal sizes on the graphite surface. Consequently, taller and narrower absorption signals than those obtained by the liquid drop injection are expected.

However, some researches [Howell and Koirtyohann (1992)] show contrary results. With the sample introduction as an aerosol, a ring of solids particles is formed around the impact point of each aerosol particle on the hot surface. The aerosol is seen as cooling the surface and thus permitting the accumulation of more un-evaporated liquid.

This liquid would be forced towards the outside of the ring through the impact of the aerosol being deposited, but when the liquid ring reaches the hottest region of the tube it would evaporate, producing a larger crystalline solid ring. Nonetheless, the sample introduction as an aerosol allows the analyte to be retained on the graphite surface for longer periods, until the furnace reaches higher temperatures in comparison with those obtained through manual injection. On the other hand, the sample introduction as an aerosol is more corrosive for a pyrolytic graphite-coated graphite surface, which causes a loss of sensitivity against manual injection. Analytical differences between the two ways of sample introduction are minimal when the solutions introduced hold no dissolved solids [Howell and Koirtyohann (1992)], but these differences are more and more noticeable as the quantity of dissolved solids increases.

If several successive injections are carried out through the liquid drop (and the aerosol) approaches, a pre-concentration of the analyte into the graphite tube occurs [Nae-wen Kuo *et al.* (1993); Ceccarini *et al.* (2001); Lanza *et al.* (1997)].

When a mass, m_{dep}, of the analyte from the original sample solution is deposited into the graphite tube, the maximum absorbance, A_{max}, and the integrated absorbance, A_{int}, due to the analyte can be expressed as a function of certain operating parameters [Lanza et al. (1997)]

$$A_{max} = S\varphi'\varphi''C_s V_{in} \tag{3.1}$$

and

$$A_{int} = \int_0^{t_i} A(t)dt = S'\varphi'\varphi''C_s V_{in}, \tag{3.2}$$

where C_s is the analyte concentration in the sample solution, V_{in} is the sample volume introduced, S and S' are the method sensitivities, t_i is the integration time, which refers to the atomization stage that follows the deposition stage, φ' is the atomization efficiency (which is controlled by the graphite interface and by the concentration and nature of the matrix), and φ'' is the deposition efficiency.

While experimental conditions remain constant, and for a given analyte, if φ', and φ'' do not vary during the pre-concentration stage, Eq. (3.2) may be rewritten as follows

$$A_{int} = S'\varphi\varphi'\varphi''v'C_s t_{dep} = k\varphi v'C_s t_{dep}, \qquad (3.3)$$

where v' is the sample aspiration rate (mL/min) and φ the aerosol efficiency reaching the graphite tube. Equation (3.3) demonstrates that pre-concentration of analytes by means of aerosol deposition in ETAAS depends linearly on the deposition or aspiration time, t_{dep}, from the sample solution. Similar equations can be written for A_{max}.

A platform-to-platform sample transfer, distribution, dilution, and dosing system has been described, which provides higher transport efficiency and allows external sample pretreatment, sampling, and handling with aerosol dosing of a primary solid organic or inorganic, bulk, powder or liquid sample [Hermann *et al.* (2004)].

3.2.1.1. *Introduction of organic solvents*

The use of organic solvents in flame atomic absorption spectrometry provides some advantages. However, advantages are not usually present in ETAAS. On the contrary, drawbacks and restrictions have been noted in respect of their use [Cassella *et al.* (2002); Saint'Pierre *et al.* (2003)]. Thus, viscosity, density, surface tension and moistening capacity of the organic solvents affect on the sampling and injection stages.

The solvent effects depend on the solvent type and the atomizer surface [Sommer *et al.* (1992)]. Organic solvents penetrate deeply into the graphite surface showing heavy background absorption. Thus, solvents containing benzene rings (xylene, toluene) usually give more background absorption than non-aromatic solvents such as methylisobutyl ketone (MIBK) and *n*-butyl acetate (*n*BA). Organic solvents containing chlorine form chemical bonds with carbon atoms from the atomizer [Tserovsky and Arpadjan (1991)], although chloroform and carbon tetrachloride have been occasionally used [Chung *et al.* (1984); Fan and Zhou (2006)]. Uncoated graphite tubes are not suitable to work with organic solvents, while pyrolytic graphite-coated graphite tubes yield better results because penetration into the graphite is probably more limited. Their elimination requires a high charring temperature (at least 1200°C) and a long pre-heating stage (30–40 seconds).

Negative effects from the physical properties of the organic solvents can be compensated for, at least in part, by injection of the solvent into a pre-heated furnace. An increased sensitivity is usually found in this way. Other alternatives involve injecting the sample as an organic-in-water micro-emulsion [Aucélio *et al.* (2000)]. In this way, calibration by direct comparison with aqueous standards is possible, while sampling is easy and repeatable. Platforms and tubes impregnated with tungsten provide high sensitivity working with organic solvents and permitting higher char and atomization temperatures. The most appropriate drying temperature is whichever coincides with the boiling point of the organic solvent. The inclusion of a cool-down step prior to atomization offers no benefits when organic solvents are used.

Absorbance-time peaks derived using organic solvents are slightly lower and wider than those obtained with aqueous solutions. These differences might be explained on the basis of several different physical processes: greater dissemination of the sample within the tube and deeper penetration by the organic solvents into the graphite, exceeding the losses by diffusion through the walls of the furnace. In an organic medium, the stabilizing effect is achieved with smaller quantities of the chemical modifier (around 50 times less).

3.2.2. Liquid Chromatography (HPLC)

The growing interest in speciation of metals has brought with coupling studies that would take advantage of the capacity of the separation techniques in combination with the high sensitivity and selectivity of various spectroscopic detectors. Various *on-line* and/or *in-line* combinations of different separation techniques and atomic absorption spectroscopy have been tried out [Tsalev (1999)]. Among all the separation techniques *on-line* coupled to ETAAS, the chromatographic techniques are the most widely used. Some of these systems include *on-line* couplings of ionic chromatography and HPLC with ETAAS (Fig. 3.3) [Brinckman *et al.* (1977); Kölbl *et al.* (1993)]. However, other

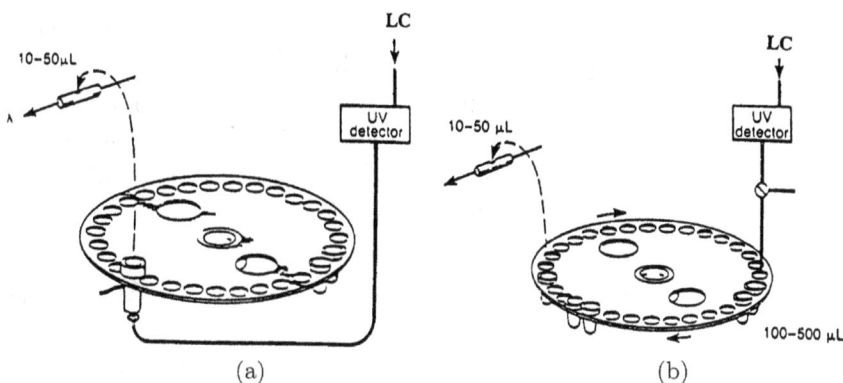

Fig. 3.3. Schematic for a sampling carrousel used to collect HPLC eluents before ETAAS in pulsed (a) and continuous (b) modes [Brinckman *et al.* (1977)].

combinations incorporate automatic sampling systems [Kuo *et al.* (1993)], trapping systems within the atomizer [Zhe-ming *et al.* (1993); Heng-bin *et al.* (1993)], and automatic *on-line* and *in-line* column pre-concentration and separation systems [Sahayam (2002); Bermejo-Barrera *et al.* (2002); Robles *et al.* (1999)]. Sample introduction systems from high-pressure hydraulic nebulizers have also been coupled to ETAAS. In this, the sample is introduced into an aerosol deposition module with a high pressure hydraulic nebulizer so that the aerosol is injected into the graphite tube [Berndt and Schaldach (1994)].

The use of liquid chromatography (LC) for this purpose offers various advantages with respect to gas chromatography (GC) [Donard and Martín (1992)]. Particularly, it permits separation of species without derivatization and the great variety of LC techniques (adsorption, ion-exchange, gel-permeation, normal and reverse phase chromatography) allows the separation of different species: ions, volatile species, high molecular mass organo-metals, complexes, and biological large size compounds. Coupling between LC techniques and atomic absorption spectrometry mostly yields instrumentation with low sensitivity; the contrary happens with cryogenic trapping methods which are much more sensitive. The interface is one of the factors conditioning sensitivity. Finally, the organic solvents are usually a source of interference in the detector.

The two main problems encountered in direct coupling between LC and ETAAS instrumentation are [Ebdon *et al.* (1987)]: (i) to keep an ET atomizer at temperatures above 1000°C for a long time, likely causing a shortening of the half-life of the furnace; and (ii) systems for atomization and detection are not designed to work continuously with the relatively large solvent volumes used in LC, although some alternative approaches have been proposed [Soldado Cabezudo *et al.* (1997)]. Transformation of the liquid eluent into a gas at 1000–2500°C is a drawback that has been solved on various occasions through using an electrically heated interface (Fig. 3.4) [Nygren *et al.* (1988); High *et al.* (1992); Bendicho (1994a)].

This thermal pulverization (thermospray) system shows great potential, also finding check applications in the introduction of discrete samples into electrothermal atomizers [Zhang *et al.* (2000)]. This thermal interface has been applied to the determination of various tin species [Nygren *et al.* (1988)] and arsenic [Bendicho (1994b)]. One of the advantages exhibited by the thermospray system is the possibility of using shorter temperature programs (Fig. 3.5) [Bendicho (1994a)].

3.2.3. Flow Injection (FI) Systems

FI techniques have been defined as the techniques which provide information from a concentration gradient formed in a well-defined zone in a fluid, dispersed within a non-segmented continuous flow of transporter. This broad definition includes both GC and HPLC. However, when the definition is established in terms of the application, a clear difference appears between chromatographic techniques and FI techniques. In this sense, chromatographic techniques are designed in the first place to achieve *on-line* separations of a given number of components of the sample based on the relative affinities between the stationary and mobile phases. However, FI techniques are designed to quantify a limited number of components of a sample based on a selective *on-line* chemical reaction. Furthermore, the usual chromatographic detectors are non-specific detectors, while FI detectors usually play an important role in achieving the method's selectivity.

The basic characteristic of the FI thechniques is that they are dynamic and the signal produced by the detector is continuously changing with time. Thus, the signal profile produced is controlled

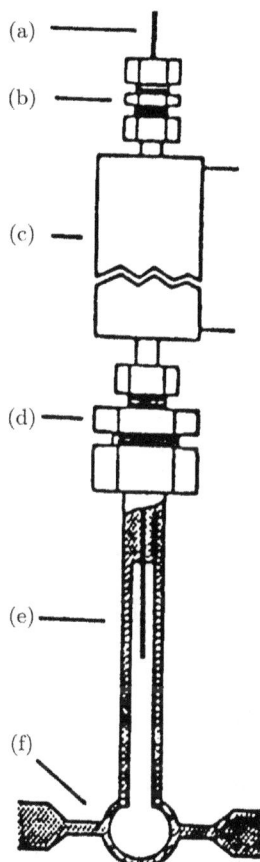

Fig. 3.4. Schematic of a thermal pulverization interface: (a) fused silicon capillary coming from the HPLC column; (b) joint; (c) upper part of the interface, consisting of a stainless steel tube covered by a heating wire insulated with porcelain and with electrical connections; (d) joint; (e) lower part of the interface consisting of a vitreous carbon tube; (f) integrated contact cuvette in the furnace [Nygren *et al.* (1988)].

by a combination of the reagent concentration profiles and thermodynamics and kinetics of the relevant chemical reactions.

The FI systems provide the most flexible and versatile automation of the different operations necessary for a chemical assay, because flows may be mixed, trapping, reversed, split, recombined and sampled, for different contact times and with great precision [Burguera and Burguera (1997)].

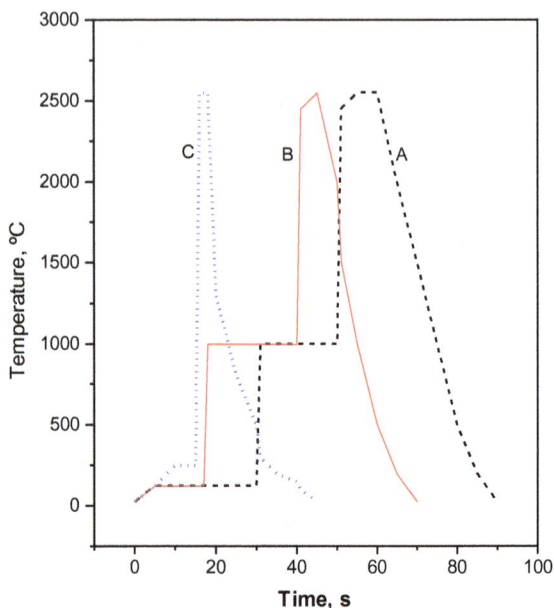

Fig. 3.5. Thermal program for: (A) conventional sample introduction; (B) deposition through thermal pulverization and introduction by flow injection; and (C) introduction through thermal pulverization in an HPLC-ETAAS coupling [Bendicho (1994a)].

The FI methods, widely utilized in FI analysis (FIA), present two important features: (i) a high pre-concentration factor, generally between one and two orders of magnitude; and (ii) excellent separation of the analyte from the interferring matrix constituents. By means of this sample introduction system, both the standards and samples reach the atomizer with the same matrix, the eluent. The FI system permits working with very small sample volumes. Furthermore, it is possible to handle samples with high contents of solids and varied viscosity. It also gives the possibility of adding reagents and of pre-concentrating the analyte, eliminating the matrix, or using organic solvents.

The FI method, seen as a micro-sample introduction system, offers some advantages when compared with the manual injection procedure, such as the totally automatic manipulation of the sample within a closed system and the possibility of an automatic *on-line* pre-concentration of the analyte (Fig. 3.6) [Bosch Ojeda *et al.* (2003); Zhen-Shan Liu and Shang-Da Huang (1993)]. Pre-concentration and

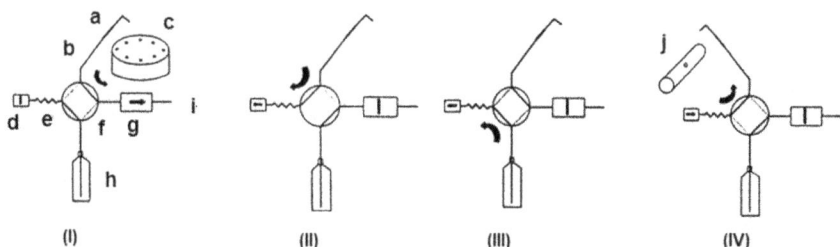

Fig. 3.6. On-line pre-concentration system. (a) Microcolumn C_{18}, (b) sampling capillary, (c) sample tray, (d) pump, (e) delay tube, (f) four-way distribution valve, (g) pomp, (h) bottle containing methanol, (i) drain, (j) graphite tube [Zhen-Shan Liu and Shang-Da Huang (1993)].

separation on solid and liquid phases by means of FI are discontinuous processes that are well suited to hyphenate with techniques like ETAAS because of their discrete and discontinuous nature [Tao and Fang (1995)]. The closed status of the FI system, together with direct coupling to the sample introduction capillary, reduces contamination problems. Since the analyte is separated efficiently from the matrix and is introduced into the graphite furnace practically in the form of a pure solution, chemical modifiers are usually not necessary. This also reduces chemical contamination. *On-line* purification of the reagent is another characteristic of FI that aids in maintaining the blank values at a low level, permitting determination in the ng/L range for a great number of elements [Fang *et al.* (1990); Sperling *et al.* (1991)] and in many different types of samples [Tyson (1999)].

Commercially available FI systems comprise a range of valves, injectors, detectors, autosamplers and a data station controlled by computer. Various FI systems using microcolumns have been coupled to ETAAS (Fig. 3.7) [Welz *et al.* (1993); Liu and Huang (1993); Tyson (1991); Arruda *et al.* (1993)]. Couplings were made of a FI system which included *on-line* pre-concentration and precipitation stages followed by collection of the analyte precipitated. This precipitate is dissolved in 20 μL of acidic solution and transported directly to the graphite atomizer [Sella *et al.* (1997)]. González *et al.* (2001) used a continuous precipitation-dissolution flow system to determine the total arsenic. Arsenic was precipitated as Ag_3AsO_4, and then dissolved

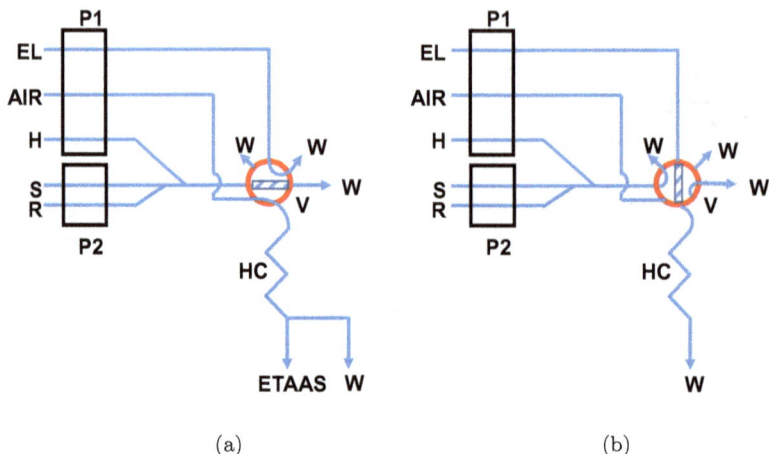

(a)   (b)

Fig. 3.7.   Solid-phase extraction *on-line* coupled with ETAAS. (P1 and P2: peristaltic pumps; W: drain; ETAAS: electrothermal atomizer). There are five steps to the procedure: (1) The sample, S, is mixed with the reagent, R, and introduced into the column through the valve, V, (placed as in the left (a) with pumps P1 and P2 running. (2) The column is washed with water, H, while the position of the valve is as in the left, and pumps P1 and P2 are switched off. (3) The column is eluted within the coil, HC, using ethanol, EL, the valve position as in the right (b) and P1 and P2 turned off. (4) The eluate is transferred to the furnace (atomizer) with the valve set as in the left and pumps P1 and P2 not running. (5) Residual analyte is eluted and the sample line is filled with fresh sample, the valve being set as in the right, pumps P1 and P2 turned on and the outlet of the coil, HC, open to the drain [Tyson (1991)].

in ammonia 6 M and transported to an autosampler cup of the ETAAS instrument. A flow injection system has also been coupled to a tungsten coil electrothermal atomizer for determination of lead in natural waters [Barbosa *et al.* (1999)].

Some variants of FIA are sequential injection analysis (SIA) and bead injection lab-on-valve (BI-LOV) introduced in the early and at the end of 1990s, respectively (Fig. 3.8) [Ruzicka and Marshall (1990); Ruzicka (2000); Hansen and Wang (2002); Grudpan (2004); van Staden and Stefan (2004)]. Relatively little work has been done involving coupling FI systems to ETAAS [Zhaolun Fang *et al.* (1996)]. The future offers considerable promise [Zhaolun Fang (1998)], because miniaturization of the manifolds reduces consumption of sample and reagents, minimizes the amounts of waste, allows complex sample

(a) **Flow Injection**

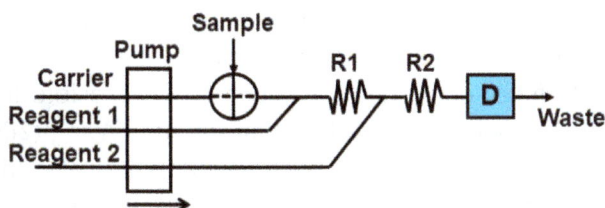

(b) **Sequential Injection**

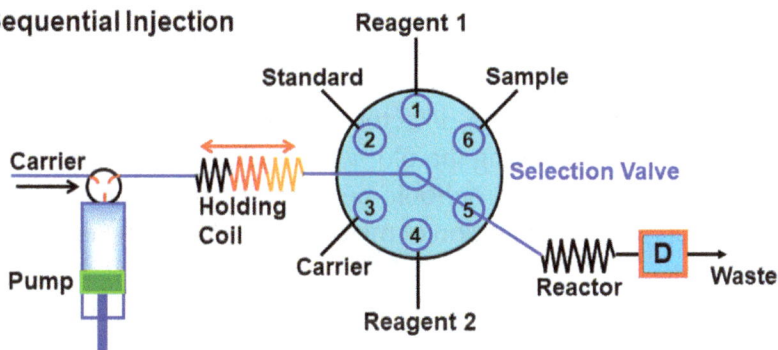

(c) **Lab-on-Valve**

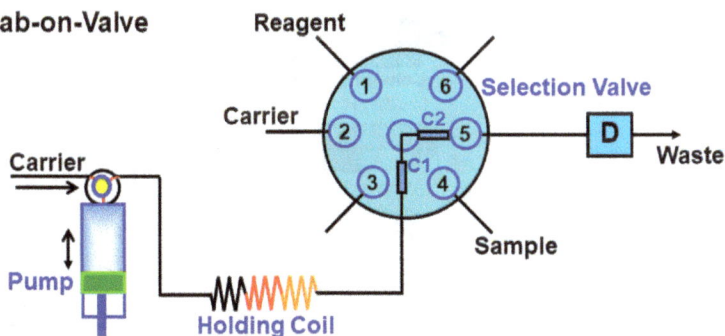

Fig. 3.8. The three generations of FIA, where (a) depicts a typical FIA-manifold, where a defined volume of sample is injected into a continuous flowing carrier stream, which is subsequently merged with two reagent streams. The ensuing transient generation of product is monitored by a suitable detector (D). (b) Typical sequential injection system, based on using a selection valve and a bi-directional syringe pump. (c) Schematic drawing of a lab-on-valve (LOV) [Hansen and Wang (2002)].

manipulation and integrated sequential unit operations [Hansen (2004); Wang and Hansen (2002); Wang and Hansen (2005)].

The use of packed columns reactors in flow systems represents a different separation/pre-concentration scheme now feasible for FI and SI [Wang and Hansen (2000, 2001); Hansen and Wang (2002); Economu (2005)]. Difficulties reported have been circumvented in different ways. One of them is the so-called renewable surface scheme, where a new packed reactor is generated for each assay following four stages (Fig. 3.9): (i) packing of the column, (ii) loading of the sample, (iii) elution of the retained analyte by an appropriate eluent and transfer to the detector, sandwiched by air segments for ETAAS, and (iv) replacing the old beads with new beads for the next assay. A number of hydrophobic bead materials have been used for separation/pre-concentration of metals by ETAAS [Wang *et al.* (2003); Miró *et al.* (2003)].

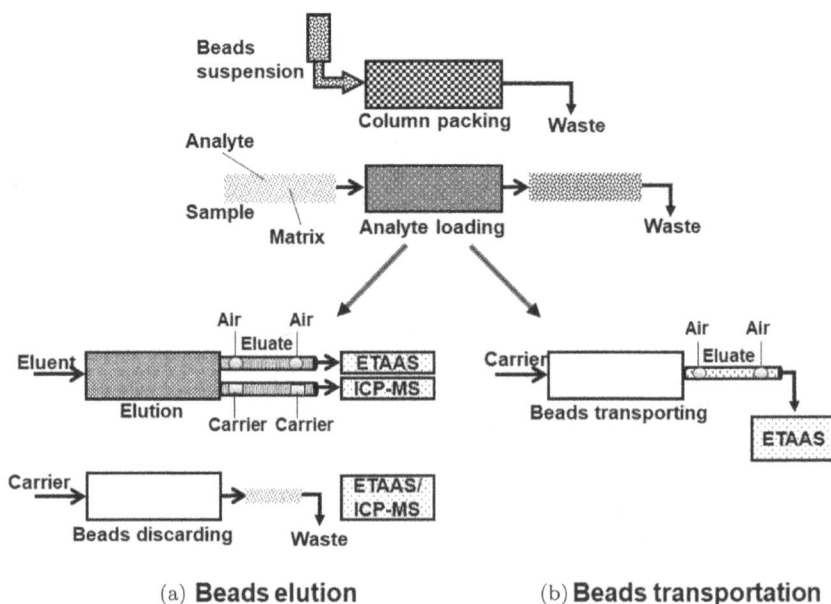

(a) **Beads elution**   (b) **Beads transportation**

Fig. 3.9. The principle of BI as realized in (a) by bead elution, and (b) bead transportation directly into the graphite furnace [Hansen and Wang (2002)].

3.3. Solids

Direct analysis of solid samples, without chemical pretreatment, is an idea extensively pursued with the aim to search for quicker, more efficient, safer, cheaper, simpler and shorter ETAAS analysis [Ali *et al.* (1989); Miller-Ihli (1989)], especially with refractory materials [Hauptkorn and Krivan (1996); Schäffer and Krivan (1996); Friese *et al.* (1996); Krivan and Mao Dong (1998)]. In this procedure, no prior chemical treatment of the sample is used, although there may be some physical changes affecting the size and shape of particles, surface polishing, and the like. In many cases, these physical changes are necessary in order to achieve greater homogeneity of the sample.

Any method for direct analysis of solid samples must provide the following features [Bendicho and Loss-Vollebregt (1991)]:

- applicable to samples with a broad range of composition,
- relatively rapid,
- simple standardization,
- capable of correctly handling large samples to avoid problems of heterogeneity,
- capability of simultaneous multi-element analyses,
- cheap analysis,
- suitable precision and accuracy.

Direct analysis of solids can be undertaken in two ways, depending on the manner of introducing the solid sample: fully dry or as a suspension (or slurry). Analysis of slurries involves previously triturating or grinding up the solid sample (so it is finely divided) and which dispersed in an appropriate liquid diluent, introducing the slurry into the tube. In contrast, introduction of a fully dry sample means that this sample is put directly into the furnace. Introduction of solid particles by means of sputtering from a surface has also been a method used as an analytical tool, although it has not shown much promise [Müller-Vogt *et al.* (1996)].

The advantages of direct solid sampling as opposed to the conventional procedures are evident [Miller-Ihli (1992)]:

- shorter sample preparation time (pretreatment of the sample is considerably reduced, so quicker analyses),
- less sample contamination (no reagents are used or at least they are used to a lesser extent),
- less analyte losses through volatilization prior to the analysis stage and through retention in the insoluble residues during the sample pretreatment,
- avoiding corrosive and dangerous chemicals,
- increased sensitivity (since the samples are not diluted),
- no random risks associated with the use of acids
- easier selective analyses of microsamples.

However, to achieve these benefits some problems need to be overcome [Belarra *et al.* (1999)]:

- great care and training is necessary to put a solid sample into the atomizer,
- different weights are used at each replication,
- many difficulties to prepare standards and blanks,
- the chemical modifiers employed in the direct atomization of solids are usually less effective,
- solid samples offer few possibilities for dilution with high analyte concentration,
- precision is poorer than using liquid samples,
- part of the matrix can remain in the furnace after the atomization stage,
- spectral interferences from a structured background are more frequent,
- the slurry needs to be kept stable and homogeneous during sampling,
- incomplete vaporization and atomization, as well as occlusion within a solid matrix, are the usual problems with refractory metals.

3.3.1. Direct Solid Introduction (Dry Sample)

Direct solid introduction has used [Docekal and Krivan (1995)]: (i) graphite supports (platforms, cups, micro-cups and probes) for weighing the sample (before and after atomization) and for carrying out the atomization itself; (ii) special injectors; and (iii) special sampling systems to gather the solid sample, filtering it through porous graphite or accumulating it electrostatically.

Preparation of the solid sample usually includes the processes of grinding, sieving, homogenization, drying and so forth. However, samples such as plastic can be directly analyzed [Resano *et al.* (2000)]. Biological samples and vegetable matter can be converted into a solid (by drying, dry charring, plasma charring or lyophilization) and this can be ground up thereafter.

Errors in direct solid atomization are, or may be, derived from the operator (inaccurate weighing of the sample, errors in the transfer operation and sample introduction) and from the sample itself (particle sizes, heterogeneity, amount used and analyte concentration).

Hence, the final result is influenced significantly both by the level as defined anatomically for biological samples (i.e. different lobes from a liver) and by the level as averaged (i.e. different sampling points) of the applied sampling procedure [Lücker (1997)]. To compensate for this lack of homogeneity and to reduce the total analytical error, it is advisable to analyze several (at least six) microsamples, each one taken from a different part of the overall sample, depending on the sample type and analysis goal [Lücker and Thorius-Ehrler (1993); Lücker *et al.* (1993a,b); Lücker (1997)]. The strong background may be corrected, at least in refractory samples with high WO_3 contents, by adding hydrogen as a purge gas during the pyrolysis stage [Hornung and Krivan (1998)] or by optimizing the analysis conditions [Lucic and Krivan (1998a,b)].

In the direct solid sampling, chemical modifiers in both liquid and solid state have been used. However, a chemical modifier added in liquid form is often not very effective for stabilization. The use of chemical modifiers in a solid state requires the modifier and the sample to be mixed and well ground up so as to achieve good contact between them.

To dilute the solid sample and ensure that the analyte mass is within the linear calibration range, it is customary to add graphite powder. The added graphite powder also inhibits formation of solid residues after atomization, decreases chemical interference and encourages desorption of the analyte when atomization occurs in a reducing atmosphere. Besides this, the half-life of the graphite tube is lengthened.

3.3.2. Introduction of Slurries

Preparation of any slurry is done by adding a liquid diluent to the solid sample (weighed, ground and sieved), keeping the slurry stable during sampling. The amount of solid sample to be weighed depends upon the analyte concentration, sample homogeneity and slurry concentration. The slurry introduction into the atomizer may be carried out manually or by using an autosampler.

The solid sample introduction as a slurry allows some of the problems of the direct solid sampling to be overcome, since slurry can be introduced using micropipettes, samplers and nebulizers [Dobrowolski and Mierzwa (1992)]. In general, the sample pretreatment is simple, and the lack of homogeneity is minimal, so that the precision obtained is acceptable. Nevertheless, the most crucial difficulty in the slurry technique is maintaining the slurry stable during the sample introduction. To get a representative aliquot into the furnace, the slurry must be stabilized and homogenized by using thyxotropic stabilizing agents such as "viscalex" and glycerol. Stability of ceramic slurries can be increased by immersion of the particles in a polar liquid to build up a double electrical layer around each particle, and thus generating a repulsion force that predominates over the London/van der Waals attraction forces. The stability of slurries containing vegetable matter may be improved by the use of 5% nitric acid as a solvent and ultrasound mixing rather than mechanical preparation of the slurry. In this case, the slurry method is comparable to the conventional solution procedure [Dobrowolski and Mierzwa (1993)], with the additional advantage that the slurry technique reduces the sample manipulation time. Nonetheless, there are several difficulties that make it advisable to homogenize the sample by means of ultrasound agitation (the approach most frequently used) [Lucic and Krivan (1998a)], although other procedures such as magnetic agitation, pre-digestion of the slurry, and electrical dispersion in a condensed medium have also been investigated [Bendicho (1994b)]. Recently, descriptions of several "vortex" and gas mixing systems, utilizing bubbles of argon to homogenize slurries prior to autosampling, have been published [López-García *et al.* (1997a)]. Such a system can be easily adapted to any commercial auto-sampler. Extraction *in situ* of metals from a slurry by the use of an acid diluent (HF) permits the analysis of volatile [López-García *et al.* (1997b)] and non-volatile [Bendicho and de Loss-Vollebregt (1990b)] metals in a refractory matrix of SiO_2, with high precision and accuracy.

The use of micro-organisms to pre-concentrate elements, followed by the sample introduction as a slurry, is an attractive area. The bacteria *Escherichia coli* and *Pseudomonas putida* have been used, both living and dead (immobilized or in suspension) for pre-concentrating several elements, Au, Be, Cd, Hg and Se. Cell suspensions have been maintained using ultrasound agitation [Robles *et al.* (1993); Robles and Aller (1994, 1995, 1996); Aller and Robles (1998)]. Organic slurries behave like organic chemical modifiers in a graphite atomizer, permitting an adequate thermal stabilization. The slurry's organic content decreases the atomization temperature (at least for Cd and Pb), due to the formation of activated carbon during the pyrolysis step. This activated carbon adsorbs isolated analyte atoms, giving rise to a very rapid atomization [Terzieva and Arpadjan (1998)].

The slurry introduction allows the use of chemical modifiers in solution [Baralkiewiez *et al.* (2005)]. Interaction between the chemical modifier and the particles of the solid sample is much more effective than for the direct solid sampling (dry sample). Moreover, for some refractory samples, it is possible to carry out quantitative analyses without having a pyrolysis stage and even without adding chemical modifiers, if a sample's particle size is under 25 μm, as the components of the matrix act as chemical modifiers [Bendicho and de Loss-Vollebregt (1990a)].

3.3.3. Electrodeposition

Electrodeposition as a means of introducing analytes into the graphite tube has not been widely used. However, V-shaped pyrolytic graphite platforms have been used as a cathode to electrodeposite the sample at a potential of 4–6 V, whilst a platinum wire served as an anode [Matousek and Powell (1993)]. In another study, a flow system incorporating flow cells with three electrodes was employed to make electrodeposition at potentials between −0.9 V and 1.2 V [Beinrohr *et al.* (1993)]. The graphite tube itself, forming part of a flow system [Komárek *et al.* (1999)], or a graphite probe [Komárek and Holý (1999)] has also been utilized as a cathode for electrolytic deposition of metal analytes.

Electrostatic precipitation has also been used for sampling solid particles from the air, being thereafter introduced into a graphite tube or deposited on a graphite rod [Bitterli *et al.* (1997); Buchkamp and Hermann (1999)]. Air sampling was carried out by using a sample flow of 2 L/min, while the voltage employed for electrostatic precipitation oscillates between +500 V and +2500 V [Bitterli *et al.* (1997)]. This technique eliminates sample preparation procedures and decreases the risk of contamination.

3.4. Gases and Vapors

A very effective and selective way to introduce the sample into the graphite tube consists of generating compounds in the vapor state. This is possible only when the vapors or gaseous compounds in the sample can be transported efficiently to the atomizer. The generation of gaseous analytes and their introduction into the atomization cells offer several advantages over the sample introduction system based on the pneumatic nebulization [Matusiewicz and Sturgeon (1996); Smichowski and Farías (2000)], particularly as the need for a nebulizer obviated, and there is increased efficiency in transporting the analyte (almost 100%), as well as the existence of a homogeneous vapor in the atomizer. Furthermore, with this sample introduction system, the relative detection limits and accuracy can be significantly improved, as long as there is: (a) separation of the analyte from the matrix (high selectivity); (b) easy pre-concentration; (c) chemical speciation; (d) simple instrumentation; and (e) procedures for automation.

On the other hand, there are some drawbacks [Matusiewicz and Sturgeon (1996)], such as: possible interference from concomitant elements (principally transition elements); effect of pH; influence of oxidation states (this can also be a positive point if used for speciation); and atomization interference in the gaseous phase (the mutual effect of other similar substances). There are several techniques for introducing samples in a gaseous state, but the most widespread is the technique based on hydride generation.

3.4.1. Hydride Generation

The determination of some elements by atomic absorption spectrometry can be improved using the chemical vapor generation technique where the analyte in the gas phase is transported from the sample solution to the atomizer. The most widely used procedure for the production of chemical vapors is the formation of volatile hydrides, principally of the following elements: As, Bi, Ge, In, Pb, Sb, Se, Sn, Te and Tl, as also of some of their organic derivatives [Donard *et al.* (1986); Camero and Sturgeon (1999); Nakahara (2005); Kumar and Riyazuddin (2005)]. The precise composition of the volatile species of Ag, Cu, Cd, and Zn generated at room temperature by reaction of the aquo-ions with sodium tetrahydroborate(III) remains to be definitely elucidated [Luna *et al.* (2002)]. To generate these hydrides, it is necessary to add a reducer, normally sodium tetrahydroborate ($NaBH_4$) and tin(II) chloride ($SnCl_2$), to an acidic sample solution. It is generally accepted that the chemical reducing agent produces nascent hydrogen, which would be involved in the analyte reduction. Nonetheless, other opinions have also been outlined [Laborda *et al.* (2002)] and the mechanism of hydride formation can be represented by the model reaction scheme shown in Fig. 3.10 [D'Ulivo *et al.* (2005)].

The hydrogen and analyte hydride gases produced are transported by means of an argon or helium flow into a quartz tube, which is electrically heated (or by a flame). The radiation beam coming from the

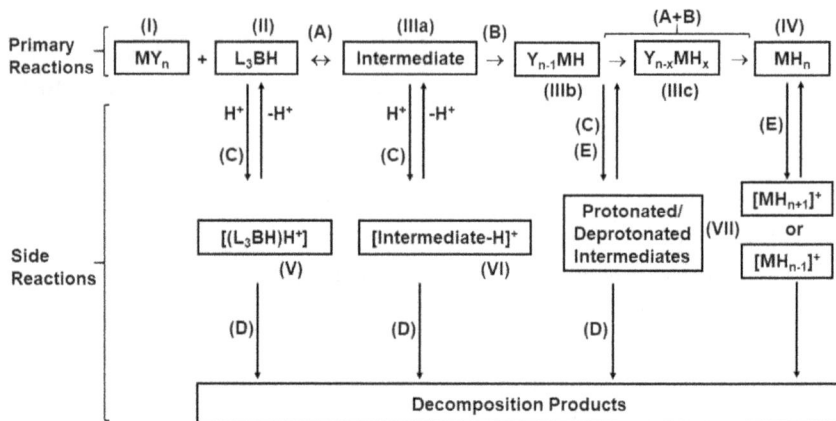

Fig. 3.10. Model reaction scheme for hydride generation using tetrahydroborate(III) reagent. The charge on complex species (I), (II), (III) and (IV) is omitted for the sake of simplicity and $1 < x < n$ [D'Ulivo *et al.* (2005)].

hollow cathode lamp is passed into this tube. Hydrogen is known to play a very important role in the hydride dissociation in the absorption cell at high temperatures. However, the occasional production of hydrogen during the analyte reduction process represents a disadvantage, since it may give rise to excessive error in the baseline signal and sporadically influence the atomization process. The results become more accurate if the hydrides produced are made to pass through a dryer and then trapped in silica-packed polytetrafluorethylene (PTFE) coil, cooled with liquid nitrogen. This coil is then rapidly warmed by use of a heating strip or through immersion in a water bath and a pulse of the analyte hydride vapor is set free and can be transported to the absorption cell by means of a helium flow containing hydrogen at a fixed concentration. It is also possible to use a system for continuous production of hydride vapors, yielding a constant AAS signal. In this case, a gas–liquid dryer is used to eliminate waste liquids and allow transport of the gases (analyte hydrides) to the absorption cell.

The presence of certain metal ions, such as Ni^{2+}, Co^{2+}, and Cu^{2+}, in the sample solution can give rise to interference in the hydride generation process [Bax *et al.* (1988)]. These interfering ions are easily reduced by sodium tetrahydroborate, and this produces a slurry with the metal particles. The analyte hydride is broken down by these metal particles, disrupting the efficiency of the hydride generation process. The addition of Fe^{3+} to the standards and samples can compensate for such interference effects. It is believed that Fe^{3+} is fundamentally reduced to Fe^{2+} and that the formation of metal particles diminishes. As an alternative to the chemical hydride generation based on the use of sodium tetrahydroborate, electrochemical hydride generation has also been accomplished in continuous or flow injection mode [Laborda *et al.* (2000)]. Through vapor generation, it is possible to pre-concentrate and eliminate *in situ* any matrix interference.

Some published works describe procedures developed for direct sampling of particulate matter from air within the graphite atomizer. The solid matter in suspension is collected on the inner surface of a porous graphite tube and air is made to pass through the wall of the tube, which is used as an electrothermal graphite atomizer [Lüdke *et al.* (1994)]. Other authors refer to direct introduction of a gaseous sample (0.1 mL) into the atomizer at a temperature of 1100°C without using an internal gas flow, being immediately atomized at 2300°C [Baaske *et al.* (1994); Baaske and Telgheder (1995)].

The most widely used atomizer for the hydride generation in AAS is a quartz T tube (normally $\varnothing = 10$ mm, and $l = 100-150$ mm). The use of continuously heated graphite tubes has been very limited. The principal obstacles encountered in conventional hydride detection and generation techniques are the following: (a) the analyte hydride is diluted by the co-formed hydrogen and the carrier gas, thus decreasing sensitivity; and (b) many hydrides are not efficiently atomized at the maximum temperature achieved with the conventionally heated quartz tubes.

Techniques for *in situ* trapping, incorporating a coupling between hydride generation and a graphite furnace, permit major increases in the detection capacity relative to conventional bath generation and continuous systems for determining ultra-traces of metal hydrides. The graphite furnace is utilized to break down the volatile hydride and trap analyte species on the tube surface. This provides a clean and effective separation from the matrix, along with good analyte concentration. This methodology offers several advantages over conventional quartz tube or furnace analytical techniques: simplicity of operation, flexibility in the sample volumes used, the mass of the modifier employed, trapping temperature, carrier gas flow, capillary long, high sensitivity and a considerable increase in the relative detection power as a result of *in situ* pre-concentration [Docekal *et al.* (1997)].

A flow injection hydride generation manifold coupled to a tungsten coil electrothermal atomizer for *in situ* collection of selenium has been developed [Barbosa *et al.* (2002)]. A continuous flow system for generating hydrides, which is cleaner and simpler to operate than using $NaBH_4$, has also been developed [Ding and Sturgeon (1996)]. This technique allows direct determination of elements like Sb, without any necessity to reduce Sb(V) to Sb(III), where the hydride generated is captured in a graphite furnace coated with Pd. Later atomization provides very good detection limits for Sb (45 pg and 0.02 μg/L using a 2 mL sample volume) with a precision better than 6%.

3.4.2. Generation of Other Gaseous Compounds

Some elements have been introduced into the graphite tube by generation of some other volatile compounds including chlorides (As, Bi, Cd, Ge, Mo, Pb, Sn, Tl and Zn), fluorides (Ge, Mo, Re, U, V and W), β-dicetonates (Al, Co, Cu, Cr, Fe, Mn, Ni, Pb and Zn), and dithiocarbamates (Co, Cr and Cu) [Smichowski and Farías (2000)]. However, other organometallic compounds, such as alkyl derivatives (propylates [Radojevic *et al.* (1986)] and particularly ethylates [Rapsomanikis *et al.* (1986); Rapsomanikis and Craig (1991); Fischer *et al.* (1993); Cai *et al.* (1993); Rapsomanikis (1994)]) of various elements: Pb, Hg, Sn, Tl, Cd, Zn, and Cu, have also been generated. Alkyl derivatives are formed by using alkylborates as a reagent, although sodium tetraethylborate dissolved in water has also been used to generate ethyl derivatives.

Chemical vapor generation (CVG) of transition and noble metals, firstly introduced in 1996 [Sturgeon *et al.* (1996)] is an interesting alternative way for the introduction of these metals into the atomizer. CVG tries to take advantage of the main well-proven capability of hydride generation and the high analyte fraction actually introduced into the absorption cell, but extended to other elements. Elucidation of the mechanisms of the chemical vapor generation has assumed two types of reactions contributing to the formation of volatile species: (i) homogeneous liquid phase reactions between aqueous analyte complexes and hydroboron species; and (ii) surface mediated reaction between reaction intermediates and hydroboron species, where the intermediate stability plays an important role in determining the nature of the final product [Feng *et al.* (2005)].

In addition to the noble metals (Ag, Au, Ir, Os, Pd, Pt, Rh, Ru), some transition metals (Co, Cr, Cu, Fe, Mn, Ni, Zn, Ti, Tl) have also been generated as volatile species using the acid-borohydride reaction. The identity of the species formed under direct reduction with tetrahydroborate is currently unknown, although the formation of metal nanoparticles small enough to be transported as a gas has been suggested. However, photo-CVG provides a powerful alternative to conventional CVG due to its simplicity, versatility, and cost effectiveness [He *et al.* (2007)]. Photo-CVG retains the advantages of conventional CVG in addition to the following unique features: (i) simple reactions, (ii) amenability to speciation analysis, (iii) greener analytical methodology, (iv) cost-effectiveness, (v) expanded number of detectable elements, and (vi) less interference or potentially interference-free analysis.

A special case of the vapor generation technique is the largely known cold vapor generation method initially developed for the determination of mercury. Basically, Hg(II) ions are reduced to elemental Hg and the gaseous atoms of Hg at room temperature are transported by an argon or helium flow to the absorption cell (quartz tube). Obviously, no electrothermal atomization is required.

3.4.3. Gas Chromatography

Electrothermal atomizers coupled to a gas chromatograph can be grouped into three categories [Ebdon *et al.* (1986)]: (i) lab-made electrothermally heated ceramic or quartz tubes; (ii) commercial graphite furnaces; and (iii) commercial cold vapor mercury analyzers. The various developments aimed at coupling ETAAS with GC normally involve the introduction of a sample coming from the chromatographic column into the furnace using the purge gas through the sample introduction hole. In any case, it is usually necessary to heat up the furnace at frequent intervals or continuously during the chromatographic separation stage. For those elements needing high temperatures for atomization, any practical coupling will result in a very short average tube-life.

References

Ali, A.M.; Smith, B.W. and Winefordner, J.D. (1989). Direct analysis of coal by electrothermal atomization atomic-absorption spectrometry. *Talanta,* **36**: 893–896.

Aller, A.J. and Robles, L.C. (1998). Determination of selenocystamine by slurry sampling electrothermal atomic absorption spectrometry after a selective preconcentration by living *Pseudomonas putida. J. Anal. At. Spectrom.,* **13**: 469–476.

Alary J.F. and Salin E.D. (1995). Rapid sample preconcentration by spray deposition for electrothermal vaporization inductively coupled plasma spectrometry. *Spectrochim. Acta., Part B,* **50**: 405–413.

Arruda, M.A.Z.; Gallego, M. and Valcarcel, M. (1993). Determination of aluminium in slurry and liquid phase of juices by flow-injection analysis-graphite-furnace atomic absorption spectrometry. *Anal. Chem.,* **65**: 3331–3335.

Aucélio, R.Q., Curtius, A.J. and Welz, B. (2000). Sequential determination of Sb and Sn in used lubricating oil by electrothermal atomic absorption spectrometry using Ru as a permanent modifier and microemulsion sample introduction. *J. Anal. At. Spectrom.,* **15**: 1389–1393.

Baaske, B.; Golloch, A. and Telgheder, U. (1994). Application of a modified atomic absorption spectrometer for the determination of iron traces in gaseous hydrogen chloride. *J. Anal. At. Spectrom.,* **9**: 867–870.

Baaske, B. and Telgheder, U. (1995). Automated sampling system for the direct determination of trace amounts of heavy metals in gaseous hydrogen chloride by atomic absorption spectrometry. *J. Anal. At. Spectrom.,* **10**: 1077–1080.

Baralkiewiez, D.; Gramowska, H.; Kózka, M. and Kanecka, A. (2005). Determination of mercury in sewage sludge by direct slurry sampling graphite furnace atomic absorption spectrometry. *Spectrochim. Acta., Part B,* **60**: 409–413.

Barbosa, F.,Jr., Krug, F.J. and Lime, E.C. (1999). *On line* coupling of electrochemical preconcentration in tungsten coil electrothermal atomic absorption spectrometry for determination of lead in natural waters. *Spectrochim. Acta., Part B,* **54**: 1155–1166.

Barbosa, F.,Jr., de Souza, S.S. and Krug, F.J. (2002). *In situ* trapping of selenium hydride in rhodium-coated tungsten coil electrothermal atomic absorption spectrometry. *J. Anal. At. Spectrom.,* **17**: 382–388.

Bax, D.; Agterdenbos, J.; Worrell, E. and Kolmer, J.B. (1988). The mechanism of transition metal interference in hydride generation atomic absorption spectrometry. *Spectrochim. Acta., Part B,* **43**: 1349–1354.

Beinrohr, E.; Lee, M.L.; Tschöpel, P. and Tölg, G. (1993). Determination of platinum in biotic and environmental samples by graphite furnace atomic absorption spectrometry after its electrodeposition into a graphite tube packed with reticulated vitreous carbon, *Fresenius' J. Anal. Chem.,* **346**: 689–692.

Belarra, M.A.; Resano, M. and Castillo, J.R. (1999). Theoretical evaluation of solid sampling-electrothermal atomic absorption spectrometry for acreening purposes. *J. Anal. At. Spectrom.,* **14**: 547–552.

Bendicho, C. (1994a). Evaluation of an automated thermospray interface for coupling electrothermal atomization atomic absorption spectrometry and liquid chromatography. *Anal. Chem.,* **66**: 4375–4381.

Bendicho, C. (1994b). Determination of metal impurities in elctrolytic iron by GF-AAS using electric dispersion in liquid medium. *Fresenius' J. Anal. Chem.,* **348**: 353–355.

Bendicho, C. and de Loss-Vollebregt, M.T.C. (1990a). The influence of pyrolysis and matrix modifiers for analysis of glass materials by GFAASZ using slurry sample introduction. *Spectrochim. Acta,* **45**: 679–693.

Bendicho, C. and de Loss-Vollebregt, M.T.C. (1990b). Metal extraction by hydrofluoric acid from slurries of glass materials in graphite furnace atomic absorption spectrometry. *Spectrochim. Acta,* **45**: 695–710.

Bendicho, C. and de Loss-Vollebregt, M.T.C. (1991). Solid sampling in electrothermal atomic absorption spectrometry using commercial atomizers. *J. Anal. At. Spectrom.,* **6**: 353–374.

Bermejo-Barrera, P.; Anllo-Sendín, R.M.; Cantelar-barbazán, M.J. and Bermejo-Barrera, A. (2002). Selective preconcentration and determination of tributyltin in fresh water by electrothermal atomic absorption spectrometry. *Anal. Bioanal. Chem.,* **372**: 837–839.

Berndt, H. and Schaldach, G. (1994). High-performance flow electrothermal atomic absorption spectrometry for *on-line* trace element preconcentration-matrix separation and trace element determination. *J. Anal. At. Spectrom.,* **9**: 39–44.

Bitterli, B.A.; Cousin, H. and Magyar, B. (1997). Determination of metals in airborne particles by electrothermal vaporization inductively coupled plasma mass spectrometry after accumulation by electrostatic precipitation. *J. Anal. At. Spectrom.,* **12**: 957–961.

Bosch Ojeda, C.; Sánchez Rojas, F.; Cano Pavón, J.M. and García de Torres, A. (2003). Automated on-line separation-preconcentration system for platinum determination by electrothermal atomic absorption spectrometry. *Anal. Chim. Acta,* **494**: 97–103.

Brinckman, F.E.; Blair, W.R.; Jewett, K.L. and Iverson, W.P. (1977). Application of a liquid chromatograph coupled with a flameless atomic absorption detector for speciation of trace organometallic compounds. *J. Chromatogr. Sci.,* **15**: 493–503.

Buchkamp, T. and Hermann, G. (1999). Solid sampling by electrothermal vaporization in combination with electrostatic particle deposition for electrothermal atomization multi-element analysis. *Spectrochim. Acta, Part B,* **54**: 657–668.

Burguera, J.L. and Burguera, M. (1997). Flow injection for automation in atomic spectrometry. *J. Anal. At. Spectrom.,* **12**: 643–651.

Cai, Y.; Rapsomanikis, S. and Andreae, M.O. (1993). Determination of butyltin compounds in river sediment samples by gas chromatography–atomic absorption spectrometry following *in situ* derivatization with sodium tetraethylborate. *J. Anal. At. Spectrom.*, **8**: 119–125.

Camero, R.M. and Sturgeon, R.E. (1999). Hydride generation-electrostatic deposition-graphite furnace atomic absorption spectrometric determination of arsenic, selenium and antimony. *Spectrochim. Acta, Part B*, **54**: 753–762.

Cassella, R.J.; de Sant'Ana, O.D.; Rangel, A.T.; de Carvalho, M.F.B. and Santelli, R.E. (2002). Selenium determination by electrothermal atomic absorption spectrometry in petroleum refinery aqueous streams containing volatile organic compounds. *Microchemical J.*, **71**: 21–28.

Castro, M.A.; Aller, A.J.; McCabe, A.; Smith, W.E. and Littlejohn, D. (2005). Spectrometric and morphological characterization of condensed phase zirconium species produced during electrothermal heating on a graphite platform. *J. Anal. At. Spectrom.*, **20**, 385–394.

Ceccarini, A.; Fuoco, R. and Vecchio, C. (2001). Capillary injection device for in situ pre-concentration of trace elements in graphite furnace-atomic absorption spectrometry. *Spectrochim. Acta, Part B*, **56**: 2439–2448.

Chung, C.-H.; Iwamoto, E.; Yamamoto, M. and Yamamoto, Y. (1984). Selective determination of arsenic(III, V), antimony(III, V), selenium(IV, VI) and tellurium(IV, VI) by extraction and graphite furnace atomic absorption spectrometry. *Spectrochim. Acta., Part B*, **39**: 459–466.

Ding, W.-W. and Sturgeon, R.E. (1996). Evaluation of electrochemical hydride generation for the determination of total antimony in natural waters by electrothermal atomic absorption spectrometry with *in situ* concentration. *J. Anal. At. Spectrom.*, **11**: 225–230.

Dobrowolski, R. and Mierzwa, J. (1992). Direct solid *versus* slurry analysis of tobacco leaves for some trace metals by graphite furnace AAS — a comparative study. *Fresenius' J. Anal. Chem.*, **344**: 340–344.

Dobrowolski, R. and Mierzwa, J. (1993). Determination of trace elements in plant materials by slurry sampling furnace AAS — some analytical problems. *Fresenius' J. Anal. Chem.*, **346**: 1058–1061.

Docekal, B. and Krivan, V. (1995). Determination of trace impurities in powdered molybdenum metal and molybdenum silicide by solid sampling GFAAS, *Spectrochim. Acta, Part B*, **50**: 517–526.

Docekal, B.; Dédina, J. and Krivan, V. (1997). Radiotracer investigation of hydride trapping efficiency within a graphite furnace. *Spectrochim. Acta, Part B*, **52**: 787–794.

Donard, O.F.X.; Rapsomanikis, S. and Weber, J.H. (1986). Speciation of inorganic tin and alkyltin compounds by atomic absorption spectrometry using an electrothermal quartz furnace after hydride generation. *Anal. Chem.*, **58**: 772–777.

Donard, O.F.X. and Martín, F.M. (1992). Hyphenated techniques applied to environmental speciation studies. *Trends in Anal. Chem.*, **11**(1): 17–26.

D'Ulivo, A.; Mester, Z. and Sturgeon, R.E. (2005). The mechanism of formation of volatile hydrides by tetrahydroborate(III) derivatization: A mass spectrometric study performed with deuterium labelled reagents. *Spectrochim. Acta, Part B*, **60**: 423–438.

Ebdon, L.; Hill, S. and Ward, R.W. (1986). Directly coupled chromatography — Atomic spectroscopy. Part 1. — Directly coupled gas chromatography — Atomic spectroscopy. A review. *Analyst*, **111**: 1113–1138.

Ebdon, L.; Hill, S. and Ward, R.W. (1987). Directly coupled chromatography — Atomic spectroscopy. Part 2. — Directly coupled liquid chromatography — Atomic spectroscopy. A review. *Analyst*, **112**: 1–16.

Economu, A. (2005). Sequential-injection analysis (SIA): A useful tool for on-line sample-handling and pre-treatment. *Trends in Anal. Chem.*, **24**(5): 416–425.

Fan, Z. and Zhou, W. (2006). Dithizone-chloroform single drop microextraction system combined with electrothermal atomic absorption spectrometry using Ir as permanent modifier for the determination of Cd in water and biological samples. *Spectrochim. Acta, Part B*, **61**: 870–874.

Fang, Z.; Sperling, M. and Welz, B. (1990). Flow injection on-line sorbent extraction pre-concentration for graphite furnace atomic absorption spectrometry. *J. Anal. At. Spectrom*, **5**: 639–646.

Feng, Y.-L.; Sturgeon, R.E.; Lam, J.W. and D'Ulivo, A. (2005). Insights into the mechanism of chemical vapour generation of transition and noble metals. *J. Anal. At. Spectrom*, **20**: 255–265.

Fischer, R.; Rapsomanikis, S. and Andreae, M.O. (1993). Determination of methylmercury in fish samples using GC-AAS and sodium tetraethylborate derivatization. *Anal. Chem.*, **65**: 763–766.

Friese, K.-C.; Krivan, V. and Schuierer, O. (1996). Electrothermal atomic absorption spectrometry using an improved solid sampling system for the analysis of high purity tantalum powders. *Spectrochim. Acta, Part B*, **51**: 1223–1233.

González, M.M.; Gallego, M. and Valcárcel, M. (2001). Determination of arsenic in wheat flour by electrothermal atomic absorption spectrometry using a continuous precipitation-dissolution flow system. *Talanta*, **55**: 135–142.

Grudpan, K. (2004). Some recent developments on cost-effective flow-based analysis. *Talanta*, **64**: 1084–1090.

Hansen, E.H. and Wang, J. (2002). Implementation of suitable flow injection/sequential injection-sample separation/preconcentration schemes for determination of trace metal concentrations using detection by electrothermal atomic absorption spectrometry and inductively coupled plasma mass spectrometry. *Analytica Chimica Acta*, **467**: 3–12.

Hansen, E.H. (2004). The impact of flow injection on modern chemical analysis: has it fulfilled our expectations? And where are we going?. *Talanta*, **64**: 1076–1083.

Hauptkorn, S. and Krivan, V. (1996). Solution and slurry sampling electrothermal atomic absorption spectrometry for the analysis of high purity quartz. *Spectrochim. Acta, Part B*, **51**: 1197–1210.

He, Y.; Hou, X.; Zheng, Ch. and Sturgeon, R.E. (2007). Critical evaluation of the application of photochemical vapour generation in analytical atomic spectrometry. *Anal. Bioanal. Chem.*, **388**: 769–774.

Heng-bin, H.; Yan-bing, L.; Shi-fen, M. and Zhe-ming, N. (1993). Speciation of arsenic by ion chromatography and off-line hydride generation electrothermal atomic absorption spectrometry. *J. Anal. At. Spectrom.*, **8**: 1085–1090.

Hermann, G.; Trenin, A.; Matz, R.; Gafurov, M.; Gilmutdinov, A.Kh.; Nagulin, K.Yu.; Frech, W.; Björn, E.; Grinshtein, I. and Vasilieva, L. (2004). Platform-to-platform sample transfer, distribution, dilution and dosing via electrothermal vaporization and electrostatic deposition. *Spectrochim. Acta., Part B*, **59**: 737–748.

High, K.A.; Azani, R.; Fazckas, A.F.; Chee, Z.A. and Blais, J.S. (1992). Thermospray-microatomizer interface for the determination of trace cadmium and cadmium-metallothioneins in biological samples with flow injection- and high-performance liquid-chromatography-atomic absorption spectrometry. *Anal. Chem.*, **64**: 3197–3201.

Hornung, M. and Krivan, V. (1998). Determination of trace impurities in high-purity tungsten by direct solid sampling electrothermal atomic absorption spectrometry using a transversely heated graphite tube. *Anal. Chem.*, **70**: 3444–3451.

Howell, A.G. and Koirtyohann, S.R. (1992). Effect of sample injection modes on residue deposits in graphite furnace atomic absorption. *Appl. Spectrosc.*, **46**(6): 953–958.

Kölbl, G.; Kalcher, K. and Irgolic, K.J. (1993). Computerized data treatment for an HPLC-GFAAS system for the identification and quantification of trace element compounds. *J. Automatic Chem.*, **15**: 37–45.

Komárek, J. and Holý, J. (1999). Determination of heavy metals by electrothermal atomic absorption spectrometry after electrodeposition on a graphite probe. *Spectrochim. Acta, Part B*, **54**: 733–738.

Komárek, J.; Krásenský, P.; Balcar, J. and Rehulka, P. (1999). Determination of palladium and platinum by electrothermal atomic absorption spectrometry after deposition on a graphite tube. *Spectrochim. Acta, Part B*, **54**: 738–743.

Krivan, V. and Mao Dong, H. (1998). Direct analysis of pieces of materials by solid sampling electrothermal atomic absorption spectrometry demonstrated using high-purity titanium. *Anal. Chem.*, **70**: 5312–5321.

Kumar, A.R. and Riyazuddin, P. (2005). Mechanism of volatile hydride formation and their atomization in hydride generation atomic absorption spectrometry. *Anal. Sci.*, **21**: 1401–1410.

Kuo, N.-W.; Lan, C.-R. and Alfassi, Z.B. (1993). *In situ* preconcentration of trace metals in high-purity water on to a graphite tube by multiple injections followed by graphite furnace atomic absorption spectrometry. *J. Radioanal. Nucl. Chem.*, **172**: 117–123.

Laborda, F.; Bolea, E.; Barabguan, M.T. and Castillo, J.R. (2002). Hydride generation in analytical chemistry and nascent hydrogen: when is it going to be over?. *Spectrochim. Acta, Part B.*, **57**: 797–802.

Laborda, F., Bolea, E. and Castillo, J.R. (2000). Tubular electrolytic hydride generator for continuous and flow injection sample introduction in atomic absorption spectrometry. *J. Anal. At. Spectrom*, **15**: 103–107.

Lanza, F.; Ceccarini, A. and Papoff, P. (1997). Preconcentration of analytes by aerosol deposition in graphite furnace atomic absorption spectrometry at the pg ml^{-1} level. *Spectrochim. Acta, Part B*, **52**: 113–123.

Liu, Z.-S. and Huang, S.-D. (1993). Automatic *on-line* preconcentration system for graphite furnace atomic absorption spectrometry for the determination of trace metals in sea-water. *Anal. Chim. Acta*, **281**: 185–190.

López-García, I.; Sánchez-Merlos, M. and Hernández-Córdoba, M. (1997a). Slurry sampling device for use in electrothermal atomic absorption spectrometry. *J. Anal. At. Spectrom.*, **12**: 777–779.

López-García, I.; Sánchez-Merlos, M. and Hernández-Córdoba, M. (1997b). Determination of mercury in soils and sediments by graphite furnace atomic absorption spectrometry with slurry sampling. *Spectrochim. Acta, Part B*, **52**: 2085–2092.

Lucic, M. and Krivan, V. (1998a). Slurry sampling electrothermal atomic absorption spectrometry for the analysis of aluminium-based ceramic powders. *Appl. Spectrosc.*, **52**: 663–672.

Lucic, M. and Krivan, V. (1998b). Solid sampling electrothermal atomic absorption spectrometry for analysis of aluminium powders. *J. Anal. At. Spectrom.*, **13**: 1133–1139.

Lücker, E.; Gerbing, C. and Kreuzer, W. (1993a). Distribution of Pb and Cd in the liver of the mallard — direct determination by means of solid sampling ZAAS. *Fresenius' J. Anal. Chem.*, **346**: 1062–1067.

Lücker, E.; Meuthen, J. and Kreuzer, W. (1993b). Distribution of Pb and Cd in equine liver — direct determination by means of solid sampling ZAAS. *Fresenius' J. Anal. Chem.*, **346**: 1068–1071.

Lücker, E. and Thorius-Ehrler, S. (1993). Solid sampling ZAAS determination of endogenous lead contamination in muscle tissue caused by calcification. *Fresenius' J. Anal. Chem.*, **346**: 1072–1076.

Lücker, E. (1997). Samplig strategy and direct solid sampling electrothermal atomization atomic absorption spectrometric analysis of trace elements in animal tissue. *Appl. Spectrosc.*, **51**: 1031–1036.

Lüdke, Ch.; Hoffmann, E. and Skole, J. (1994). Studies on the determination of the metal content of airborne particulates by furnace atomization non-thermal excitation spectrometry. *J. Anal. At. Spectrom.*, **9**: 685–689.

Luna, A.S.; Pereira, H.B.; Takase, I.; Gonçalves, R.A.; Sturgeon, R.E. and de Campos, R.C. (2002). Chemical vapour generation-electrothermal atomic absorption spectrometry: new perspectives. *Spectrochim. Acta, Part B*, **57**: 2047–2056.

Matousek, J.P. and Powell, H.K.J. (1993): Analyte preconcentration and separation from small volumes by electrodeposition for electrothermal atomic absorption spectroscopy. *Talanta*, **40**: 1829–1831.

Matusiewicz, H. and Sturgeon, R.E. (1996). Atomic spectrometric detection of hydride forming elements following *in situ* trapping within a graphite furnace. *Spectrochim. Acta, Part B*, **51**: 377–397.

Miller-Ihli, N.J. (1989). Graphite furnace atomic absorption spectrometry for the analysis of biological materials. *Spectrochim. Acta, Part B*, **44**: 1221–1227.

Miller-Ihli, N.J. (1992). Solid analysis by GFAAS. *Anal. Chem.*, **64**: 964A–968A.

Miró, M.; Jończyk, S.; Wang, J.-H. and Hansen, E.H. (2003). Exploiting the bead-injection approach in the integrated sequential injection lab-on-valve format using hydrophobic packing materials for on-line matrix removal and preconcentration of trace levels of cadmium in environmental and biological samples *via* formation of non-charged chelates prior to ETAAS detection. *J. Anal. At. Spectrom.*, **18**: 89–98.

Müller-Vogt, G.; Huwe, A. and Wendl, W. (1996). The use of sputtering as a solid sampling technique in graphite furnace atomic absorption spectrometry. *Spectrochim. Acta, Part B*, **51**: 1191–1196.

Nae-wen Kuo; Chi-ren Lan and Alfassi, Z.B. (1993). *In situ* preconcentration of trace metals in high purity water onto graphite tube by multiple injections followed by graphite furnace atomic absorption spectrometry. *J. Radioanal. Nucl. Chem.*, **172**: 117–123.

Nakahara, T. (2005). Development of gas-phase sample-introduction techniques for analytical atomic spectrometry. *Anal. Sci.*, **21**: 477–484.

Nygren, O.; Nilsson, C.-A. and Frech, W. (1988). On-line interfacing of a liquid chromatograph to a continuously heated graphite furnace atomic absorption spectrophotometer for element-specific detection. *Anal. Chem.*, **60**: 2204–2208.

Radojevic, M.; Allen, A.; Rapsomanikis, S. and Harrison, R.M. (1986). Propylation techniques for the simultaneous determination of tetraalkyllead and ionic alkyllead species by gas chromatography-atomic absorption spectrometry. *Anal. Chem.*, **58**: 658–661.

Rapsomanikis, S.; Donard, O.F.X. and Weber, J.H. (1986). Speciation of lead and methyllead ions in water by chromatography-atomic absorption spectrometry after ethylation with sodium tetraethylborate. *Anal. Chem.*, **58**: 35–38.

Rapsomanikis, S. and Craig, P.J. (1991). Speciation of mercury and methylmercury compounds in aqueous samples by chromatography-atomic absorption spectrometry after ethylation with sodium tetraethylborate. *Anal. Chim. Acta*, **248**: 563–567.

Rapsomanikis, S. (1994). Derivatization by ethylation with sodium tetraethylborate for the speciation of metals and organometallics in environmental samples. A review. *Analyst*, **119**: 1429–1439.

Resano, M.; Belarra, M.A.; Castillo, J.R. and Vanhaecke, F. (2000). Direct determination of phosphorus in two different plastic materials (PET and PP) by solid sampling-graphite furnace atomic absorption spectrometry. *J. Anal. At. Spectrom.*, **15**: 1383–1388.

Robles, L.C.; Feo, J.C.; de Celis, B.; Lumbreras, J.M.; Garcia-Olalla, C. and Aller, A.J. (1999). Speciation of selenite and elenate using living bacteria. *Talanta*, **50**: 307–325.

Robles, L.C.; Garcia-Olalla, C. and Aller, A.J. (1993). Determination of gold by slurry electrothermal atomic absorption spectrometry after preconcentration by *Escherichia coli* and *Pseudomonas putida*. *J. Anal. At. Spectrom.*, **8**: 1015–1022.

Robles, L.C. and Aller, A.J. (1994). Preconcentration of beryllium on the outer membrane of *Escherichia coli* and *Pseudomonas putida* prior to determination by electrothermal atomic absorption spectrometry. *J. Anal. At. Spectrom.*, **9**: 871–879.

Robles, L.C. and Aller, A.J. (1995). Determination of cadmium in biological and environmental samples by slurry electrothermal atomic absorption spectrometry. *Talanta*, **42**: 1731–1744.

Robles, L.C. and Aller, A.J. (1996). Solid extraction — Electrothermal atomic absorption spectrometric determination of selenium in environmental samples. *Anal. Sci.*, **12**: 783–787.

Ruzicka, J. and Marshall, G.D. (1990). Sequential injection: a new concept for chemical sensors, process analysis and laboratory assays. *Anal. Chim. Acta,* **237**: 329–343.

Ruzicka, J. (2000). Lab-on-valve: universal microflow analyzer based on sequential and bead injection. *Analyst,* **125**: 1053–1060.

Sahayam, A.C. (2002). Speciation of Cr(III) and Cr(VI) in potable waters by using activated neutral alumina as collector and ET-AAS for determination. *Anal. Bioanal. Chem.,* **372**: 840–842.

Saint'Pierre, T.; Aucélio, R.Q. and Curtius, A.J. (2003). Trace elemental determination in alcohol automotive fuel by electrothermal atomic absorption spectrometry. *Microchem. J.,* **75**: 59–67.

Schäffer, U. and Krivan, V. (1996). Slurry sampling electrothermal atomic absorption spectrometry for the analysis of graphite powders. *Spectrochim. Acta, Part B,* **51**: 1211–1222.

Sella, S.; Sturgeon, R.E.; Willie, S.N. and Campos, R.C. (1997). Flow injection on-line reductive precipitation preconcentration with magnetic collection for electrothermal atomic absorption spectrometry. *J. Anal. At. Spectrom.,* **12**: 1281–1285.

Smichowski, P. and Farías, S. (2000). Advantages and analytical applications of chloride generation. A review on vapour generation methods in atomic spectrometry. *Microchem. J.,* **67**: 147–155.

Soldado Cabezuelo, A.B.; Blanco González, E. and Sanz-Medel, A. (1997). Quantitative studies of aluminium binding species in human uremic serum by fast protein liquid chromatography coupled with electrothermal atomic absorption spectrometry. *Analyst,* **122**: 573–577.

Sommer, L.; Komarek, J. and Burns, D.Th. (1992). Organic analytical reagents in atomic absorption spectrophotometry of metals. *Pure & Appl. Chem.,* **64**: 213–226.

Sperling, M.; Yin, X-F. and Welz, B. (1991). Flow injection on-line separation and preconcentration for electrothermal atomic absorption spectrometry. Part 1. Determination of ultratrace amounts of cadmium, copper, lead and nickel in water samples. *J. Anal. At. Spectrom.,* **6**: 295–300.

Sturgeon, R.E.; Liu, J.; Boyko V.J. and Luong, V.T. (1996). Determination of copper in environmental matrices following vapor generation. *Anal. Chem.,* **68**: 1883–1887.

Tao, G. and Fang, Z. (1995). On-line flow injection solvent extraction for electrothermal atomic absorption spectrometry: determination of nickel in biological samples. *Spectrochim. Acta, Part B,* **50**: 1747–1755.

Terzieva, V. and Arpadjan, S. (1998). Determination of metals in atmospheric particulates by electrothermal atomic absorption spectrometry with organic slurry sample introduction. *J. Anal. At. Spectrom.,* **13**: 815–817.

Tsalev, D.L. (1999). Hyphenated vapour generation atomic absorption spectrometric techniques. *J. Anal. At. Spectrom.,* **14**: 147–162.

Tserovsky, E. and Arpadjan, S. (1991). Behaviour of various organic solvents and analytes in electrothermal atomic absorption spectrometry, *J. Anal. At. Spectrom.,* **6**: 487–491.

Tyson, J.F. (1991). Flow injection atomic spectrometry. *Spectrochim. Acta Rev.,* **14**: 169–233.

Tyson, J.F. (1999). High-performance, flow-based, sample pre-treatment and introduction procedures for analytical atomic spectrometry. *J. Anal. At. Spectrom.,* **14**: 169–178.

van Staden, J.F. and Stefan, R.I. (2004). Chemical speciation by sequential injection analysis: an overview. *Talanta,* **64**: 1109–1113.

Wang, J.-H. and Hansen, E.H. (2000). Coupling on-line preconcentration by ion-exchange with ETAAS: a novel flow injection approach based on the use of a renewable microcolumn as demonstrated for the determination of nickel in environmental and biological samples. *Anal. Chim. Acta,* **424**: 223–232.

Wang, J.-H. and Hansen, E.H. (2001). Coupling sequential injection on-line preconcentration by means of a renewable microcolumn with ion-exchange beads with detection by electrothermal atomic absorption spectrometry: Comparing the performance of eluting the loaded beads with transporting them directly into the graphite tube, as demonstrated for the determination of nickel in environmental and biological samples. *Anal. Chim. Acta,* **435**: 331–342.

Wang, J.-H. and Hansen, E.H. (2002). Development of an automated sequential injection on-line solvent extraction-back extraction procedure as demonstrated for the determination of cadmium with detection by electrothermal atomic absorption spectrometry. *Anal. Chim. Acta,* **456**: 283–292.

Wang, J.-H.; Hansen, E.H. and Miró, M. (2003). Sequential injection–bead injection–lab-on-valve schemes for on-line solid phase extraction and preconcentration of ultra-trace levels of heavy metals with determination by electrothermal atomic absorption spectrometry and inductively coupled plasma mass spectrometry. *Anal. Chim. Acta,* **499**: 139–147.

Wang, J. and Hansen, E.H. (2005). Trends and perspectives of flow injection/sequential injection on-line sample-pretreatment schemes coupled to ETAAS. *Trends in Anal. Chem.,* **24**: 1–8.

Welz, B.; Sperling, M. and Sun, X. (1993). Analysis of high-purity reagents using automatic on-line column preconcentration-separation and electrothermal atomic absorption spectrometry. *Fresenius' J. Anal. Chem.,* **346**: 550–555.

Zhang, X.; Chen, D.; Marquardt, R. and Koropchak, J.A. (2000). Thermospray sample introduction to atomic spectrometry. *Microchem. J.,* **66**: 17–53.

Zhaolun Fang (1998). Trends and potential in flow injection on-line separation and preconcentratrion techniques for electrothermal atomic absorption spectrometry. *Spectrochim. Acta, Part B,* **53**: 1371–1379.

Zhaolun Fang, Shukun Xu and Guanhong Tao (1996). Developments and trends in flow injection atomic absorption spectrometry. *J. Anal. At. Spectrom.,* **11**: 1–24.

Zhe-ming, N.; Bin, H. and Heng-Bin, H. (1993). *In situ* concentration of selenium and tellurium hydride in a silver-coated graphite atomizer. *J. Anal. At. Spectrom.,* **8**: 995–998.

Zhen-Shan Liu and Shang-Da Huang (1993). Automatic on-line preconcentration system for graphite furnace atomic absorption spectrometry for the determination of trace metals in sea water. *Anal. Chim. Acta,* **281**: 185–190.

Chapter 4

Radiation Sources, Spectral Dispersion, Isolation and Detection of Radiation

4.1. Radiation Sources

Two types of radiation sources are usually employed in AAS: lines sources and continuum sources.

4.1.1. Line Sources

The two most usual lines sources employed in AAS are hollow-cathode lamps (HCLs) and electrodeless discharge lamps.

4.1.1.1. *Hollow-cathode lamps*

The electromagnetic radiation source normally used in atomic absorption spectrometry is the HCL [Lowe and Sullivan (1999)] (Fig. 4.1).

There has also been intense interest in other luminescent discharge lamps [Leis and Steers (1994); Schnürer-Patschan *et al.* (1993)], as they have proved to be of great use as electromagnetic radiation sources. In HCLs, the first step is excitation of the filler gas through high-voltage between the anode and cathode. The first discharges produce metastable forms of the filler gas atoms, the half-life time of which is several orders of magnitude longer than that of the commonest excited states, so playing an important role in maintaining the discharge. The metastable filler gas atoms transmit their high energy to the analyte atoms present in the hollow cathode, likewise through electron collisions, and bring about their excitation or even ionization, with subsequent emission of specific radiation.

The current intensity passing through the lamp needs to be controlled below a limiting value in order to avoid self-reversal of the emission lines [Freeman *et al.* (1980)] (Fig. 4.2). Other parameters,

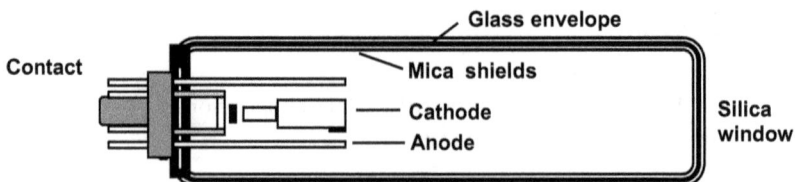

Fig. 4.1.   Scheme for a HCL.

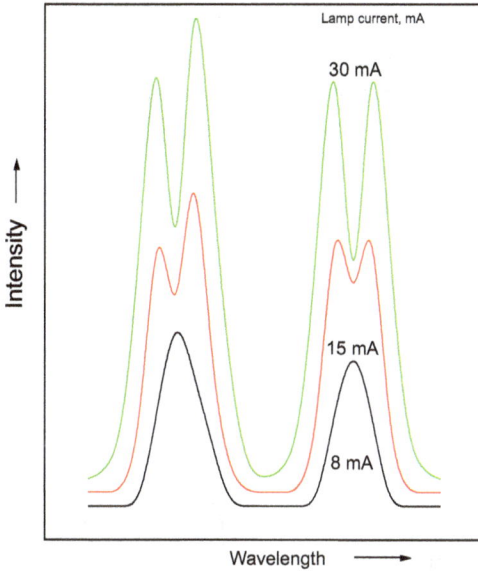

Fig. 4.2. Form of two spectral lines for copper as a function of the current intensity passing through the hollow cathode.

not controllable by the analyst, but which also influence the radiation emitted, are the geometry of the hollow cathode, the type of the filler gas and pressure [Oliver and Finlayson (1998)].

When the internal diameter, d, of the hollow cathode is less than or equal to 2 mm, the intensity of the emitted radiation, I, is inversely proportional to that diameter, $Id = K$ (where K takes a constant value). It is advisable that the length of the hollow cathode should be at least three times greater than the cathode diameter. The optimum distance between the cathode and anode is usually around 20 mm to 25 mm, although it may vary in accordance with the design and experimental conditions. The shorter the distance, the more difficult it is to achieve ionization of the filler gas. On the other hand, longer distances will cause prolonged flashing before a stable or stationary discharge is attained.

The rare gases most frequently used as fillers are argon and helium. These gases have metastable energy levels of 11.55 eV ($4S_{3/2}$) and 11.72 eV ($4S_{1/2}$) for argon, and 19.82 eV ($^3S_1$) and 20.62 eV ($^1S_0$)

for helium. Neon has metastable levels intermediate between helium and argon, but its high price drastically restricts its use. On the other hand, the 193.7 nm line in the neon filled arsenic HCL suffers from several spectral interferences, where the cause is the presence of the NeII lines at 193.89 nm and 193.01 nm [Freeman *et al.* (1980)]. The radiation intensity increases as the pressure decreases [Freeman *et al.* (1980)].

According to Beer's law, analytical sensitivity increases logarithmically with the electromagnetic radiation intensity employed to excite the analyte atoms. For this reason, attempts to improve the intensity of the analytical signal given off when HCLs are used have been carried out by:

- coupling the luminescent discharge to some other form of external energy (for example, a source of microwaves or radio frequency (*r.f.*), or a magnetic field) [Pavlovic and Dobrosavijevic (1992)],
- modifying the geometry of the hollow cathode: cones, cylinders, microcavities [$\varnothing \leq 2$ mm, $l \leq 6$ mm]) [Morgan *et al.* (1994)], and
- pulsing the flux emitted [Smith *et al.* (1994)].

Not only the radiation intensity, but also the radiation intensity profile or spatial distribution from HCLs depends strongly on the lamp current, lamp power, carrier gas pressure and cathode geometry, particularly on its diameter [Freeman *et al.* (1980); Larkins (1985)]. Thus, for cathodes with 3-mm inner diameters, the radiation intensity emitted shows a symmetrical bell curve distribution independent on the lamp current. For large cathodes, the radiation intensity distribution is not parabolic, in some cases having a minimum along the cathode axis or a non-uniform transverse distribution (Fig. 4.3) [Gilmutdinov *et al.* (1994, 1995)]. The emission-time profiles coming from spherical cathodes are similar to those observed for cylindrical cathodes, where the emission-time profile changes with both the cavity diameter [Williams *et al.* (1995)], and the longitudinally or radially non-uniform distributions [Gilmutdinov *et al.* (1996a,b,c)]. The lamp emission profile has a considerable effect on the analytical absorbance [Piepmeier (1989)].

The pulsed radiation lamps provide some improvement over the continuum and line radiation sources. Thus, instability and

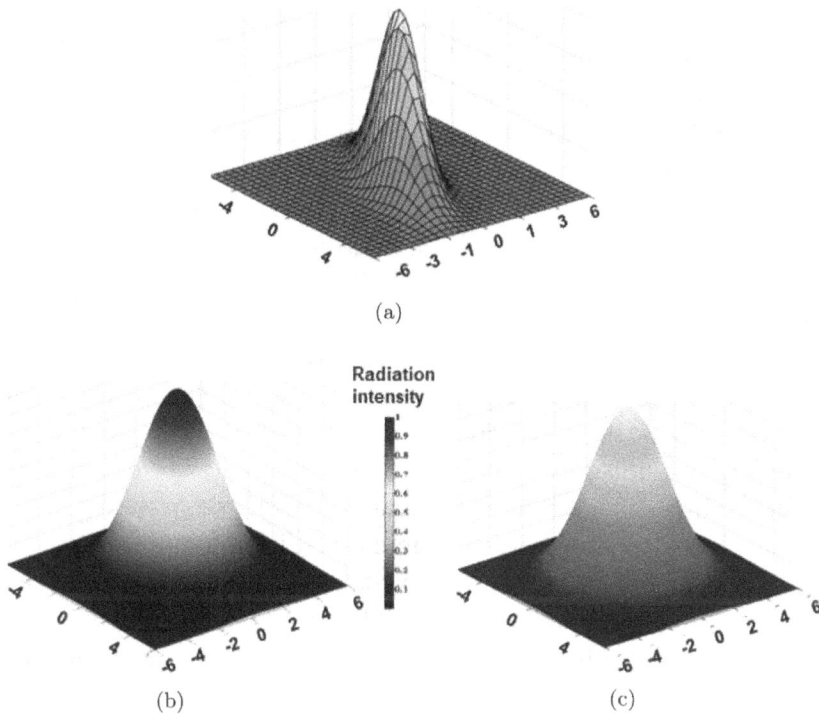

Fig. 4.3. Schematic representation of an asymmetrical emission profile (a) and two symmetrical emission profiles before (b) and after (c) the atomizer containing an atomic vapor of analyte.

reproducibility can generally be reduced by operating in the pulsed mode. The overall intensity diminishes rapidly with the number of pulses. So, after 5.10^3 pulses this intensity is approximately 95% less than its original value. After 5.10^3 and 5.10^4 pulses, the overall intensity remains stable in lamps operating on direct current ($d.c.$). An increase in the overall intensity produces a proportional decrease in the absorbance noise at the photodiode detector (LPDA or linear photodiode array). In this way detection limits will improve proportionally as the characteristic mass remains constant since it is not affected by intensity [Smith *et al.* (1994)]. In both pulsed and *d.c.* lamps, changes in the position of the cathode relative to the anode govern the average effective life of the lamps [Smith *et al.* (1994)].

Radiation intensity from a pulsed HCL is superior to what can be obtained from *d.c.* lamps, and the difference is even greater if N_2 is added along with the filler gas [Niemczyk *et al.* (1994)]. Pulsed HCLs use short pulses (a few μs) at a high current (several hundred mA). There are four parameters directly affecting the emited radiation intensity: pulse amplitude (I_p), pulse width (t), pulse frequency (f), and the *d.c.* component ($I_{d.c.}$). The intensity gain (I/I_0) is maximized as a function of the four parameters. The parameter I is the spectral line intensity of the cathode material under pulsed excitation, and I_0 is the intensity for *d.c.* excitation at the same average current value. Optimization of the pulse parameters to achieve the maximum intensity gain (I/I_0) does not necessarily give the best analytical conditions using HCL as emission sources. The *d.c.* discharge current must be greater than 5 mA in order to maintain a discharge in the HCL. For currents less than 5 mA, the discharge is restricted to the cathode surface and will not have any part in excitation of the analytes which are within the cathode [Mixon *et al.* (1994)].

Incorporation of lasers as radiation sources in analytical AAS, would surely improve analytical characteristics [Omenetto (1998)]. However, the cost and complexity of tunable lasers (primarily colorant lasers) has rendered their general use difficult, despite their excellent analytical characteristics. Diode lasers have several advantages: small, inexpensive and easy to operate (simply tuned)) over other tunable lasers. Atomic absorption spectrometry in graphite tubes using diode lasers with wavelength modulation has been employed to determine metals in complex samples with good analytical results (10 fg for Rb) [Ljung and Axner (1997)]. Nonetheless, the drawbacks derived are the lack of lasers able to operate in the visible and/or ultraviolet spectral zones.

4.1.1.2. *Electrodeless discharge lamps*

For many of the elements of interest in atomic absorption, HCLs offer adequate luminosity and stability, and emit a spectrum consisting of lines narrow enough to obtain good sensitivity. Nevertheless, there are elements (As, Hg, Se, Te and some others) for which better results are obtained by utilizing electrodeless discharge lamps (EDLs),

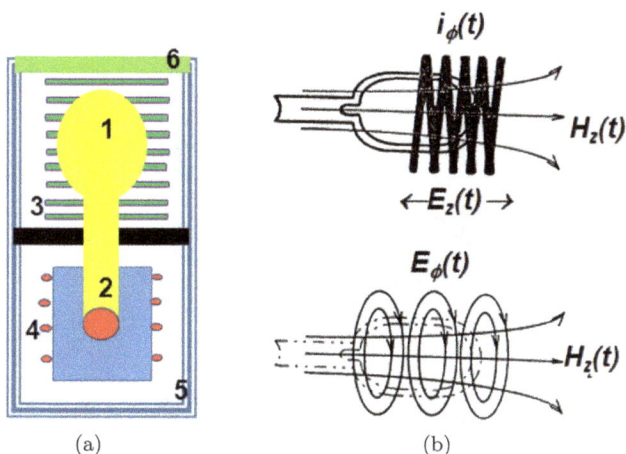

Fig. 4.4. (A) Schematic diagram of a high frequency (HF) electrodeless discharge lamp EDL: (1) Lamp bulb, (2) side-arm containing the working element, (3) HF-induction coil, (4) temperature stabilization, (5) insulation, (6) window [Ganeev *et al.* (2003)], and (B) EDL and structure of the magnetic field within the lamp [Gilmutdinov *et al.* (1996a)].

particularly if high-frequency (HF) EDLs are used (Fig. 4.4) [Ganeev *et al.* (2003)].

The EDLs are constituted by a quartz bulb with the analyte surrounded by a radio frequency (RF) generator inside a glass tube filled with an inert gas, usually argon, under low pressure. The short half-life time and the high temperature of operation of HCL make them inadequate for volatile species [Cooke *et al.* (1972)]. However, in EDLs, a HF electromagnetic field (RF or microwaves (MW): 300–3000 MHz) accelerates free electrons of the fill material colliding with the gas atoms and ionizing them to release more electrons (the "avalanche" effect).

The energetic electrons collide with the analyte-atom particles present in the quartz bulb, thus exciting them from the ground state to higher energy levels. Then, the excitation energy is released as a characteristic electromagnetic radiation of the fill material. The EDL performance is affected by temperature, the nature and pressure of the fill gas, fill material, dimensions and quality of the lamp envelope, the nature and characteristics of the MW energy coupling device, and the frequency and intensity of the MW energy [Müller *et al.* (2005)].

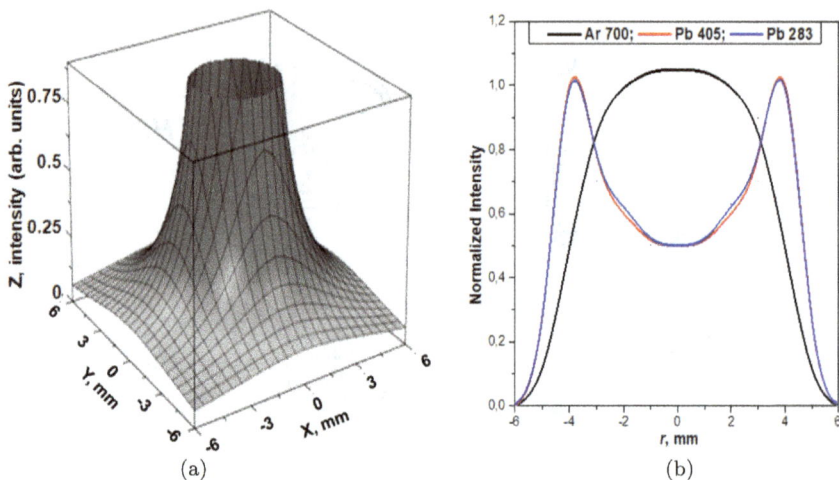

Fig. 4.5. Schematic representation of the spatial distribution of the radiation emitted by a lead (Pb) electrodeless discharge lamp (EDL) operating at a current intensity of 440 mA. (a) Pb atomic line at 405.7 nm. (b) Transversal distributions superimposed across the diameter of the lamp with normalized intensities for different spectral lines and the edges of the graph (b) coincide with the inner radius of the bulb (5.75 mm) [Gilmutdinov *et al.* (1996b)].

However, the use of these EDLs in AAS has been restricted by the large size and weight of HF generators. The plasma formed in EDLs is highly structured, with excited species both from the filler gas and analyte. For example, the excited atoms of Pb detected by both resonance and non-resonance lines take the shape of a thin layer concentric to the walls of the bulb and located near the bulb surface (the optical sink effect) (Fig. 4.5a). In contrast, the distribution of the emitted radiation for Ar atomic lines is bell-shaped with a maximum at the center of the plasma (Fig. 4.5b) [Gilmutdinov *et al.* (1996b)]. Self-absorption (and self-reversal) can be reduced by using high-power RF fields and much better by thermostating the side-arm of the lamp volume. In this case, the partial pressure of the element inside the lamp is controlled [Spietz *et al.* (2001)].

For mercury EDL, the difference between the spatial intensity profiles for the analyte lines and the filler gas (Ar) is less pronounced, owing to the use of a higher lamp filler gas pressure [Gilmutdinov *et al.* (1996b)].

4.1.2. Continuum Radiation Sources

Continuum radiation lamps show several drawbacks to be used as an ideal excitation sources in atomic absorption measurements. The electromagnetic radiation emitted by these spectral sources is not monochromatic, which is in contradiction to the requirements of Beer's law. Moreover, there is not any reference point for selecting and checking accurately the analytical wavelength. Thus, the establishment of the analytical wavelength and the spectral bandwidth would not be generally reproducible from one instrument to another. The principal difficulties associated with non-monochromatic radiation are well known: *non-linear calibration lines* and *poor sensitivity* (Fig. 4.6).

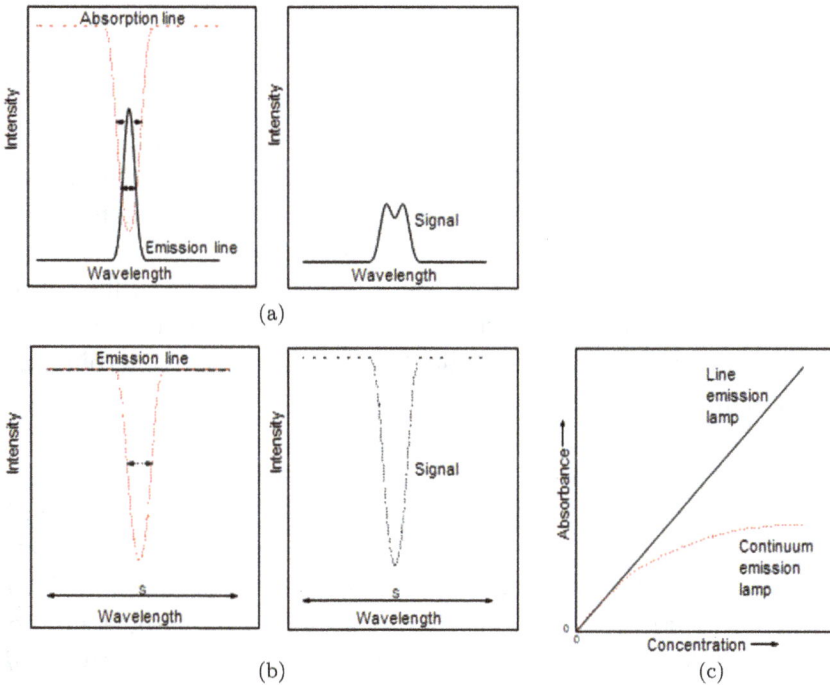

Fig. 4.6.   Comparison between atomic absorption processes using (a) line radiation and (b) continuum sources, also showing the net signal. (c) Two calibration graphs with line radiation (straight line) and continuum (curved line) sources.

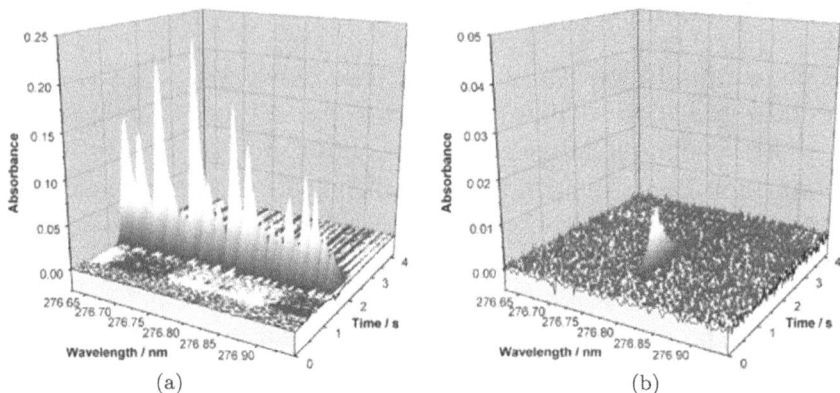

Fig. 4.7. Least squares background correction for structured background; atomization temperature 1650°C; (a) molecular absorption spectrum recorded during the vaporization of 10 μg KHSO$_4$ at 276.787 $\pm$ 0.16 nm; (b) residual absorbance spectrum for PACS-2 (marine sediment reference material) around the resonance line for Tl at 276.787 nm, after subtracting the reference spectrum [Welz *et al.* (2003)].

These drawbacks become more crucial when measurements are made using a narrow absorption band, where slight variations in the wavelength position or in the spectral bandwidth trigger sizeable changes in the absorbance observed. Furthermore, since the monochromator transmits a narrow but finite radiation bandwidth, the detector will indicate the average intensity over the whole selected spectral range, which is not necessarily the true absorbance at the exact center of the analytical wavelength profile. Despite this, continuum radiation sources (pulsed or not) have also been utilized to carry out multi-element atomic absorption measurements, but combining high resolution spectrometers with solid state detectors, photodiode arrays, and charge coupled devices (CCD) (Fig. 4.7) [Harnly (1986); Schmidt *et al.* (1990); Fernando and Jones (1994); Harnly *et al.* (1997); Schuetz *et al.* (2000); Welz *et al.* (2003); Rust *et al.* (2005a,b)]. Investigations of spectral interferences in Zeeman AAS are also possible for these situations [Heitmann *et al.* (1996)].

The most important advantages of using continuum radiation sources (CS) in AAS would be the following [Welz *et al.* (2003)]: (i) only one single radiation source is used for all elements and lines; (ii) CS makes AAS a truly simultaneous multi-element technique; (iii) for most elements, the radiation intensity of the xenon short arc lamp is at least one or two orders of magnitude higher than that of the corresponding line sources; (iv) the problem of "weak" lines disappears using CS, allowing secondary lines to be used; (v) it is possible to determine elements with CSAAS for which no line sources are available; and (vi) the determination of isotope ratios is also possible.

4.2. Spectral Apparatus or Instrument

Any optical system or any instrument dispersing optical (electromagnetic) radiation in a spectrum and/or isolating a specific spectral band is termed spectral apparatus or spectral instrument. If the entry slit forms a bi-dimensional image, whose width and length coincide on the same focal plane, the spectral instrument is said to have a stigmatic ordering. On the other hand, if the focal planes do not coincide in two dimensions, it is termed astigmatic ordering. When radiation passes through the same optical components before and after being dispersed, the spectral system is selfcollimating [Butler and Laqua (1996)].

4.2.1. Dispersive Systems: Monochromators and Polychromators

For isolating adequately each wavelength and gathering the greatest possible amount of light from the selected spectral zone, the optical systems are equipped with a monochromator in single channel spectral apparatus (normally in AAS) and a polychromator in multichannel ones (normally in AES). Monochromators and polychromators consist of the following parts (Fig. 4.8): (i) an input slit; (ii) a collimator to produce a parallel beam; (iii) a light-dispersive element (a prism, a diffraction grating, or a multiple-beam interferometer); (iv) a focusing element to reform the specific broad bands of the radiation dispersed; and (v) one (in the case of a monochromator) or more (in the case of a polychromator) output slits (in the Rowland's circle) to isolate the spectral band or bands desired.

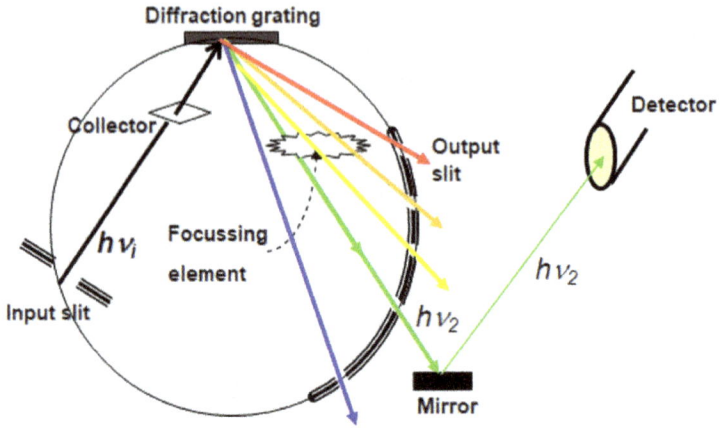

Fig. 4.8.   Main elements of any spectral instrument.

A Rowland circle is a mounting where a spherical concave grating with a radius of curvature R is mounted on the perimeter of a circle (real or imaginary) with a diameter equal to R. The lines of the grating are normal to the plane of the circle and the radius of the grating sphere passes through the center of the circle. An input slit positioned at an arbitrary point on the Rowland circle guides the light to be diffracted in the concave grating, producing a focused aberration-free spectrum on a different position of the Rowland circle.

The mounting or layout of the elements composing a spectral instrument can be very varied, but the most widely used in AAS are the following (Fig. 4.9).

- *Wadsworth mounting.* This concave grating mounting containts a concave mirror (the imaging element of the entrance collimator) and a concave grating (the dispersive element and at the same time the imaging element at normal angle of diffraction of the exit collimator). This mounting is used because of its stigmatic imaging properties.
- *Seya-Namioka mounting.* This is also a concave grating mounting and contains input and output collimators fixed at an angle ($\sim70°$ 30′) and wavelength variation is effected by rotation of the grating. It is mainly used in the vacuum UV region.

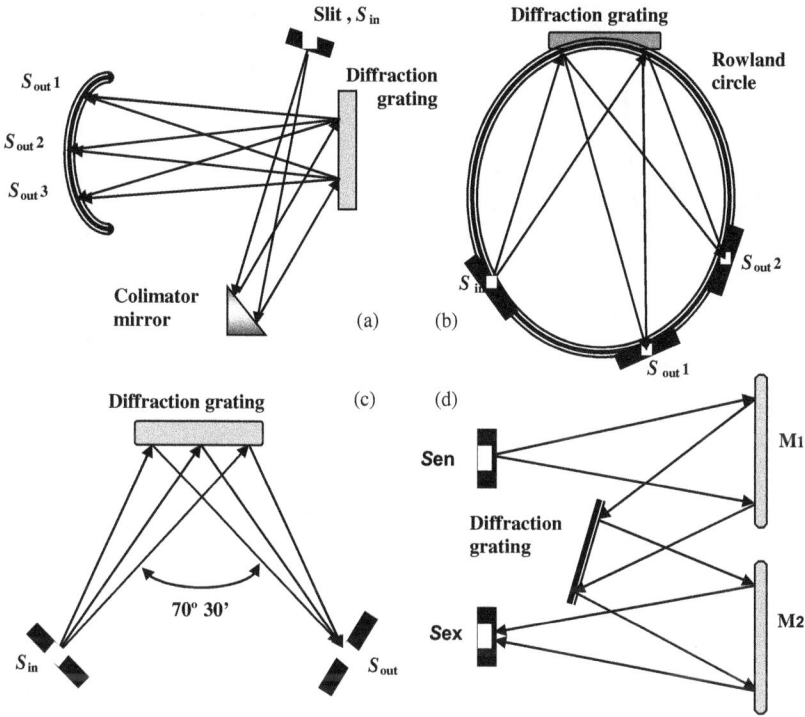

Fig. 4.9. Various assembly layouts for the components of a spectral instrument: (a) Wadsworth layout, (b) Paschen–Runge layout, (c) Seya–Namioka layout, and (d) Czerny–Turner layout.

- *Paschen–Runge mounting.* This is a Rowland circle mounting with an input slit and a concave grating fixed on the Rowland circle. The spectral lines are astigmatic.
- *Czerny–Turner mounting.* This is a plane grating mounting with two concave mirrors for input and output collimators. This mounting is similar to the in-plane Ebert mounting where one concave mirror acts as an imaging element symmetrically for both the input and the output collimators.

Some important characteristic parameters of the dispersive elements are as follows (Fig. 4.10):

- *Total deviation angle,* φ, of the radiation beam after refraction or diffraction.

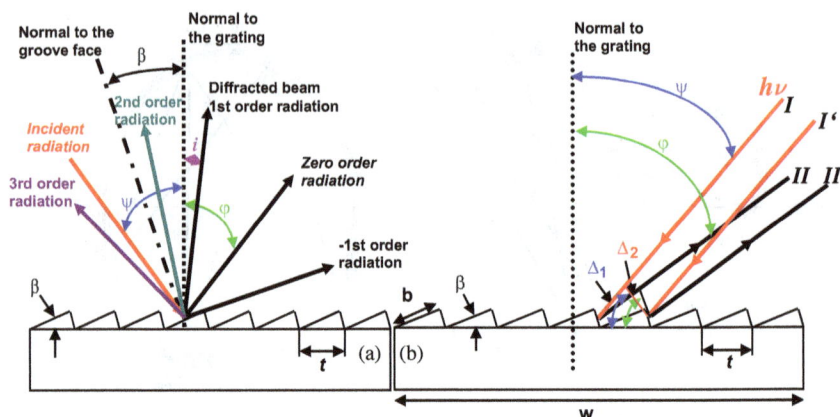

Fig. 4.10.   (a) Conventional, and (b) echelle diffraction gratings.

- *Angular dispersion, $\frac{d\varphi}{d\lambda}$, with respect to the wavelength, λ*

$$D_\varphi = \frac{d\varphi}{d\lambda} = \frac{K}{t\,\mathrm{Cos}_0\,\varphi} \tag{4.1}$$

$$\frac{d\varphi}{d\lambda} = \frac{R_0}{B_w}. \tag{4.2}$$

where B_w is the width of the light beam refracted on the refraction plane and R_0 is the theoretical resolving power.

- *Theoretical resolving power*

$$R_0 = \frac{\lambda}{\delta\lambda}. \tag{4.3}$$

where λ is the average wavelength of two consecutive lines and $\delta\lambda$ is the difference in wavelength between the two lines.

- *Wavelength limits (upper, λ_s, and lower, λ_y) between which the transmission (or reflection) factor goes above a specific fraction of the maximum.*

4.2.1.1. *Diffraction gratings*

In atomic spectrometry, the most widely used dispersive element is the diffraction grating, which may be of two types: (i) conventional diffraction grating, and (ii) echelle grating (Fig. 4.10).

Although previously used diffraction gratings were termed transparent, because they were made up of a series of slits, currently used diffraction gratings are composed of a grooved metal surface (Fig. 4.10). The main characteristic parameters of the grating include the following.

- *Grating width, W*, over the grooved area (measured in direction perpendicular to the grooves, on the plane of the grating).
- *Total number of grooves, N*

$$N = n W, \tag{4.4}$$

where n is the number of grooves per unit length (groove density) across W.

- *Grating constant, t*, which is the inverse of n.
- *Grating function*, the function relating the incidence angle, ψ, with the diffraction angle, φ. It is possible to distinguish two situations:

  (i) the incident beam and the reflected beam are on the same side of the grating normal

$$Sin\,\psi + Sin\,\varphi = K\frac{\lambda}{t}, \text{ and} \tag{4.5}$$

  (ii) both beams are on the opposite side of the grating normal

$$Sin\,\varphi - Sin\,\psi = K\frac{\lambda}{t}, \tag{4.6}$$

  where K is the diffraction order.

- *Usable free spectral range* (without overlap with other spectral orders)

$$\Delta\lambda = \frac{\lambda}{K}. \tag{4.7}$$

The position of the principal maxima for the reflective grating is given by Eq. (4.5). The energy distribution for this case shows its principal maximum for the null order ($K = 0$), if the angle φ is equal to the specular reflection angle. Occasionally, this diffraction grating is called a reflection grating. The position of the maxima produced by interferences will be observed when the difference in the optical path travelled by the beams, given by $\Delta_2 - \Delta_1$, is equal to an integer, K, of wavelengths, λ

$$\Delta_2 - \Delta_1 = K\lambda \tag{4.8}$$

Thus, Eq. (4.6) can be reordered as follows

$$t(\text{Sin } \varphi - \text{Sin } \psi) = K\lambda = K_i\lambda_i \tag{4.9}$$

As K is any whole number, this means that the condition set by Eq. (4.6) is fulfilled, or may be fulfilled, simultaneously for several values of K. Hence, a diffraction grating produces a great number of spectra, each of which corresponds to a value of K. When $K = 0$, the spectrum is of zero order (Fig. 4.10), in which Sin φ = Sin ψ independently on the wavelength. This spectrum corresponds to the incident radiation or otherwise to the transmitted radiation. When $K = 1, 2, 3, \ldots$, etc., the spectra are of first, second, third order and so forth (Fig. 4.10). In this way, if for the maximum wavelength observed, λ, the first order spectrum emerges with an angle φ with respect to the normal, from that the same angle there will also be observed diffracted radiation at the wavelength $\lambda/2$ in the second order spectrum, $\lambda/3$ in the third order spectrum, and so on. The uppermost value of K is limited by the condition

$$|\text{Sin } \varphi - \text{Sin } \psi| \leq 2. \tag{4.10}$$

Thus, from Eq. (4.6) it results that

$$K\lambda \leq 2t \tag{4.11}$$

or alternatively

$$K_{max} \leq 2t/\lambda \tag{4.11'}$$

which represents the maximum value for K. For the grating to produce a spectrum, even of the first order, the condition $t > \lambda/2$ must be fulfilled. That is, the grating constant must be small for the shortest wavelengths, and larger for a longer λ.

When relatively low spectral orders are used, these overlapping emissions can be separated fairly easily through the use of filters. However, as the spectral order increases, it becomes more difficult

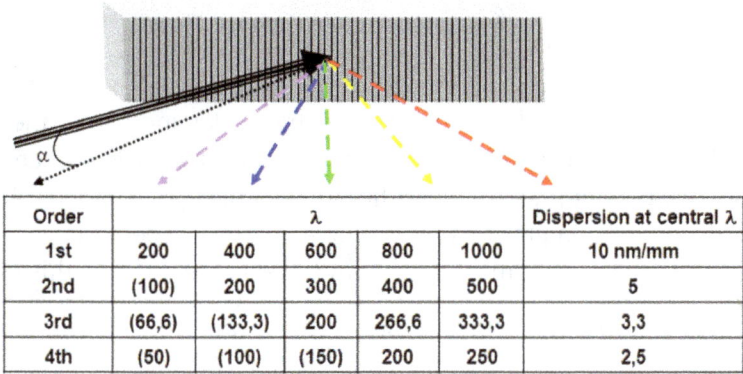

Order	λ					Dispersion at central λ
1st	200	400	600	800	1000	10 nm/mm
2nd	(100)	200	300	400	500	5
3rd	(66,6)	(133,3)	200	266,6	333,3	3,3
4th	(50)	(100)	(150)	200	250	2,5

Fig. 4.11. Two-dimensional spectrum set obtained with an echelle spectrometer. The orders are classified vertically and dispersed horizontally within each order. Wavelengths for emission spectra of differing orders and hence observed at different angles.

to separate the first radiation, because (Fig. 4.11): (1) with an increase in the spectral order, the wavelengths get closer one to another; and (2) intensity diminishes with the spectral order. Hence, if a diffraction grating is used to observe higher order spectra, the overlapping free region becomes smaller. In this case, it is necessary to take special care to separate spectra from neighboring orders.

The distance ($\Delta\lambda$) between the maxima of two spectra from neighboring orders diffracted at the same angle, φ, may be worked out as follows. As the incidence angle and the diffraction angle for overlapping neighboring orders are equal, it holds that:

- for a given order

$$t(\mathrm{Sin}\,\varphi - \mathrm{Sin}\,\psi) = K\lambda_1 \qquad (4.9')$$

- for a neighboring order

$$t(\mathrm{Sin}\,\varphi - \mathrm{Sin}\,\psi) = (K+1)\lambda_2 \qquad (4.9'')$$

that is, λ_2 diffracts along the direction of λ_1, but in order $(K+1)$, with

$$\Delta\lambda = \lambda_1 - \lambda_2 \qquad (4.12)$$
$$\lambda_2 = \lambda_1 - \Delta\lambda \qquad (4.12')$$

$$K\lambda_1 = (K+1)(\lambda_1 - \Delta\lambda) \qquad (4.13)$$

$$\Delta\lambda = \lambda_1/(K+1). \qquad (4.14)$$

Thus, it may be noted that as the spectral order increases, the distance between two wavelengths from neighboring orders decreases (Fig. 4.11). Equation (4.14) is a different way of expressing Eq. (4.7).

For a diffraction grating to produce a first order spectrum ($K = 1$), the condition, $t > \lambda/2$ must hold (why?, see Eq. (4.11′)). In this way

- for $\lambda = 800$ nm, it holds that $t > 400$ nm, and the number of grooves on the grating surface (N) < 2500 ($N\,t/10^6 = 1$ mm),
- for $\lambda = 200$ nm, it is the case that $t > 100$ nm, and the number of grooves on the grating surface (N) < 10000.

This means that to observe the radiation dispersion for large wavelengths, it is not necessary to use gratings with many grooves. However, there are two reasons to increase the number of grooves on the grating, or to decrease the grating constant.

(i) The first reason is the theoretical resolving power of the grating, given by

$$R_0 = \frac{\lambda}{\delta\lambda} = KN = KnW_g, \qquad (4.3')$$

where K is the diffraction order, N the total number of grooves, W_g is the illuminated grating width, and n is the groove density.

It will be observed that the resolving power of the grating is determined not just by the spectral order, but also by the number of grooves on the grating (N). It is evident that an increase in the number of grooves on the grating is possible only at the cost of diminishing the grating constant ($t = 1/n$). If this were not so, it would be necessary to construct gratings of large size. However, to increase the resolving power without upping the number of grooves, it is possible to increase the spectral order, as in the echelle gratings.

(ii) The second reason for an increase in the total grooves on the grating is the tendency for angular dispersion to grow (see Section 4.2.1.4. *Dispersion*).

4.2.1.2. *Apparatus profile*

Owing to the permanent presence of all, or at any rate of many, of the various causes for the broadening of the spectral lines (see Section 1.4. *Broadening of the Spectral Lines*), when a monochromatic radiation is recorded, a line showing a finite width contour or profile is always observed. Each elemental part of this profile is related to the signal in the following way

$$d\phi = \phi f(\lambda) d\lambda, \tag{4.15}$$

where ϕ is the complete signal corresponding to the total flow recorded, and the function $f(\lambda)$ is determined by the properties of the spectral instrument and is called the *apparatus function* or *instrumental profile*. This function must fulfil the uniformity condition

$$\int_0^\infty f(\lambda) d\lambda = 1. \tag{4.16}$$

Each wavelength, λ, deviates in the spectrometer with the angle, ϕ, focusing at the point x, in the focal coordinate (Fig .4.12).

The instrumental profile can present different forms.

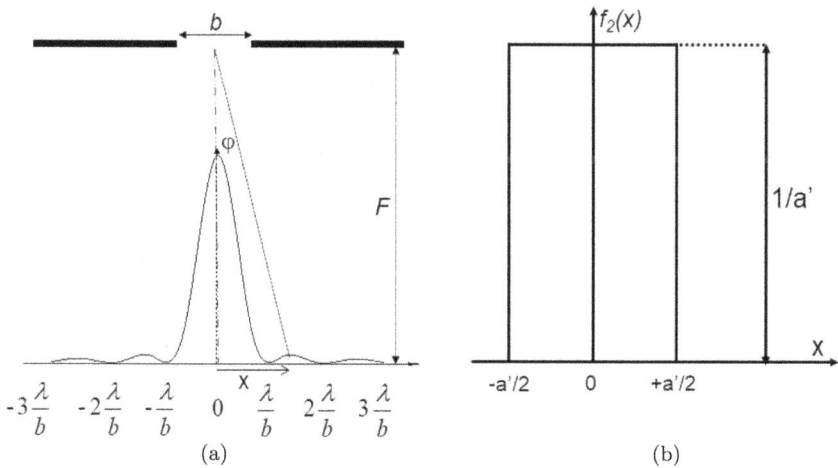

Fig. 4.12. Diffraction (a), and rectangular (b) instrumental profiles.

(i) An infinitely narrow slit with the image broadening solely due to diffraction on the edges (Fig. 4.12a). The illumination distribution on the focal surface is the following

$$f(x) = \frac{\text{Sin}^2\left(\frac{\pi x}{\delta l}\right)}{\left(\frac{\pi x}{\delta l}\right)^2}, \tag{4.17}$$

where $\delta l = \lambda F/b$, F is the focal length (see F in Fig. 4.12a) of the lens producing the image of the line on the focal plane; and $\varphi = x/F$, x is the distance from the line center in the direction of dispersion. The minimum and maximum intensities appear for $\text{Sin }\phi = K\lambda/b$ and $\text{Sin }\phi = (K+1/2)\lambda/b$, respectively.

(ii) A wide slit illuminated with monochromatic light (Fig. 4.12b). The geometrical image width is much greater than the image width of the null maximum of diffraction. In this situation, the illumination on the focal plane is constant throughout the image from the slit if no account is taken of the diffraction phenomena. In this case, the instrumental profile is described by the function

$$f(x) = \begin{cases} 1/a' & \forall |x| \leq a'/2 \\ 0 & \forall |x| \geq a'/2 \end{cases}, \tag{4.18}$$

where a' is the image width from the slit.

(iii) A wide slit, but not sufficiently so for it to be possible to ignore diffraction phenomena on the edges. In this case, the slit is considered a light source of finite width that may be divided into a series of narrow sources of width dy, situated at distance y from its center. Thus, if the *instrument's slit is wide,* keeping in mind the diffraction phenomena on the edges, and if there are n independent causes for broadening and each of them is described by function $f_k(x)$, the resulting contour $F_n(x)$ may be obtained through the following convolution function (Fig. 4.13)

$$F_n(x) = \int_{-\infty}^{+\infty} f_n(x - x')F_{n-1}(x')dx'. \tag{4.19}$$

where $f_n(x - x')$ represents the instrumental profile.

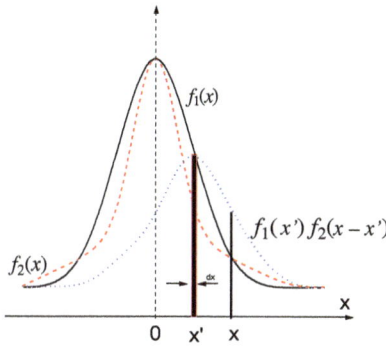

Fig. 4.13. Convolution of two functions.

4.2.1.3. *Resolving power*

The width and form of the instrumental profile in general determine the possibility of investigating a spectrum in more or less detail. This feature of the instrument is expressed as a magnitude called resolving power. This magnitude represents the minimum interval in wavelength terms for two monochromatic spectral lines to be observed as separated one from the other. However, this definition is not fully sufficient, as the possibility of observing two monochromatic lines as distinct (or separated) depends on the precision with which the instrumental profile is known.

The quantitative criterion of the instrument's resolving power must be expressed on the assumption that a given level of precision is applied to energy measurements. This criterion was established by Rayleigh (Fig. 4.14), who took the spectral range for minimum resolution, $\delta\lambda$ (or resolution limit), as the distance between the principal maximum and the first minimum of the function that describes a contour for which two monochromatic lines of the same intensity are considered distinct if the drop in brightness between them is 20%. Resolving power, R_0, is usually represented by the following dimensionless magnitude,

$$R_0 = \frac{\lambda}{\delta\lambda} = bD_\varphi, \tag{4.3''}$$

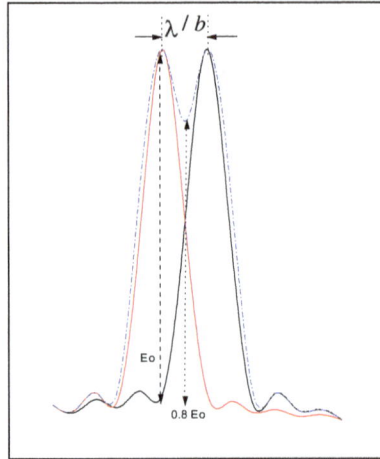

Fig. 4.14.   Rayleigh's criterion.

where b is the linear dimension of the diaphragm limiting the beam width, and D_φ is the angular dispersion of the instrument. This magnitude, R_0, is termed Rayleigh's resolving power, or theoretical resolving power.

The theoretical resolving power, determined by the phenomenon of diffraction, is contrasted with the practical resolving power, determined by other factors (finite slitwidth, defects in optics and focusing, light dispersion), which cause a broadening of the instrumental contour. The practical resolving power is defined as

$$R_p = \frac{\lambda}{\delta\lambda_p},\tag{4.3'''}$$

where $\delta\lambda_p$ is the resolution limit. It is always the case that $R_p < R_0$.

Logically, the slitwidth influences the resolving power. The width of the instrumental contour grows very slowly for slitwidth values equal to the normal width, so that the diminution of resolving power is very gradual. For slitwidths greater than double the normal value, the instrumental contour expands proportionally to the slitwidth, and the resolving power diminishes sharply. However, the practical resolving power is a more important characteristic than linear dispersion, since resolving power is precisely what characterizes the capacity of the apparatus to differentiate two close-by lines. If the instrumental contour is wider than the distance between the lines, these will be poorly separated, even if they are a good way one from the other. The practical resolving power is determined by the width of the real instrumental contour, which cannot be less that the width of the geometrical image of the slit, so that

$$R_p = \frac{\lambda}{\delta\lambda_p} = \frac{\lambda}{\Delta l}D_l. \tag{4.3''''}$$

If $\Delta l = 10^{-3}$ cm

$$R_p = 10^3 \lambda D_l. \tag{4.20}$$

This means that the practical resolving power is directly proportional to the linear dispersion, D_l, of the apparatus. However, this resolution is not always attained, as the linear dispersion is also proportional to, or dependent on, the angular dispersion and the parameters of the focusing optics.

In general, it is impossible to study absolutely all the factors influencing the practical resolving power, so that habitually this is experimentally determined. The simplest way to determine the resolving power consists of finding lines which are narrow and very close to each other but which can be separated. Depending on the resolving power of the apparatus, the lines chosen may be:

- multiplets from known spectra or superfine structure lines, and
- components from Seeman's decomposition, whose distance may be gently varied by changing the magnetic field value.

It is preferable to select components of equal intensity and polarization state. The resolving power is determined by measuring the width of the instrumental profile at the point where it drops to a value equal to 0.4 times the maximum. This width corresponds to Rayleigh's separation limit.

Resolutions of the order of 0.02 nm (FWHM) or even better [Florek and Becker-Ross (1995)] are now standard characteristics. To improve the resolution of any spectrometer, it is essential to be able to fix precisely the position of the wavelengths. To fix this wavelength position various methods may be used, including automatic calibration controlled by the instrument's software. Automatic fixing of the wavelength must be reproducible and precise in order for the resolution and speed of the technique to be adequate.

All spectral instruments can suffer slight shifts in the wavelength position as analyses progress, owing to changes in temperature, vibrations and other factors. For this reason, instruments usually incorporate an algorithm to control factors which affect the wavelength shifts. Nonetheless, errors due to thermal, mechanical and mathematical variations are usually very small and generally lie between 0.003 nm and 0.005 nm, or even lower values, depending on the type of monochromator used [Grosser and Collins (1991)].

4.2.1.4. *Dispersion*

Isolation of different wavelengths is called dispersion. Two types of dispersion may be distinguished: angular dispersion, D_φ, and linear dispersion, D_l.

Angular dispersion, D_φ, is given by the relationship obtained through differentiating with respect to λ Eq. (4.6)

$$D_\varphi = \frac{d\varphi}{d\lambda} = \frac{K}{t\,\mathrm{Cos}\,\varphi} = \frac{\mathrm{Sin}\,\varphi - \mathrm{Sin}\,\psi}{\lambda\,\mathrm{Cos}\,\varphi} \text{ (see Eq. (4.6))}, \qquad (4.21)$$

where $d\varphi$ is the angular separation between two wavelengths (radians), and $d\lambda$ is the differential separation between two wavelengths (nm). Consequently, a decrease in the grating constant, t, or an increase in the spectral order, K, will increase the angular dispersion of the apparatus.

Linear dispersion, D_l, provides the distance, l, along the focal plane for any wavelength. The angular dispersion and the focal distance, F, of the spectrometer determine the reciprocal **linear dispersion**, $1/D_l = d\lambda/dl$. However, for small diffraction angles, $\mathrm{Cos}\,\varphi \cong 1$, so that, the linear dispersion of a diffraction grating monochromator is constant. **Linear dispersion** is given by

$$D_l = dl/d\lambda = KF/t\,\mathrm{Cos}\,\varphi. \qquad (4.22)$$

while, considering Eqs. (4.3'), (4.20) and (4.22), the focal distance is given by

$$F = Nt \, \mathrm{Cos}\,\phi/1000\lambda. \tag{4.23}$$

The precise focal distance of the objective depends on the grating width and not on the spectral order and the grating constant. This is the result of an increase in the number of grooves per millimeter or in the spectral order; both equally increase dispersion and the resolving power of the grating. The resolving power grows as the number of grooves per millimetre (n) increases, since Nt is constant, and the total number of grooves, N, is proportional to n.

The reciprocal linear dispersion is related to the spectral bandwidth (bandpass), $\Delta\lambda$, and the geometric width of the monochromator slit, s, as follows:

$$\Delta\lambda = s\left(\frac{d\lambda}{dl}\right). \tag{4.24}$$

4.2.2. Non-Dispersive Systems

In non-dispersive spectral apparatus, isolation of a spectral band is achieved without dispersion of the wavelength through the use of *absorption, fluorescence, reflection*, or *optical dispersion* [Pogue *et al.* (2000)]. It may also be isolated by using an interference filter based on the interference from multiple beams. Such filters are examples of *spectral filters* (see Section 4.4.4. *Filter Spectrometer*). A *double-beam interferometer* may also be included in non-dispersive spectral instruments.

4.3. Detection Devices: Reading and Recording Analytical Signals

Detection of electromagnetic radiation is carried out by a detection device in which the incident radiation produces a measurable effect. Detectors can be grouped as a function of this effect.

- Thermal detector, variation in temperature.
- Photoacoustic detector, variation in pressure.

- Photoelectrical detector, electrical signal.
- Photochemical detector, the radiation produces a chemical reaction.

The responsivity of a detector, that represents a special case of the general term sensitivity, refers to the detector output and the detector input ratio.

4.3.1. Photomultiplier Tubes

The photomultiplier tubes are the detectors most widely used in AAS. A photomultiplier tube is a vacuum phototube (a photoemissive detector) with an additional electron multiplication. It is made up of a series of electrodes (or dynodes), each one with a photoemissive surface and a positive potential relative to the previous electrode. When a photon hits the first photoemissive surface, an electron is released, and this is attracted by the next dynode. During its trajectory, the electron is accelerated, so that when it hits the second dynode it expels several electrons. These in their turn are attracted and accelerated by the third dynode, which in its turn expels a greater number of electrons than its predecessor. This process continues at each dynode, so that the exit from the photomultiplier tube is reached by an extraordinarily large number of electrons. In this way, a single photon generates a current of electrons that produces a measurable signal. The sensitivity of this system of amplification depends on the voltage between the dynodes. The larger the potential difference between dynodes, the greater will be the amplification. However, if the voltage is amplified a great deal, the output signal starts to become erratic and noisy.

The gain, G, of the multiplier tube is expressed as $G = \varepsilon\sigma^m$, where ε is the photoelectron collection efficiency on the first dynode, σ is the secondary emission ratio, and m is the number of dynodes.

The type of photomultipliers used depends on the radiation wavelength measured, as the intensity of the output signal varies considerably with wavelength and so with the photomultiplier used (Fig. 4.15) [Sweedler *et al.* (1988)]. It is essential to choose a suitable photomultiplier, because if the resonance line is outside the optimum wavelength

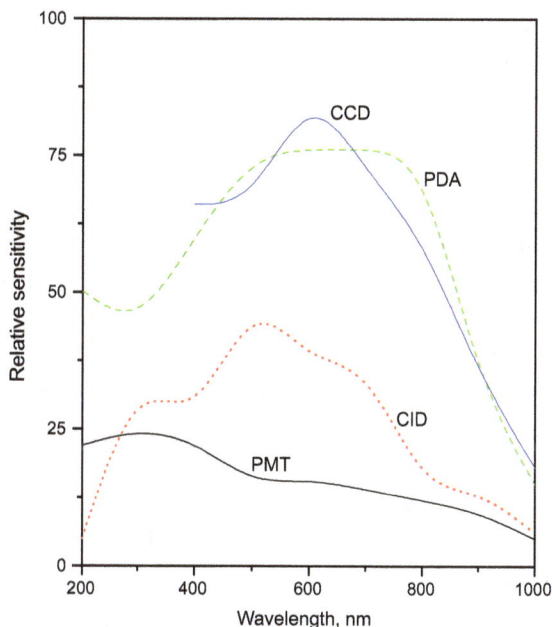

Fig. 4.15. Sensitivity graphs for photomultiplier tubes (PMT) and other detectors: charge-coupled device (CCD), charge-injection device (CID) and photodiode array (PDA) [Sweedler *et al.* (1988).]

range of the photomultiplier, the analytical signals obtained will be erratic and lacking in sensitivity and precision.

4.3.2. Charge-Transfer Devices

Charge-transfer devices (CTD) are solid-state detectors in a linear or two-dimensional array which gather and store photogenerated charges, and then measure the amount of charge present. A charge-transfer device consists of a metal oxide semiconductor structure composed of many independent pixels where charge is stored in such a way that the charge pattern corresponds to the irradiation pattern. In CTDs, the charge detection is accompanied by a prior process of charge-*transfer*, from a collection zone to the detection zone (Fig. 4.16) [Sweedler *et al.* (1988)]. The charge-transfer process takes place as variation occurs in the voltage applied to a series of electrodes covering the whole collection zone.

The CTDs accumulate charges when light hits them (as in photographic films). This process is distinct from what happens in photomultiplier tubes, which produce a single current proportional to the *instantaneous* flow of photons reaching the photocathode. Almost all the CTDs manufactured today are based on silicon, although there are alternative substrates built with other materials, such as InGaAs, MCT (HgCdTe), and PtSi. The CTDs are made up of a multilayer structure of silicon consisting of a substrate, an *epitaxi*, an insulating layer, and various electrodes (Fig. 4.16). An individual detector in a CTD array consists of several conductive electrodes set into an insulating layer that form a series of metal oxide semiconductor (MOS) capacitors. The insulator separates the electrodes from the doped silicon zone used to store the photogenerated charge. The amount of this photogenerated charge is proportional to the number of photons incident upon the detector. When a photon is absorbed into the epitaxi, an electron is shifted from the valency band to the conduction band of silicon, creating an electron-hole pair. The electron or hole displaced is trapped at a potential much lower than that created in the electrodes by the voltage applied (Fig. 4.16).

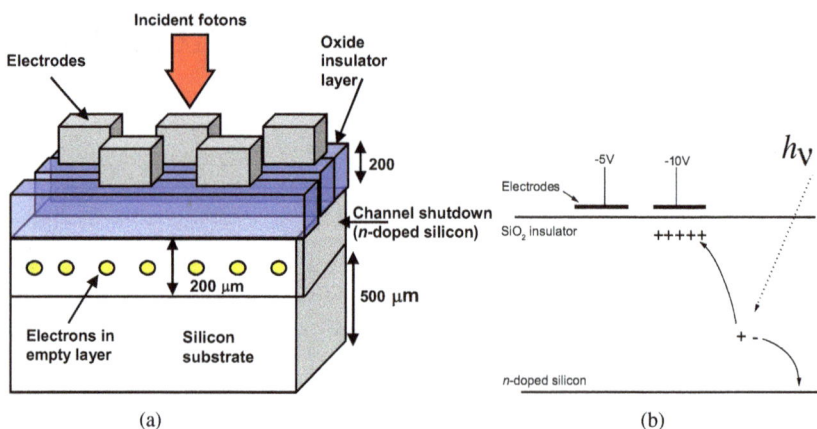

Fig. 4.16. (a) Representative scheme of an ideal CTD detector with the electrodes ready for charge integration. (b) Transverse section of a CTD detector showing how the absorption of a photon causes the formation of an electron-hole pair, and the positively-charged hole is collected under the negatively charged electrodes [Sweedler *et al.* (1988)].

The most important features of the single-channel detectors are their electro-optical characteristics: reading noise, dark current, efficiency, dynamic range, and response time. However, the bidimensional nature of CTDs adds other additional parameters, such as linear format, specialized reading modes, charge-transfer efficiency, and the intra- and inter-scenic dynamic range.

The CTDs are nowadays the commonest multichannel electronic detectors. The CTDs offer high sensitivity, can reach up to 90% peak quantum efficiency, are sensitive to photons coming from the region of soft X-rays to near infrared, i.e. from 0.12 nm (UV) up to 1100 nm (NIR) [Epperson *et al.* (1988); Sweedler *et al.* (1994)], and have count speeds for darkness permitting integrations running over many hours with very low reading noise (equivalent to less than 1 *e*). Thus, the CTDs are also the first image detectors to combine the characteristics of integration and high spatial resolution of photographic films with a sensitivity that rivals, and in some cases exceeds, that offered by a (single-channel) photomultiplier tube [Earle *et al.* (1993)].

Solid state detectors provide better analytical capabilities [better detection limits, possibilities for studying the spectral region where spectral interferences occur, and multi-element determinations] than the photomultiplier tubes, even for carrying out single-element measurements [Gilmutdinov *et al.* (1994); Hanley *et al.* (1996); Haisch and Becker-Ross (2003)]. Thus, in ETAAS the transverse distribution of both the radiant intensity and the analyte atoms may not be uniform within the atomizer. Consequently, the absorbance measured by a detection system based on the use of photomultipliers tubes depends both on the number and gradients of free atoms and on the gradient of the radiation beam. These drawbacks can be overcome by using solid state detectors. In this case, the absorbance recorded depends solely on the number of absorbing atoms and may be integrated not only with respect to time but also with respect to the monochromator slit, which can be used as a measure of the analyte concentration in the sample (Table 4.1).

According to the method used to detect the charge pattern, two types of charge-transfer devices can be distinguished: charge-coupled devices (CCDs), and charge-injection devices (CIDs). Although the two

Table 4.1. Various Limit Expressions Applicable to a CTD Detector used for Measurements of Absorbance [Hanley *et al.* (1996)]

Parameter	Without Background	With Background[a]
Minimum detectable absorbance	$A_{min} = -\log\left(1 - k\sqrt{\dfrac{2}{Q_{sat}}}\right)$	$A_{min} = -\log\left(1 - \dfrac{k\sqrt{2(Q_{sat} + B)}}{Q_{sat} - B}\right)$
Maximum absorbance	$A_{max} = -\log\left(\dfrac{k\sigma_r}{Q_{sat}}\right)$	$A_{max} = -\log\left(\dfrac{k\sqrt{B}}{Q_{sat} - B}\right)$
Maximum dynamic range	$DR = \dfrac{A_{max}}{A_{min}}$	$DR = \dfrac{A_{max}}{A_{min}}$

[a]The noise in the background signal is assumed to be equal to $\sqrt{B}$. B is the magnitude of the background signal. The value of k is selected in accordance with the level of confidence required. Q_{sat} is the total capacity of the system. σ_r is the read noise. DR is the dynamic range. A is the absorbance.

main subclasses of CTDs are much alike in some ways, their behavior and modes of operation are very different. The charge stored at each pixel may be transferred to the reading amplifier system in either of the two following ways. Charges from each pixel are systematically sent along the line and presented sequentially to the reading amplifier (the reading mode in a CCD), or the reading amplifier is connected sequentially to each pixel through a multiplex system (the reading mode in a CID). In the CCD system, the electrodes (capacitors) collecting the electrons are also used to shift them systematically along the line to the reading amplifier. The CID mode is a random access mode, which permits any given pixel to be controlled independently [Harnly and Fields (1997)]. Chemists were the first to demonstrate the utility of CTDs for carrying out spectroscopic measurements. Hence, Ratzlaff and Paul (1979) used a linear CCD to detect molecular absorption, while in other applications CCDs were used in ETAAS [Harnly *et al.* (1997); Radziuk *et al.* (1995a,b)].

4.3.2.1. CCD

CCD detector has a metal-oxide-semiconductor structure which houses carriers for photogenerated charges (Fig. 4.17) [Bilborn *et al.* (1987); Sweedler (1993)].

Above the silicon there is a series of conductor electrodes. When an incident photon penetrates into the silicon substrate and form a hole-electron pair, the electrons migrate to one of the electrodes under the influence of the potential applied, and are stored there until a reading of the charge is taken. By correctly biasing the numerous inset electrodes, the photogenerated charge can be kept in the element (*pixel*) where it is generated. In a charge-coupled device, the signal charge is transferred to the edge of the array for readout. Alternatively, multiplexing can be used. The charge packets are transferred in discrete time increments by the controlled movement of potential wells. In a linear CCD, the charge is moved in a stepwise way from element to element and is detected at the end of the line. A "two-dimensional

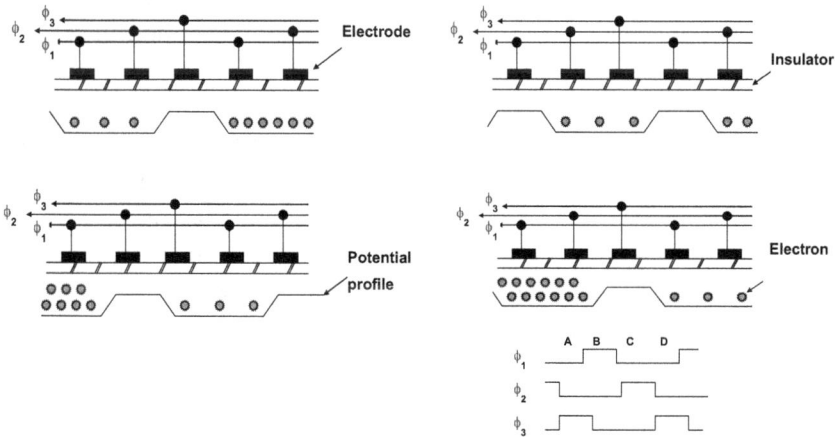

Fig. 4.17. Diagram of a three-phase CCD showing the processes of integration and charge transfer for 4 different time states. The voltages applied to the 3 independently controllable phases are connected between a high level and a low level, forcing the charge to migrate towards the right. (The voltages applied in each phase for each time are shown in the lower part of the figure) [Bilhorn *et al.* (1987)].

array" CCD consists of a two-dimensional assembly of interconnected linear CCDs. The on-chip summing of charges in adjacent pixels along rows or columns is called "binning".

The CCDs are built with two, three, or four electrodes per pixel (systems for two, three or four phases, respectively).

After the integration period has been completed, the charge is shifted from the photoactive area to the charge-sensitive amplifier located at one corner of the array. On the edge of the array, there is a single output node (various on many occasions) and the charge is shifted sequentially to the amplifier. Figure 4.17 shows the charge-transfer process for a three-phase CCD; at least one of the three regions of background potential in the detector element is always kept low (collapsed) so that there will be a barrier preventing the charge from falling into the adjacent element. Displacement of the location of this barrier causes migration of the charge. What distinguishes CCDs from CIDs, photodiode arrays (PDAs) and other array detectors is the ability to transfer the charge from the sensitive element to a low-capacitance output node, and which is thus specially designed to eliminate high-capacitance multiplexed architectures. The extremely small capacitance at the entry to the output amplifier permits CCDs to attain ultra-low read noise levels.

Quantum efficiency, a measure of the system sensitivity, is defined as the number of detectable electrons created per photon of incident light. It is generally expressed in percentage terms. For photons with wavelengths greater than 250 nm, absorption is the result of the participation of a single electron, and the maximum possible efficiency is 100%. The quantum efficiency varies with the wavelength and with each CCD system. Thus, for other regions of the electromagnetic spectrum (λ < UV and even in the region of soft X-rays), creation of multiple electrons can lead to quantum efficiencies greater than 100%.

One reason for the limitation of quantum efficiency is that silicon has a reflectivity of about 30% at 1090 nm, reaching 5% below 400 nm. Another factor limiting quantum efficiency is that the depth to which a photon penetrates the *epi* before creating an electron-hole pair depends on the wavelength. At 275 nm, the penetration depth (the depth by which 90% of the photons have been absorbed) is only 20 Å,

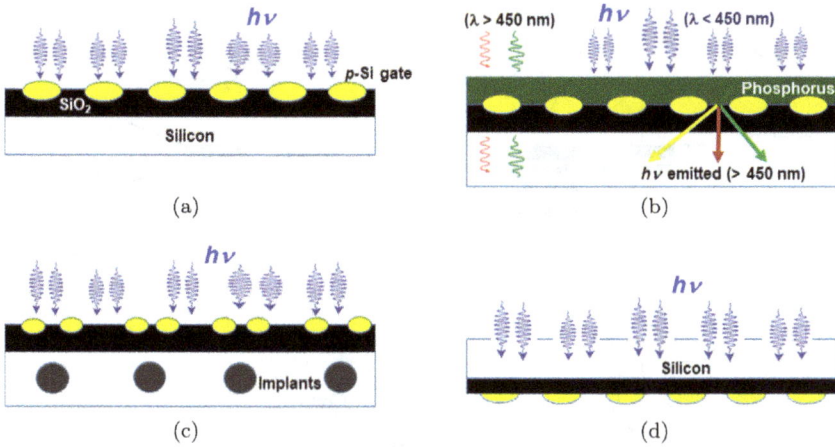

Fig. 4.18. Various assembly layouts of a CCD to illustrate the methods used to increase quantum efficiency in the UV region. (a) Normal CCD; (b) phosphor-coated CCD; (c) open phase CCD; and (d) rear-light CCD [Sweedler (1993)].

but this grows exponentially up to ~30 μm at 800 nm, and reaches several hundred micrometres at 1000 nm [Giles *et al.* (1998)]. There are nonetheless, various methods for increasing quantum efficiency in the ultraviolet region (Fig. 4.18) [Sweedler (1993)].

4.3.2.2. *CID*

In a charge-injection device (CID), the accumulated charge is not transferred serially out of the array, but is shifted between two adjacent capacitors. In non-destructive readout, the output is derived from the electric potentials on these two capacitors, which retain the information. Alternatively, the output can be derived from the stored charge after it has been injected into the substrate, thus destroying the original information.

As is the case for CCDs, in CID systems the individual detector element comprises several electrically conductor electrodes set into a thin insulator made of silicon nitride or oxide. CIDs are manufactured using an *n*-doped epitaxial region grown on a *p*-doped substrate, and so a CID collects photogenerated holes. The inserted electrodes are adapted to permit collection and integration of the charge in

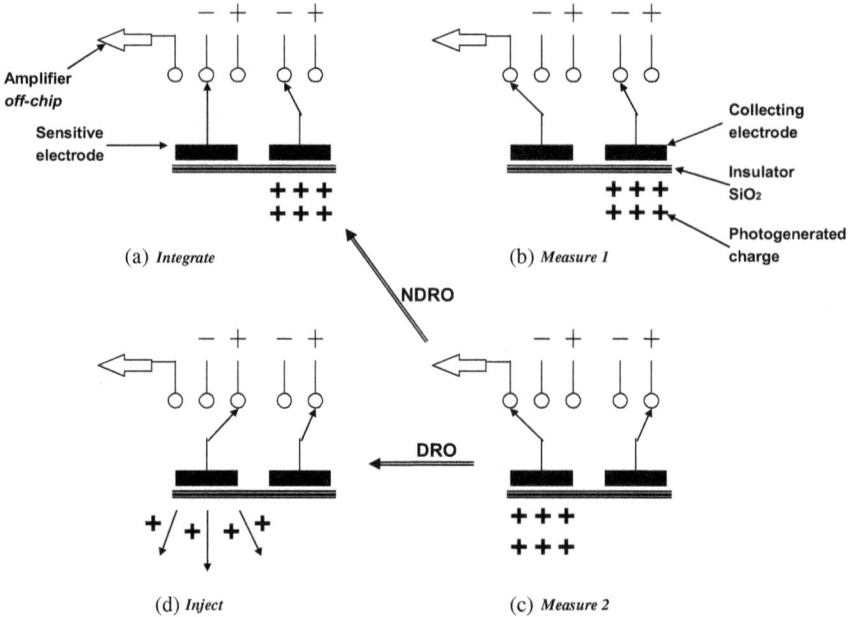

Fig. 4.19. Reading from a CID to show the non-destructive measurement process. (a) The CID is in the integration mode. (b) The first of the two measurements of potential is carried out at the sensitive electrode. (c) The charge moves under the sensitive electrode and a second measurement takes place. The process of reading can be completed by rearward displacement of the charge back under the collector electrode (non-destructive read-out mode, NDRO) (a) or by injecting the charge (destructive read-out mode, DRO) (d) [Bilhorn *et al.* (1987)].

a similar way to CCDs. In a CID, each pixel contains a pair of electrodes.

The charge gathered is measured in the detector element itself, where it is collected, rather than being transferred to an on-chip amplifier. In CIDs, the charge collected is measured by recording a voltage change when the accumulated charge moves between the two electrodes. Figure 4.19 shows a diagram of a simple CID detector element with the two electrodes used for reading the charge and its elimination [Bilhorn *et al.* (1987)]. Since the charge never leaves the detector element where it is collected, the CCD modes of reading that permit combination of charges are not applicable to CIDs. On the other hand, as CIDs allow a non-destructive reading of the information about the charge contained in a detector element (NDRO), they make possible an increased inter-scene dynamic range using random access integration (RAI), and are extremely resistant to charge blooming.

A CID covers with one single structure a wide wavelength range, from the near IR up to UV, with high resolution and good multi-element sensitivity. A CID can simultaneously see more lines with greater quantum efficiency than a multichannel spectrometer with detection using photomultiplier tubes [True *et al.* (1999)]. Pixels can be integrated individually for variable times. The longest integration times are especially critical for photons with short wavelengths (<190 nm).

4.4. Spectrometers

Any spectrometer combines a spectral apparatus with one or more detectors for measuring the intensity of spectral bands. Spectrometers may be of various types.

4.4.1. Sequential Spectrometer

With a sequential spectrometer, the intensity of spectral bands is measured sequentially (that is, one after another). This spectrometer is not currently used in AAS.

4.4.2. Simultaneous Spectrometer

A simultaneous spectrometer has more than one detector and permits the intensity of various spectral bands to be detected at the same time. Attempts are being made to incorporate this spectrometer in ETAAS by combining the use of continuous emission lamps with photodiode detectors to carry out multielement analysis [Harnly and Kane (1984); Tittarelli *et al.* (1985); Lewis *et al.* (1985); Lundberg *et al.* (1988); Jones *et al.* (1989); Ratliff and Majidi (1992); Harnly (1993); Chakrabarti *et al.* (1993); Edel *et al.* (1995); Radziuk *et al.* (1995a,b); Harnly *et al.* (1997); Wagner *et al.* (1998); Harnly (1999)].

4.4.3. Multiplex Spectrometer

In a multiplex spectrometer, a single photodetector simultaneously receives signals from different coded spectral bands. In this case of multiple frequencies, each spectral band is modulated at a specific frequency [Kluczynski *et al.* (2001)]. Decoding is achieved by means of the electronic filtering of the appropriate signals.

The Michelson interferometer, considered as an amplitude division interferometer (ADI) (Fig. 4.20a), forms the central component of almost all modern Fourier Transform (FT) spectrometers. To study the polarization properties, it is very common to place a polarizer in the optical path of the light coming out of the interferometer. However, better results are achieved by using a linear polarizer and a photoelastic modulator (PEM) (Fig. 4.20b), which provides modulation of the light polarization between two orthogonal polarizations [Polavarapu (1997)]. The characteristic oscillation frequency of the PEM typically lies between 30 KHz and 100 KHz, and is selected in such a way as to be greater than the Fourier frequencies resulting from the movement of the mirror. This modulation of the polarization is transformed into an intensity modulation when a sample is placed immediately after the PEM, which transmits the two orthogonal polarizations differentially. This system constitutes a double modulation spectrometer (the interferometer provides modulation of the intensity and the PEM provides modulation of the polarization).

There are, nonetheless, other alternatives, such as the polarization division interferometer (PDI) (Fig. 4.20c), which gives greater efficiency for the study of polarized light. The unpolarized light emerging from the emission source is linearly polarized, through the use of a polarizer **P**. A beam splitter (**BS**) is used to resolve the polarized light entering BS into two components, one parallel (P_1) and the other perpendicular (S_2) to the direction of the entry beam at the beam splitter **BS**. Component P_1 is rotated 90° through reflection at M_1, yielding a perpendicular polarization component (S_1), which is now transmitted totally through the beam splitter **BS**. The polarization component S_2 is likewise rotated 90° by mirror M_2 to give rise to a parallel polarization component (P_2) that is now reflected through the beam splitter.

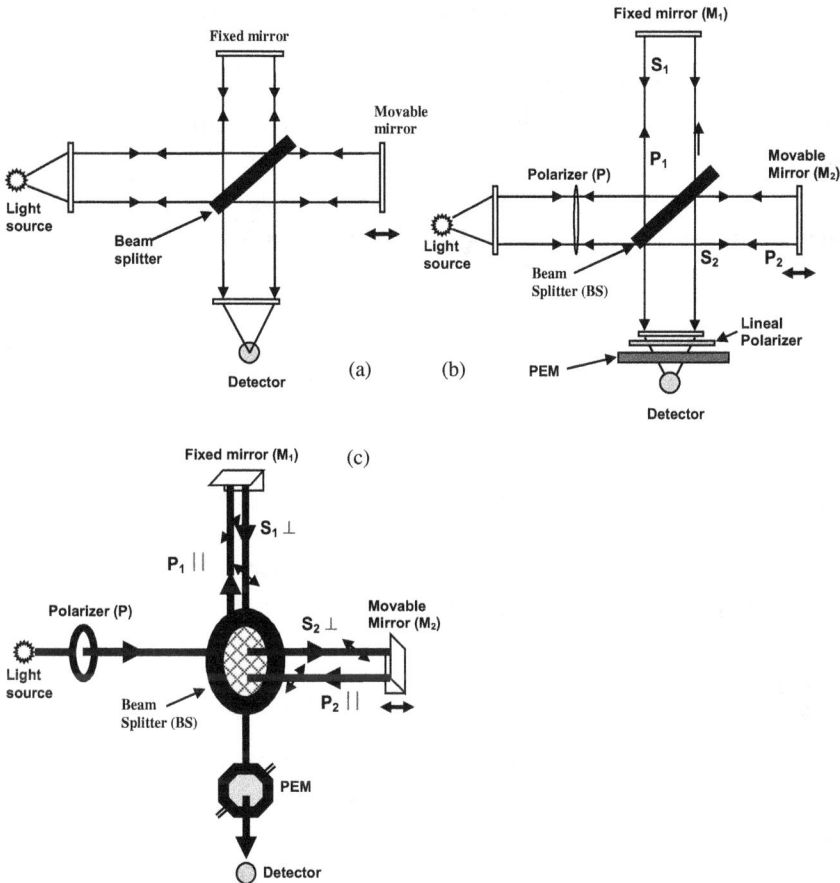

Fig. 4.20. Optical diagrams for various interferometers with division of amplitude and polarization offering modulation of intensity and polarization: (a) Amplitude division interferometer with modulation of intensity; (b) Interferometer with amplitude division offering modulation of polarization and intensity; (c) Polarization division interferometer with modulation of polarization [Polavarapu (1997)].

Components S_1 and P_2 together emerge, forming an angle of 90° with respect to the direction of entry into BS of the initial beam. At the point of zero path difference of the interferometer, the polarization of the exit beam from the beam splitter BS is identical to the beam entering BS from polarizer **P**. As mirror M_2 is moving, the polarization of the exit beam changes, depending on the distance covered by mirror M_2, because the polarization component $\mathbf{P}_2$ suffers a phase shift relative to the polarization component $\mathbf{S}_1$.

This phase shift is produced at a wavenumber $\bar{\nu}_i$ equal to $2\pi\delta\bar{\nu}_i$, where δ represents the difference in the optical path. In continuous scan interferometers, $\delta = 2vt$, where v is the mobile mirror speed, and t is time; in step scan interferometers, $\delta = 2x$, where x is the displacement of the mobile mirror from the position where the difference in path is zero. For input monochromatic light with wavelength λ, the distances of the mirror movement at $\lambda/8$, $2\lambda/8$, $3\lambda/8$, and $4\lambda/8$ correspond, respectively, with the differences in phase between $\mathbf{S}_1$ and $\mathbf{P}_2$ corresponding to $\lambda/4$, $\lambda/2$, $3\lambda/4$, and λ, or in other words to states of polarization relating to the left circularly, vertically, right circularly, and horizontally polarization. When the mirror moves greater distances, this modulation cycle is repeated. For polychromatic light, there are various wavelengths which go through the states of polarization listed above, as for a longer wavelength, the mirror would need to move a greater distance to complete a modulation cycle [Polavarapu (1997)].

4.4.4. Filter Spectrometer

A filter spectrometer contains one or more spectral filters so as to isolate one or more spectral bands. These systems contain narrowband light filters, electronically tunable and having no moving parts. These tunable filter spectrometers named acousto-optic tunable filters (AOTF) are being largely applied in combination with several molecular spectrometries and imaging techniques [Bei *et al.* (2004)], but a few of them have recently been used for atomic absorption [Fulton and Horlick (1996)] and atomic emission [Baldwin and Zamzow (1997); Gillespie and Carnahan (2001)] spectrometries.

Essentially, an AOTF is based on the diffraction of light by an acoustic wave, which can be explained in terms of wave interactions (or of particles collisions). In the first case, when an acoustic wave is propagated through a transparent material medium, it produces a change in the refractive index of the optical medium and generates a perturbation in pressure which moves and involves regions of compression and decompression in the crystal. This effect can be considered as

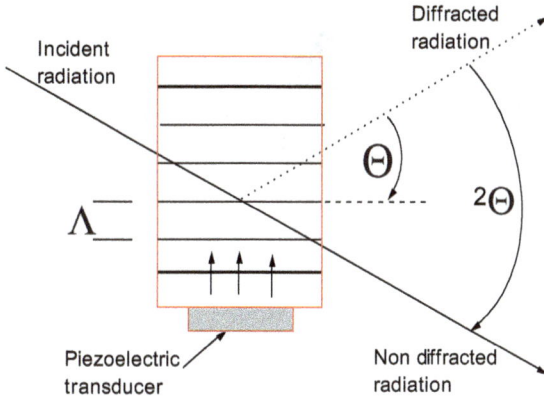

Fig. 4.21. Diffraction with an acousto-optic system. Light is incident on the crystal at Bragg's angle. By means of the piezoelectric transducer a sound wave is sent into the crystal. Radiation is diffracted at an angle of 2θ [Spudich *et al.* (1997)].

generating a diffraction grating that moves at the sonic speed and produces periodical changes in the optical phase.

Light diffraction by acoustic waves can trigger changes in the amplitude, frequency, direction and wavelength of the incident light. The optical properties of the medium control the transformation type supported by the diffracted light. If the medium is optically isotropic, a change occurs in the frequency, direction and amplitude of the light and the diffraction can be explained by the classic principle of Bragg diffraction. On the other hand, if the medium is optically anisotropic, changes in the frequency and direction of the incident beam result, but also in the wavelength. This process, named abnormal Bragg diffraction, has been used to develop acousto-optic tunable filters (AOTF) (Fig. 4.21) [Spudich *et al.* (1997)].

Depending on the interaction length, l, between the acoustic wave and the light wave, the diffraction process can be divided into two different types (Fig. 4.22) [Tran (1992)]: Raman–Nath diffraction (sometimes called Debye-Sears diffraction) and Bragg diffraction.

The first diffraction type occurs when the interaction length, l, is too short or when the Raman–Nath parameter, Q, defined by

$$Q = \frac{4l\lambda}{\Lambda^2} \tag{4.25}$$

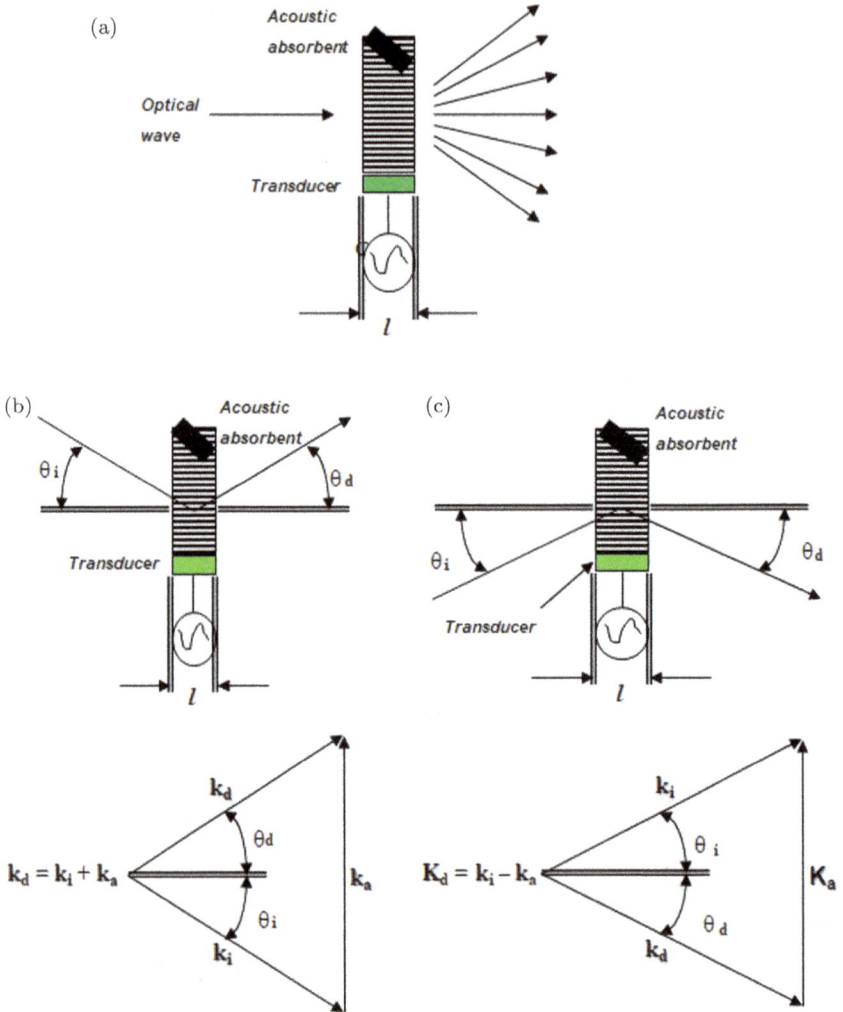

Fig. 4.22. Light diffraction by a sound wave in an isotropic medium. (a) Raman–Nath (or Debye–Sears) diffraction originating multiple orders; (b) Bragg diffraction with shift upwards, and (c) Bragg diffraction with shift downwards [Tran (1992)].

(where λ and Λ are the wavelengths of the optical and sound waves, respectively), is less than unity. In this case, the diffraction has multiple orders (Fig. 4.22a). In contrast, when the interaction length, l, is large (or when $Q \gg 1$), diffraction takes place in the Bragg region.

In this case, first order diffraction is only observed, as higher orders of diffraction produce interferences which are completely destructive if the light beam is incident normally on the sound wave. To obtain constructive interferences, the incidence angle must be divergent from the direction of the sound wave (Fig. 4.22b and 4.22c). Owing to the fact that the energy of the sound waves is much smaller than that of photons, it may be assumed that the incidence, θ_i, and diffraction, θ_d angles are equal (in an optically isotropic medium), being related to the light and sound wavelengths by Bragg's equation

$$Sin\,\theta = \frac{k_a}{2k_d} = \frac{\lambda}{2\Lambda}, \tag{4.26}$$

where k_a and k_i refer to the wave vectors related to the sound and light waves, respectively, while k_d represents the resulting vector of the colliding particles (or wave interactions). Each vector has its corresponding momentum (hk_a, hk_i and hk_d) and energy ($h\omega_a$, $h\omega_i$ and $h\omega_d$) with $\omega_d = \omega_i \pm \omega_a$.

When the medium is optically anisotropic, the angles of incidence and of diffraction are not equal ($\theta_i \neq \theta_d$), and there is also a change in the light polarization plane. As crystal is birefringent, the rotation of the light polarization plane causes a change in the refractive indices ($n_i \neq n_d$), and so the momentum and the wavelength are modified. The angles of incidence and diffraction are related to the acoustic frequency, f_a, the refractive indices for the polarizations of the incident and diffracted light, n_i and n_d, and the acoustic velocity, v_a, in the following way

$$Sin\,\theta_i = \left(\frac{1}{2n_i}\right)\left(\frac{\lambda f_a}{v_a}\right)\left[1 + \left(\frac{v_a}{\lambda f_a}\right)^2 (n_i^2\, n_d^2)\right] \tag{4.27}$$

and

$$Sin\,\theta_d = \left(\frac{1}{2n_d}\right)\left(\frac{\lambda f_a}{v_a}\right)\left[1 + \left(\frac{v_a}{\lambda f_a}\right)^2 (n_i^2\, n_d^2)\right]. \tag{4.28}$$

The interaction of light with a sound wave in an anisotropic medium has permitted the development of AOTFs. Taking into account the direction of the wave vectors of the sound and light beams, AOTFs

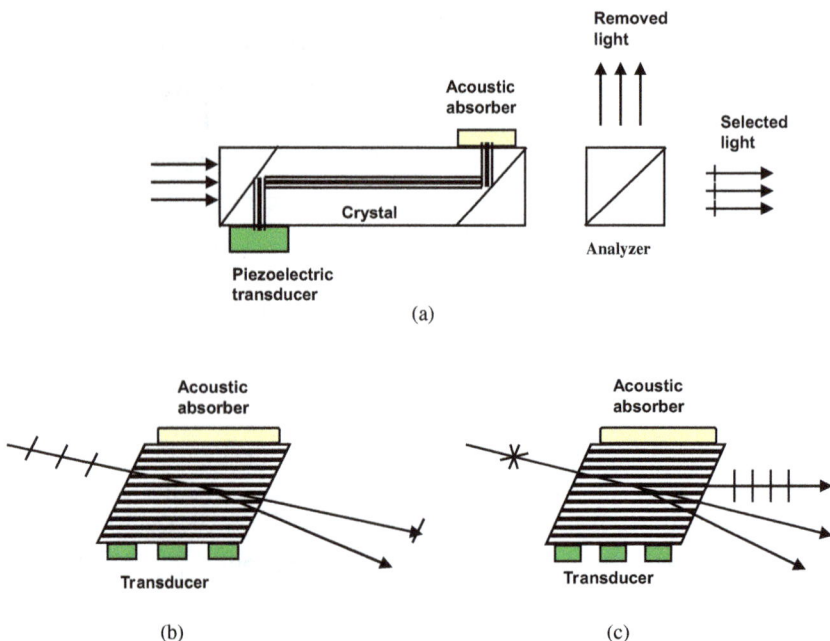

Fig. 4.23. Types of tunable acousto-optic filters. (a) Co-linear; (b) non-co-linear with incident light polarized linearly, and (c) non-co-linear with incident light not polarized [Tran (1992)].

can be divided into two types: colinear and non-colinear (Fig. 4.23) [Tran (1992)].

When a beam of light falls on a crystal at Bragg's angle, the two orthogonal polarizations of the beam move through the crystal at different speeds. As an acoustic pressure wave (created when an *RF* signal is applied to a piezoelectric transducer coupled to the crystal) (Fig. 4.23), is also being propagated through the crystal, under appropriate conditions the acoustic wave will interact with a specific wavelength of light. The mechanical compression of the medium causes an increase in the refractive index at the compression point. This pressure wave moves through the medium at a speed governed by the solid nature. The speeds typical for such pressure waves, called acoustic speed (v), are usually around 5000 m/s. If the piezoelectric transducer is used to produce pressure waves at a rate significantly higher than the timing needed for the sound wave to travel through material, then a series of pressure waves will come out of the material. If the pressure waves are induced at a selected frequency, these pressure waves will be equally spaced. When this happens, the light of the wavelength that interacts is tuned within the orthogonal polarization. The result is that when all the light leaves the crystal, three rays become visible. Two of them (monochromatics and with opposite polarities) are diffracted and appear at equal distances on either side of a third (polychromatic) central ray.

The characteristics of the solid state materials provide such modulation speeds that it would be impossible to obtain with mechanical systems [Tran (1992)]. AOTF systems can cover a broad wavelength range (350 nm–5 μm). Nevertheless, the wavelength range from 350 nm to 700 nm can be covered using a single system [Bucher and Carnahan (1999)], this being possible to scan the whole spectrum in a very short time range (around 1 s), and accessing to any wavelength in 0.1 μs. Systems designed to operate in the visible and near infrared regions are usually built with a birefringent crystal, to which are linked a series of piezoelectric transducers {LiNbO$_3$, LiTaO$_3$, LiIO$_3$, p-tellurite (TeO$_2$)} (Table 4.2) [Bucher and Carnahan (1999)]. Transmission characteristics of the AOTF operating with ultrashort light pulses (2 ps) are strongly affected by dispersion and non-linearity when the length of the AOTF increases [Sobrinho *et al.* (2002)].

To sum up, the advantages of an AOTF acting like a monochromator as against a prism or dispersion grating are the following: the AOTF system is a solid state device (compact, with no moving parts), has a

Table 4.2. Spectral Ranges Covered by Different AOTFs [Bucher and Carnahan (1999)]

Material	Type of AOTF	Wavelength Region
Quartz	Co-linear	240 nm–400 nm
MgF$_2$	Non-co-linear	200 nm–700 nm
CaMoO$_4$	Co-linear	510 nm–670 nm
TeO$_2$	Non-co-linear	370 nm–4.5 μm
Tl$_3$AsSe$_3$	Non-co-linear	1.23 μm–17 μm

broad angular field, wide range of tuning, high spectral resolution (1–6 Å), possibility of a rapid scan (in μs), gives high-speed access to a wavelength (sequentially or at random), and allows the creation of images.

4.4.5. Spectrometers In-Practice: Practical Configurations

In practice, the most widely used configurations include: single beam and double beam. In the single beam spectrometer, the relative light intensity of the beam is measured before and after insertion of the sample. In other words, the transmittance value for each wavelength of the sample is compared with the transmission value from the reference standard. That is, making transmission measurements, the spectrometer quantitatively compares the fraction of light that passes through a reference (standard or even a blank) solution and a sample solution. In the double beam spectrometer, the light intensity between two light paths, the reference solution and the sample solution is compared synchronously. Measurements from double beam instruments are easier and more stable, but single beam instruments usually have a larger dynamic range and are optically simpler and more compact.

The usual spectrometers include a monochromator containing a movable or fixed diffraction grating to produce the analytical signal. Light from the source lamp is passed through a monochromator, which diffracts the light into a "rainbow" of wavelengths and outputs narrow bandwidths of this diffracted spectrum. Discrete frequencies are transmitted through the sample. Then the photon flux density of the transmitted light is measured with a photodiode, charge-coupled device or any other light sensor. If a single detector, such as a photomultiplier tube or photodiode is used, the grating can be scanned stepwise so that the detector can measure the light intensity at each wavelength (which will correspond to each "step"). Arrays of detectors, such as charge-coupled devices (CCD) or photodiode arrays (PDA) can also be used, where the grating is fixed and the intensity of each wavelength is measured by a different detector in the array.

4.5. Acquisition and Processing of Data

The ETAAS signal is transient and usually requires correction for the background signal, baseline instabilities and variation in the generation of the atomic vapor and light scattering [Whiteside *et al.* (1980); Barnett, *et al.* (1985)]. The use of an optimal signal processing procedure can reduce the variation coefficient (better precision) and detection limit of analysis. The quality of measurements is mainly influenced by the presence of noise. In general, peak area (integrated signals) measurements are preferable to peak height measurements [Kale and Voigtman (1995)]. Integrated absorption signals are usually measured, compensating for drift in lamp intensity and the absorption-time profile [Aldous *et al.* (1973)]. Peak absorbance measurements are more susceptible to variations in physical factors than integrated absorbance measurements. For ETAAS, two noise sources may be identified which limit the precision. The first is $1/f$ noise from the light source and the electronic components, while the second is from the atomization conditions and imprecision in the sample dosing. Dawson *et al.* (1988) used white noise and random fluctuation to simulate experimental fluctuations. The data were processed using linear sliding means, inverse exponential functions and a correlation function derived from the atomization curve.

Different systems for data acquisition with scan cycles higher than 300 Hz have been described. Data processing are improved by the use of computer programs that collect and store data in digital form. Once the data have been collected, they can be manipulated in a variety of ways: handling of the analog data, selection of Baseline Offset Correction (BOC) times, definition of integration windows, data smoothing techniques, etc. In AAS, it is possible to take advantage of commercially available chemometric programs to collect and interpret data, from which the analytical characteristics of any analysis can be established on the basis of the calibration data [García Campaña *et al.* (1997)]. By means of these programs, an estimate may be made of the maximum level of interfering factors causing a bias equal to or greater than the random error of the determination.

Several signal generation and processing factors, such as the transverse *a.c.* Zeeman technique with the magnet on *off* for background correction, signal integrals at line frequency to accurately represent the shape of the peak, interpolation techniques to better correct for rapidly changing background levels, integrated areas rather than peak heights for quantitative measurements, baseline correction (BOC time) to improve limits of detection, the accuracy and precision of integrated areas, and use of graphical techniques to facilitate data interpretation and methods developments, have been used.

The program AAS-TOOLS allows [Yan-Zhong and Zhe-Ming (1994)]: (i) adjustment and filtering of analytical data; (ii) *on-line* collection of data for a given analysis; (iii) determination of kinetic parameters for a specific element through the use of the Smets method (see Chapter 7); and (iv) theoretical simulations of an AAS signal based on an exponentially modified Gaussian function. However, other programs including methods for calibration, simulations, and even intelligent programs are commercially available as well [Yan-Zhong and Zhe-Ming (1994); Pennincks *et al.* (1995); Yan-Zhong *et al.* (1995)]. Among these programs, the expert systems play an important role in achieving a greater degree of automation in the laboratory. The employment of computer programs also permits the calculation of theoretical characteristic masses [Berglund and Baxter (1992)].

The self-controlling systems have been used in ETAAS [Wienke *et al.* (1994)] by means of the development of a method based on estimation of the *on-line* state using an extended Kalman filter through quality control sampling. The advantages of this method in contrast with conventional approaches are the following: (i) calibration and re-calibration can be carried out simultaneously; (ii) drift is detected and corrected; (iii) outliers are detected and fixed; and (iv) only a minimum number of quality control samples are required to adapt parameters.

References

Aldous, K.M.; Mitchell, D.G. and Ryan, F.J. (1973). Computer-controlled atomic absorption spectrometer for measurement of transient atom populations. *Anal. Chem.*, **45**: 1990–1993.

Baldwin, D.P. and Zamzow, D.S. (1997). Limits of detection for an AOTF-FFP spectrometer in ICP atomic emission spectroscopy. *Talanta*, **45**: 229–235.

Barnett, W.B.; Bohler, W.; Carnrick, G.R. and Slavin, W. (1985). Signal processing and detection limits for graphite furnace atomic absorption with Zeeman background correction. *Spectrochim. Acta, Part B*, **40**: 1689–1703.

Berglund, M. and Baxter, D.C. (1992). Computer program (CHMASS) for calculating theoretical characteristic mass values in electrothermal atomic absorption spectrometry. *J. Anal. At. Spectrom.*, **7**: 461–470.

Bei, L.; Dennis, G.I.; Miller, H.M.; Spaine, T.W. and Carnahan, J.W. (2004). Acousto-optic tunable filters: fundamental and applications as applied to chemical analysis techniques. *Progress in Quantum Electronics*, **28**: 67–87.

Bilhorn, R.B.; Sweedler, J.V.; Epperson, P.M. and Denton, M.B. (1987). Charge transfer device detectors for analytical optical spectroscopy — Operation and characteristics. *Appl. Spectrosc.*, **41**: 1114–1125.

Bucher, E.G. and Carnahan, J.W. (1999). Characterization of an acousto-optic tunable filter and use in visible spectrophotometry. *Appl. Spectrosc.*, **53**: 603–611.

Butler, L.R.P. and Laqua, K. (1996). Instrumentation for the spectral dispersion and isolation of optical radiation. *Spectrochim. Acta, Part B*, **51**: 645–664.

Chakrabarti, C.L.; Gilmutdinov, A.Kh. and Hutton, J.C. (1993). Digital imaging of atomization processes in electrothermal atomizers for atomic absorption spectrometry. *Anal. Chem.*, **65**: 716–723.

Cooke, D.O.; Dagnall, R.M. and West, T.S. (1972). Some considerations on spectral line profiles of microwave-excited electrodeless discharge lamps. *Talanta*, **19**: 1309–1320.

Dawson, J.B.; Duffield, R.J.; King, P.R.; Hajizadeh-Saffar, M. and Fisher, G.W. (1988). Signal processing in electrothermal atomization atomic absorption spectroscopy. *Spectrochim. Acta, Part B*, **43**: 1133–1140.

Earle, C.W.; Baker, M.E.; Denton, M.B. and Pomeroy, R.S. (1993). Imaging applications for chemical analysis utilizing charge coupled device array detectors. *Trends in Anal. Chem.*, **12**: 395–403.

Edel, H.; Quick, L. and Cammann, K. (1995). Frequency-modulated simultaneous multielement atomic absorption spectrometry using electrothermal atomizer and deuterium background correction. *Fresenius' J. Anal. Chem.*, **351**: 479–483.

Epperson, P.M.; Sweedler, J.V.; Bilhorn, R.S.; Sims, G.R. and Denton, M.B. (1988). Applications of charge transfer devices in spectroscopy. *Anal. Chem.*, **60**: 327A–335A.

Fernando, R. and Jones, B.T. (1994). Continuum-source graphite-furnace atomic absorption spectrometry with photodiode array detection. *Spectrochim. Acta, Part B*, **49**: 615–626.

Florek, S. and Becker-Ross, H. (1995). High-resolution spectrometer for atomic spectrometry. *J. Anal. At. Spectrom.*, **10**: 145–147.

Freeman, G.H.; Outred, M. and Morris, L.R. (1980). A line profile study of the 193.76 nm arsenic emission line from lamps used in atomic absorption spectrometry. *Spectrochim. Acta, Part B*, **39**: 687–699.

Fulton, G. and Horlick, G. (1996). AOTFs as Atomic Spectrometers: Basic Characteristics. *Appl. Spectrosc.*, **50**: 885–892.

Ganeev, A.; Khutorshikov, V.I.; Khutorshikov, S.V.; Revalde, G.; Skudra, A.; Smirnova, G.M. and Stankov, N.R. (2003). High-frequency electrodeless discharge lamps for atomic absorption spectrometry. *Spectrochim. Acta, Part B*, **58**: 879–889.

García Campaña, A.M.; Cuadros Rodriguez, L.; Alés Barrero, F.; Román Ceba, M. and Sierra Fernández, J.L. (1997). ALAMIN: a chemometric program to check analytical method performance and to assess the trueness by standard addition methodology. *Trends in Anal. Chem.*, **16**: 381–385.

Giles, J.H.; Ridder, T.D.; Williams, R.H.; Jones, D.A. and Denton, M.B. (1998). Selecting a CCD camera. *Anal. Chem.*, **70**: 663A–668A.

Gilmutdinov, A.Kh.; Nagulin, K.Y. and Sperling, M. (2000). Spatially resolved atomic absorption analysis. *J. Anal. At. Spectrom.*, **15**: 1375–1382.

Gilmutdinov, A.Kh.; Nagulin, K.Yu. and Zakharov, Yu.A. (1994). Analytical measurement in electrothermal atomic absorption spectrometry — How correct is it?. *J. Anal. At. Spectrom.*, **9**: 643–650.

Gilmutdinov, A.Kh.; Radziuk, B.; Sperling, M. and Welz, B. (1996c). Spatially and temporally resolved detection of analytical signals in graphite furnace atomic absorption spectrometry. *Spectrochim. Acta, Part B*, **51**: 1023–1044.

Gilmutdinov, A.Kh.; Radziuk, B.; Sperling, M.; Welz, B. and Nagulin, K.Yu. (1995). Spatial distribution of radiant intensity from primary sources for atomic absorption spectrometry. Part I: Hollow cathode lamps. *Appl. Spectrosc.*, **49**: 413–424.

Gilmutdinov, A.Kh.; Radziuk, B.; Sperling, M.; Welz, B. and Nagulin, K.Yu. (1996a). Three-dimensional structure of the radiation beam in atomic spectrometry. *Spectrochim. Acta, Part B*, **51**: 931–940.

Gilmutdinov, A.Kh.; Radziuk, B.; Sperling, M.; Welz, B. and Nagulin, K.Yu. (1996b). Spatial distribution of radiant intensity from primary sources for atomic absorption spectrometry. Part II: Electrodeless discharge lamps. *Appl. Spectrosc.*, **50**: 483–497.

Gillespie S. R. and Carnahan J. W. (2001). Ultraviolet Quartz Acousto-optic Tunable Filter Wavelength Selection for Inductively Coupled Plasma Atomic Emission Spectrometry, **55**: 730–738.

Grosser, Z.A. and Collins, J.B. (1991). Identification of several wavelengths useful for ICP that are incorrectly listed in the literature. *Appl. Spectrosc.*, **45**: 367–369.

Haisch, C. and Becker-Ross, H. (2003). An electron bombardment CCD-camera as detection system for an echelle spectrometer. *Spectrochim. Acta, Part B*, **58**: 1351–1357.

Hanley, Q.S.; Earle, C.W.; Pennebaker, F.M.; Madden, S.P. and Denton, M.B. (1996). Charge-transfer devices in analytical instrumentation. *Anal. Chem.*, **68**: 661A–667A.

Harnly, J.M. (1986). Multielement atomic absorption with a continuum source. *Anal. Chem.*, **58**: 933A–943A.

Harnly, J.M. and Kane, J.S. (1984). Optimization of electrothermal atomization parameters for simultaneous multielement atomic absorption spectrometry. *Anal. Chem.*, **56**: 48–54.

Harnly, J.M. (1993). Graphite furnace atomic absorption spectrometry using a linear photodiode array and a continuum source. *J. Anal. At. Spectrom.*, **8**: 317–324.

Harnly, J.M. and Fields, R.E. (1997). Solid-state array detectors for analytical spectrometry.*Appl. Spectrosc.,* **51**: 334A–351A.

Harnly, J.M.; Smith, C.M.M.; Wichems, D.N.; Ivaldi, J.C.; Lundberg, P.L. and Radziuk, B. (1997). Use of a segmented array charge coupled device detector for continuum source atomic absorption spectrometry with graphite furnace atomization. *J. Anal. At. Spectrom.,* **12**: 617–627.

Harnly, J.M. (1999). The future of atomic absorption spectrometry: a continuum source with a charge coupled array detector. *J. Anal. At. Spectrom.,* **14**: 137–146.

Heitman, U.; Schütz, M.; Becker-Roß, H. and Florek, S. (1996): Measurements on the Zeeman-splitting of analytical lines by means of a continuum source graphite furnace atomic absorption spectrometer with a linear charge coupled device array. *Spectrochim. Acta, Part B,* **51**: 1095–1105.

Jones, B.J.; Smith, B.W. and Winefordner, J.D. (1989). Continuum source atomic absorption spectrometry in a graphite furnace with photodiode array detection. *Anal. Chem.,* **61**: 1670–1674.

Kale, U. and Voigtman, E. (1995). Signal processing of transient atomic absorption signals. *Spectrochim. Acta, Part B,* **50**: 1531–1541.

Kluczynski, P.; Gustafsson, J.; Lindberg, Å.M. and Axner, O. (2001). Wavelength modulation absorption spectrometry — an extensive scrutiny of the generation of signals. *Spectrochim. Acta, Part B,* **56**: 1277–1354.

Larkins, P.L. (1985). Atomic line profile measurements on hollow-cathode and electrodeless discharge lamps using a high-resolution echelle monochromator. *Spectrochim. Acta, Part B,* **40**: 1585–1598.

Leis, F. and Steers, E.B.M. (1994). Boosted glow discharges for atomic spectroscopy — analytical and fundamental properties. *Spectrochim. Acta, Part B,* **49**: 289–325.

Lewis, S.A.; O'Haver, T.C. and Harnly, J.M. (1985). Determination of metals at the microgram-per-lier level in blood serum by simultaneous multielement atomic absorption spectrometry with graphite furnace atomization. *Anal. Chem.,* **57**: 2–5.

Ljung, P. and Axner, O. (1997). Measurements of rubidium in standard reference samples by wavelength-modulation diode laser absorption spectrometry in a graphite furnace. *Spectrochim. Acta, Part B,* **52**: 305–319.

Lowe, R.M. and Sullivan, J.V. (1999). Developments in light sources and detectors for atomic absorption spectroscopy. *Spectrochim. Acta, Part B,* **54**: 2031–2039.

Lundberg, E.; Frech, W. and Harnly, J.M. (1988). Simultaneous multielement analysis by continuum source atomic absorption spectrometry with a spatially and temporally isothermal graphite furnace. *J. Anal. At. Spectrom.,* **3**: 1115–1119.

Mixon, P.D.; Griffin, S.T.; Williams, Jr., J.C.; Cai, X.J. and Williams, J.C. (1994). Pulse optimization criteria for the microcavity hollow cathode discharge emission source. *J. Anal. At. Spectrom.,* **9**: 697–700.

Morgan, C.A.; Davis, C.L.; Smith, B.W. and Winefordner, J.D. (1994). Evaluation of a microcavity hollow cathode discharge emission source. *Appl. Spectrosc.,* **48**: 261–264.

Müller, P.; Klán, P. and Církva, V. (2005). The electrodeless discharge lamp: a prospective tool for photochemistry. Part 5: Fill material-dependent emission characteristics. *J. Photochem. Photobiol. A: Chem.,* **171**: 51–57.

Niemczyk, T.M.; Thompson, B.D. and Angus, J.E. (1994). Intensity enhancements in hollow cathode lamps due to the addition of nitrogen to the fill gas. *Appl. Spectrosc.,* **48**: 896–899.

Oliver, D.R. and Finlayson, T.R. (1998). Effect of cathode-bore geometry and filler-gas pressure on the observed distribution of sputtered copper atom densities in a hollow-cathode lamp. *J. Anal. At. Spectrom.,* **13**: 443–446.

Omenetto, N. (1998). Role of lasers in analytical atomic spectroscopy: where, when and why. *J. Anal. At. Spectrom.,* **13**: 385–399.

Pavlovic, B.V. and Dobrosavijevic, J.S. (1992). Electric characteristics of a hollow cathode discharge in a rotating magnetic field. *Spectrochim. Acta, Part B,* **47**: 297–302.

Penninckx, W.; Vankeerberghen, P.; Massart, D.L. and Smeyers-Verbeke, J. (1995). Knowledge-based computer system for the detection of matrix interferences in atomic absorption spectrometric methods. *J. Anal. At. Spectrom.,* **10**: 207–214.

Piepmeier, E.H. (1989). The influence of the spectral profile of the source lamp on the analytical results for low pressure atomic absorption cells. *Spectrochim. Acta, Part B,* **44**: 609–616.

Pogue, R.T.; Sutherland, R.L.; Schmitt, M.G.; Natarajan, L.V.; Siwecki, S.A.; Tondiglia, V.P. and Bunning, T.J. (2000). Electrically switchable Bragg gratings from liquid crystal/polymer composites. *Appl. Spectrosc.* **54**: 12A–28A.

Polavarapu, P.L. (1997). Double polarization modulation interferometry. *Appl. Spectrosc.* **51**: 770–777.

Radziuk, B.; Rödel, G.; Stenz, H.; Backer-Ross, H. and Florek, S. (1995a). Spectrometer system for simultaneous multi-element electrothermal using line sources and Zeeman-effect background correction. *J. Anal. At. Spectrom.,* **10**: 127–136.

Radziuk, B.; Rödel, G.; Zeiher, M.; Mizuno, S. and Yamamoto, K. (1995b). Solid state detector for simultaneous multi-element electrothermal atomic absorption spectrometry with Zeeman-effect background correlation. *J. Anal. At. Spectrom.,* **10**: 415–422.

Ratliff, J. and Majidi, V. (1992). Simultaneous measurement of the atomic and molecular absorption of aluminium, copper and lead nitrate in an electrothermal atomizer. *Anal. Chem.,* **64**: 2743–2750.

Ratzlaff, K.L. and Paul, S.L. (1979). Characterization of a charge-coupled device photoarray as a molecular absorption spectrophotometric detector. *Appl. Spectrosc.,* **33**: 240–245.

Rust, J.A.; Nóbrega, J.A.; Calloway, C.P. Jr. and Jones, B.T. (2005a). Advances with tungsten coil atomizers: Continuum source atomic absorption and emission spectrometry. *Spectrochim. Acta, Part B,* **60**: 589–598.

Rust, J.A.; Nóbrega, J.A.; Calloway, C.P. Jr. and Jones, B.T. (2005b). Analytical characteristics of a continuum-source tungsten coil atomic absorption spectrometer. *Analytical Sciences,* **21**: 1009–1013.

Schmidt, K.P.; Becker-Ross, H. and Florek, S. (1990). A combination of a pulsed continuum light source, a high resolution spectrometer and a charge coupled device detector for multielement atomic absorption spectrometry. *Spectrochim. Acta, Part B,* **45**: 1203–1210.

Schnürer-Patschan, C.; Zybin, A.; Groll, H. and Niemax, K. (1993). Improvement in detection limits in graphite furnace diode laser atomic absorption spectrometry by wavelength modulation technique. *J. Anal. At. Spectrom.*, **8**: 1103–1107.

Schuetz, M.; Murphy, J.; Fields, R.E. and Harnly, J.M. (2000). Continuum source-atomic absorption spectrometry using a two-dimensional charge coupled device. *Spectrochim. Acta, Part B*, **55**: 1895–1912.

Smith, C.M.M.; Harnly, J.M.; Moulton, G.P. and O'Haver, T.C. (1994). High current pulsing of a xenon arc lamp for electrothermal atomic absorption spectrometry using a linear photodiode array. *J. Anal. At. Spectrom.*, **9**: 419–425.

Sobrinho, C.S.; Lima, J.L.S.; Almeida, E.F. de and Sombra, A.S.B. (2002). Acousto-optic tunable filter (AOTF) with increasing non-linearity and loss. *Opt. Commun.*, **208**: 415–426.

Spietz, P.; Gross, U.; Smalins, E.; Orphal, J. and Burrows, J.P. (2001). Estimation of the emission temperature of an electrodeless discharge lamp and determination of the oscillator strength for the $I(^2P_{3/2})$ 183.038 nm resonance transition. *Spectrochim. Acta, Part B*, **56**: 2465–2478.

Spudich, T.M.; Pelz, B.A. and Carnahan, J.W. (1997). Acousto-optic signal modulation for atomic emission spectrometry background correction. *Appl. Spectrosc.* **51**: 765–769.

Sweedler, J.V.; Bilhorn, R.B.; Epperson, P.M.; Sims, G.R. and Denton, M.B. (1988). High-performance charge transfer device detectors. *Anal. Chem.*, **60**: 282–291.

Sweedler, J.V. (1993). Charge transfer device detectors and their applications to chemical analysis. *Critical Rev. Anal. Chem.*, **24(1)**: 59–98.

Sweedler, J.V.; Ratzlaff, K.L. and Denton, M.B. (1994). *Charge-transfer Devices in Spectroscopy*. VCH Publishers, Inc., New York.

Tittarelli, P.; Lancia, R. and Zerlia, T. (1985). Simultaneous molecular and atomic spectrometry with electrothermal atomization and diode array detection. *Anal. Chem.*, **57**: 2002–2005.

Tran, C.D. (1992). Acousto-optic devices. Optical elements for spectroscopy. *Anal. Chem.*, **64**: 971A–981A.

True, J.B.; Williams, R.H. and Denton, M.B. (1999). On the implementation of multielement continuum source graphite furnace atomic absorption spectrometry utilizing an echelle/CID detection system. *Appl. Spectrosc.*, **53**: 1102–1110.

Wagner, K.A.; Batchelor, J.D. and Jones, B.T. (1998). A Rowland circle, multielement graphite furnace atomic absorption spectrometer. *Spectrochim. Acta, Part B*, **53**: 1805–1813.

Welz, B.; Becker-Ross, H.; Florek, S. Heitmann, U. and Vale, M.G.R. (2003). High-resolution continuum-source atomic absorption spectrometry — What can we expect? *J. Braz. Chem. Soc.*, **14**: 220–229.

Whiteside, P.J.; Stockdale, T.J. and Price, W.J. (1980). Signal and data processing for atomic absorption spectrophotometry. *Spectrochim. Acta, Part B*, **35**: 795–806.

Wienke, D.; Vijn, T. and Buydens, L. (1994). Quality self-monitoring of intelligent analysers and sensors based on an extended Kalman filter: application to graphite furnace atomic absorption spectrometry. *Anal. Chem.*, **66**: 841–849.

Williams, J.C.; Jan-Yurn Kung, Yixin Chen, Xiangjun Cai and Griffin, S.T. (1995). Some characteristics of the hollow cathode discharge source from hollow cathodes of different sizes and shapes. *Appl. Spectrosc.*, **49**: 1705–1714.

Yan-Zhong, L.; Zhe-Ming, N. and Peng-Yuan, Y. (1995). Software development for atomic absorption spectrometry — A decade of progress. *J. Anal. At. Spectrom.*, **10**: 699–702.

Yan-Zhong, L. and Zhe-Ming, N. (1994). General computer program (AAS-TOOLS) for theoretical studies in electrothermal atomic absorption spectrometry. *J. Anal. At. Spectrom.*, **9**: 669–673.

Chapter 5

Interferences: *Types and Correction*

5.1. Types of Interferences

Different criteria can be used for classification of the interferences found in ETAAS. However, none of the ensuing classifications can be seen as universally valid, or even over time, since many interference effects are the outcome of simultaneous contributions from several physical and/or chemical processes. Nonetheless, with the aim of having some form of order, despite clear awareness of the limitations just mentioned, it does seem acceptable to split the different interferences found in ETAAS into two large groups: spectral and non-spectral interferences.

Any spectral interference would involve electromagnetic radiation at the analytical wavelength, so that there can be absorption, emission and scattering processes. In contrast, non-spectral interferences do not involve electromagnetic radiation. Such interferences would be triggered by chemical and/or physical processes (including thermal ionization). Although ionization may be caused by absorption of electromagnetic radiation, it will be included among the non-spectral interference for two reasons: (i) in ETAAS it normally originates from a thermal process; and (ii) this is a conscious attempt to emphasize the diminished concentration of the analyte atoms. Subdivision of non-spectral interferences into physical and chemical would only appear advisable from a purely academic perspective, as on many, if not all, occasions there is a simultaneous contribution from both physical and chemical processes.

5.1.1. Spectral Interferences

In ETAAS, new spectral interferences have been noted in recent years (Tables 5.1, 5.2, 5.3, and 5.4) [Fassel *et al.* (1968); Pritchard and Reeves (1976); Tsunoda *et al.* (1980, 1985); Epstein *et al.* (1994); Ohlsson and Frech (1989); Daminelli *et al.* (1998)]. The spectral interferences found in ETAAS may in turn be grouped into emission and absorption categories.

Table 5.1. Some Molecular Spectral Bands [Daminelli *et al.* (1998); Ohlsson and Frech (1989)]

Molecule	Wavelength, nm
AlF	227.45
CaCl	311; 316; 320; 369; 373; 378; 383; 389, 621.3
$CaCl_2$	216
InF	233.73
MgCl	269–270; 337.9; 369; 376; 382
$MgCl_2$	220
NaCl	<200; 237
NaO_x	<200; 240; 330
NO	205; 215; 226
NO_2	<250; >270
SO; SO_2	<200; 278
Other sulfur compounds	210; 253; 259; 269

Table 5.2. Molecular Spectral Bands and some Coincident Atomic Wavelengths [Daminelli *et al.* (1998); Ohlsson and Frech (1989)]

Molecule (Band Wavelength, nm)	Element	Wavelength, nm
CuH (400–446)	Ca	422.7
	Cr	425.4
	Ga	403.3
	In	410.5
	K	404.4; 404.7
γPO (234–257)	Au	242.8
	Be	234.8
	Co	240.7; 242.5; 252.1
	Fe	246.3; 248.3; 250.1; 252.3
	Ga	245.0
	Hg	253.7
	In	256.0
	Pd	247.6; 244.8
	Si	250.7; 251.4; 251.9; 252.4
	Sn	235.4
βPO (323–328)	Ag	328.1
	Cd	326.1

Table 5.3. Molecular Interferences for some Elements in Zeeman ETAAS

Molecule (Band Wavelength, nm)	Element	Wavelength, nm
AlBr	Mn	279.5
AlCl (261.44)	Pb	261.44
AlO	Pb	217.0
BaO	Au	267.6
InCl (267.23)		
OH	Bi	306.8
CN (381–388)	Co	384.6/388.2
	Cr	357.9
	Fe	358.1/386.0
	Mg	383.8
	Ni	378.4
	Os	378.2
CS	Tl	276.8
InBr	Sn	286.3
NO, PO, Fe	Se	196.0/204.0
P_2	As	193.7
PO	In	325.8
	Sb	206.8
PO, NO, Fe	Zn	213.8
S_2	Pb	283.3

5.1.1.1. *Emission*

This type of spectral interference occurs when the detector receives radiation from some point other than the lamp. It may originate in any of the following ways.

- *Emission from the tube wall, platform or graphite particles.* Emission from the tube wall may easily be recognized by the presence of negative deviations in the baseline of the background signal, sometimes observed at the end of the atomization stage. Emission proceeding from the tube wall is more pronounced in the region of longer wavelengths ($\lambda > 400$ nm) and for high atomization temperatures. Approximately 8% of the radiation emitted by the tube wall is made up of *a.c.* components, independently

Table 5.4.   **Some Matrix Elements Causing Spectral Interferences**

Analyte	Wavelength, nm	Interferent	Wavelength, nm
Al	308.2155	V	308.2111
AS	193.6960	Pb	193.6962
			193.6911
	197.197		197.1895
	228.812	Cd	228.802
B	249.6778	Co	249.671
Bi	227.6578		227.6532
Co	243.6657	Pt	243.6689
	252.136	In	252.137
Cr	360.5333	Co	360.5356
Cu	324.754	Eu	324.753
Eu	459.402	V	459.4108
	459.3243	Cs	459.3177
	271.9025	Pt	271.9038
	279.470	Mn	279.482
	285.213	Mg	285.213
	287.417	Ga	287.424
Fe	324.728	Cu	324.754
	327.445		327.396
	338.241	Ag	338.289
	352.424	Ni	352.454
	396.114	Al	396.153
	460.765	Sr	460.733
Ga	403.298	Mn	403.307
Ge	422.657	Ca	422.673
Hg	253.652	Co	253.649
	285.242	Mg	285.213
	359.348	Cr	359.349
I	206.163	Bi	206.170
Ne	359.352	Cr	359.349
Ni	341.4765	Co	341.4736
	305.0819	V	305.0890
Pb	241.173	Co	241.162
	261.4178		261.4128
Pd	247.6418	Pb	247.6379
Pt	265.9454	Eu	265.942
	306.4712	Ni	306.4623
	273.3961	Fe	273.4004

(Continued)

Table 5.4.   (*Continued*)

Analyte	Wavelength, nm	Interferent	Wavelength, nm
Sb	217.023	Pb	216.999
	231.147	Ni	231.095
	323.252	Li	323.261
Si	250.6899	Co	250.6877
	250.6899	V	250.6905
Sn	303.4121	Cr	303.4190
	300.9147	Ca	300.9205
Zn	213.856	Fe	213.859

of the total radiation emitted by the tube which reaches the detector. The *a.c.* component of the radiation emitted by the tube wall causes systematic errors in the Zeeman AAS signal, contributing to the background signal [Loos-Vollebregt and Ochten (1990)].

- *Scattering.* Scattering of the *a.c.* radiation emitted from the tube by some of the matrix components triggers oscillations in the background absorbance signal. Although the integrated absorbance is not too badly affected by the oscillations, precision is degraded [Loos-Vollebregt and Wrouwe (1997)].
- *Molecular emission* proceeding from the matrix components.
- *Molecular emission* from the substances formed by the analyte.
- *Atomic emission* from free atoms other than analyte generated into the graphite tube.

Nonetheless, these interferences do not constitute any serious problem for ETAAS.

5.1.1.2. *Absorption*

Spectral interferences of this type may be the result of various absorption processes, but all may be put under the umbrella term of *background absorption*. Spectral interferences caused by absorption as a consequence of the overlapping of atomic lines tend to be rare in ETAAS, but molecular absorption due to the spectral bands of oxides and halides can get to be a problem in ETAAS (Tables 5.1, 5.2, 5.3,

Table 5.5. Other Line Spectral Interferences in ETAAS

Analyte	λ (nm)	Interferent	λ (nm)
Ag	328.1	Rh	328.1
Co	252.1	Fe	252.3
Cr	357.9	Fe	358.1
Cu	324.7	Pd	324.3
Cu	217.9	Fe	217.8
Cu	216.5	Fe	216.7
Mn	279.5	Mg	279.5
Sb	217.6	Fe	217.8
Tl	276.8	Pd	276.3

5.4 and 5.5) [Kurfürst and Pauwels (1994); Daminelli *et al.* (1998); Ohlsson and Frech (1989); Aller and García-Olalla (1992)]. In general, levels of background absorption are much higher in a graphite tube than in a flame, and greater in the UV region than in the visible region.

There are distinct forms of background absorption.

- *Absorption bands (molecular).* Molecular interferences arise as a consequence of the absorption of electromagnetic radiation by molecules, usually diatomic, produced during the atomization stage through dissociation processes (thermal-dissociation) of more complex molecules. These chemicals have absorption bands (either continuum or structured) (Tables 5.1, 5.2 and 5.3), which causes over- or under-compensation in the background correction, probably due to splitting of the rotational lines through the Zeeman effect [Wibetoe and Langmyhr (1987)]. Molecular species usually present in graphite atomizers likely triggering interferences of this type are PO, CN and C_2 [Ohlsson and Frech (1991)]. The use of N_2 as a purge gas can cause shift of the baseline in Zeeman ETAAS, due to overlapping of the lines from the HCL and the rotational bands of the CN molecule [Doidge (1991)]. Metal halides are frequently generated in electrothermal atomizers [Pritchard and Reeves (1976); Aller and García-Olalla (1992); Shepard *et al.* (1998); Daminelli *et al.* (1999a, b)]. The NO molecule arising from the decomposition of metal nitrites shows errors in the Zeeman effect background correction [Doidge (1992)].

- *Structured spectral background*, arising out of a molecular spectrum rich in electronic excitations. One of the commonest systematic errors due to a structured background is produced by the presence of the OH band in the wavelength region near the resonance line for bismuth (306.8 nm).

- *Atomic lines from a matrix element* (Table 5.4) [Wibetoe and Langmyhr (1984, 1985, 1986); Carnrick *et al.* (1986); Frigge and Jackwerth (1992)]. The presence of a doublet due to an Al self-ionization process can set up spectral interferences in the ETAAS determination of this element [Doidge (1992)]. An excess of Pb can also cause spectral interferences when lead itself is being determined [Siemer (1984)].

In addition, *scattering of light*, arising from the presence of solid particles of the sample in the atomizer optical path during the integration stage, is also possible, producing a very wide band [L'vov *et al.* (1991)].

5.1.2. Non-Spectral Interferences

Non-spectral interferences in ETAAS have been widely studied. Such interferences originate in the occurrence of physical and/or chemical processes that alter the quantity of the analyte atomic vapor in the atomizer. These processes can occur before and after the analyte atoms would be generated. For this reason, they can be grouped as a function of the stage in the thermal program in which they appear in the graphite tube [Frech (1997)].

One of the most important forms of non-spectral interference is that arising as the outcome of ionization of the analyte atoms. There are, nevertheless, other forms of interference originating in physical and chemical processes, distinct from this.

5.1.2.1. *Ionization*

Ionization interference occurs when the analyte (with low ionization potential) is partially ionized, decreasing the number of neutral atoms of the analyte in the vapor phase, and hence the atomic absorption signal. The analyte ionization may be expressed as follows

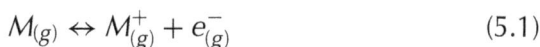

$$M_{(g)} \leftrightarrow M^+_{(g)} + e^-_{(g)} \tag{5.1}$$

which will be characterized by its corresponding equilibrium (ionization) constant, $k_i(T)$

$$k_i(T) = \frac{n_{M^+} n_{e^-}}{n_M}. \tag{5.2}$$

The ionization constant depends on the temperature and the element concerned, but not on its concentration. However, the ionization degree, α_i, depends on temperature and the analyte concentration. If a simple ionization process is assumed, then for $n_{e^-} = n_{M^+}$, the following relationship holds [Sturgeon and Berman (1983)]

$$\left(\frac{\alpha_i^2}{1 - \alpha_i} \right)_T = \frac{K_i(T)}{[n_M + n_{M^+}]_T}, \tag{5.3}$$

where α_i, $\alpha_i = \frac{n_{M^+}}{n_{M^+} + n_M}$, is the degree of ionization, n the density of species indicated by the subscript, and T, the absolute temperature.

5.1.2.2. *Physico-chemical interferences*

Non-spectral interferences originate in some physical and/or chemical processes, and may be classified as a function of the physical and chemical characteristics of the process involved.

Physical interferences arise as a consequence of any physical process, although some physical properties (such as viscosity, surface tension, and the like) of the sample do not have as marked an effect as in flame AAS [Todolí and Mermet (1999)]. The most noteworthy physical processes that may be picked out as responsible for physical interferences would be the following.

- *Losses of analyte* during the pyrolysis stage (i.e., prior to the atomization stage).
- *Incomplete vaporization of the analyte species* due to phenomena such as: (i) occlusion; (ii) incomplete cleaning of the atomizer (triggering memory effects); (iii) trapping on the solid particle surface; and (iv) migration through the atomizer surface or particles being formed.
- *Changes in the analyte elimination,* as a consequence of the following processes: (i) thermal expulsion of the analyte during the atomization stage; (ii) variation in the diffusion rate of the analyte atoms or molecules; (iii) reduction in the residence half-life time; and (iv) heterogeneous spatial distribution of the analyte (which may arise in various ways) [Panichev *et al.* (1999)].

Chemical interferences are the outcome of shifts in chemical equilibria (and probably also in ionization) involved during the analyte atomization. These changes originate in the formation of various compounds (oxides, chlorides, sulphides, refractory compounds, and so forth). The interference produced by certain metals can be interpreted as due to the formation of intermetallic compounds [García-Olalla and Aller (1992); Otha *et al.* (1992)]. Formation of salts, such as phosphates and iodides also contributes to interference [Majidi *et al.* (1991)].

Interferences based on some physical and/or chemical processes may also be classified as a function of the aggregation state of the chemicals involved in them. According to this criterion there would be interferences in the condensed phase and the vapor phase.

- *Interferences in the condensed phase.* These can take place through [Koshino and Narukawa (1993)]: (i) formation of low volatile analyte compounds; (ii) incomplete vaporization of the analyte due to occlusion; (iii) formation of the analyte refractory compounds; and (iv) changes in the generation rate of the analyte atoms. A memory effect arises from the incomplete atomization or a later failure to clean the atomizer fully, which causes an increase in signals thereafter. A matrix effect can originate as a consequence of the analyte retention within the graphite tube, this depending on the sample type [Krasowski and Copeland (1979)]. The drying and charring stages can also have a dramatic effect on the form of the atomic signal from the analyte.

- *Interferences in the vapor phase*: These interferences come from the analyte atomization in a cold atmosphere. This happens as a consequence of analyte losses (vaporization) (whether atoms or molecules) before (and while) reaching the atomization temperature. It is caused by: (i) shifts in the balance of formation of relatively stable volatile compounds [Yudelevich *et al.* (1989)]; (ii) shifts in the ionization equilibrium (as described in the previous section); and (iii) changes in the removing processes of the analyte. Interfering vapor phase absorbing species are formed from the matrix components, where aluminium is an example [L'vov *et al.* (1991); Lamourex *et al.* (1995); Castro and Aller (2003)]. Vapor phase interferences from a biological matrix seem to be less severe in probe atomization [Fernández *et al.* (2000)].

Nonetheless, many of the interferences known occur as a result of the simultaneous participation of several processes which can happen either in the vapor phase or in the condensed phase. For instance:

- Formation in the gaseous phase of the analyte species which can diffuse away from the optical path, causing losses prior to dissociation and atomization. A typical example of this sort of interference is due to the presence of chlorides.
- Reactions in the condensed phase giving rise to an analyte compound, which is later volatized and expelled from the tube before atomization.
- The co-volatization or thermal expulsion of the analyte (in a solid, liquid and/or vapor phase) together with gases from the matrix which expand rapidly, or through a transporter (or an occlusion) mechanism before (or after) the atomization temperature is reached. Sulphate interference in the determination of selenium might be included here, although it is not totally clear that this is so.
- Changes in the atomization mechanism of the analyte. These changes are usually triggered by the presence in the matrix of chemicals that act upon the analyte in a direct or indirect way.

- Other physical interferences (migration of analyte atoms or molecules).

Condensation of gas phase analyte atoms in the cooler regions of the graphite tube reduces their presence in the vapor phase, causing interferences. This condensation process occurs on the walls at the ends of the tube and near the injection hole [Frech *et al.* (1992); Frech and L'vov (1993); L'vov and Frech (1993)].

Potential interferences can be evaluated by using quantitative descriptors generated from the wavelets transform to describe electrothermal absorbance-time profiles. The descriptors used, obtained for various scales of the wavelets transform of the absorbance signal, are termed Lipschitz regularities. The number, time-position within the absorbance profile, and the value of the Lipschitz regularity provide a unique description of the shape of the absorption profile. Changes noted in the Lipschitz regularities for the absorption profiles of a standard and a sample indicate differences in the absorption profiles. Further, this is indicative of the existence of a variation in the physical or chemical conditions under which atomization is taking place. In addition, the contrary is also true; i.e., if no change is detected in the number, position in time or value of the Lipschitz regularities between the absorbance profiles of standards and samples, this means that the shape of the absorbance profile is unchanged by factors such as, for example, the presence of the matrix, differences in the way the sample is introduced, and the like [Sadler *et al.* (1998a,b)].

5.2. Effects Produced by Some Interfering Compounds

Normally, interference effects do not occur in isolation, but appear in combination with the simultaneous impact of several being visible. An example of this would be the interference effects of two typical interfering substances: chlorides and sulphates. The interference effects due to chlorides and sulphates are probably those most generally noted, since they can be present in a multitude of matrices.

5.2.1. Effect of Chlorides

The interferences produced by chlorides in the ETAAS determination of many elements are well known [Frech and Cedergren (1976a); Erspamer and Niemczyk (1982); Slavin *et al.* (1984); Kantor and Bezúr (1986); Brumbaugh and Koirtyohann (1988); Shekiro *et al.* (1988); Welz *et al.* (1985, 1988); Bektas and Akman (1990); Imai *et al.* (1991); Aller and García-Olalla (1992); Frigge and Jackwerth (1992); Grotti and Frache (1997); Akman and Tekgül (1999); Özcan and Akman (2000); Voloshin *et al.* (2003, 2004); Castro *et al.* (2004a,b)]. The mechanisms for such interferences vary with the analyte, matrix type and operational parameters. Among the different effects that chlorides may trigger are the following.

- Losses of the analyte as a chloride during the drying and pyrolysis stages.
- Occlusion of the analyte as a chloride inside the matrix microcrystals which are expelled from the tube during atomization.
- Expulsion of the analyte as a chloride as the outcome of an expansion phenomenon of the gaseous substances set free in the decomposition of the matrix during atomization.
- Formation of stable analyte chlorides (in the vapor or condensed phases) which are not completely dissociated during atomization,
- Formation in the gaseous phase of metal chlorides which absorb at the wavelength of the analyte.

The behavior of the metal chlorides in a graphite tube depends to a large degree on their capacity for hydrolization and their reactivity with the graphite surface. Thus, the broad absorption bands that appeared in the atomization of $BeCl_2$ and $MgCl_2$ have been explained as the result of the formation of the more stable hydroxyl-chlorides due to hydrolysis and thermal pretreatment. Decomposition of hydroxyl-chlorides induces the appearance of mono-chlorides and oxides (Fig. 5.1) [Katskov *et al.* (2001); Castro and Aller (2003, 2013); Castro *et al.* (2003, 2004a,b); Raseleka and Human (2004)]. A similar behavior is shown by alkaline earth fluorides [Katskov *et al.* (2000)].

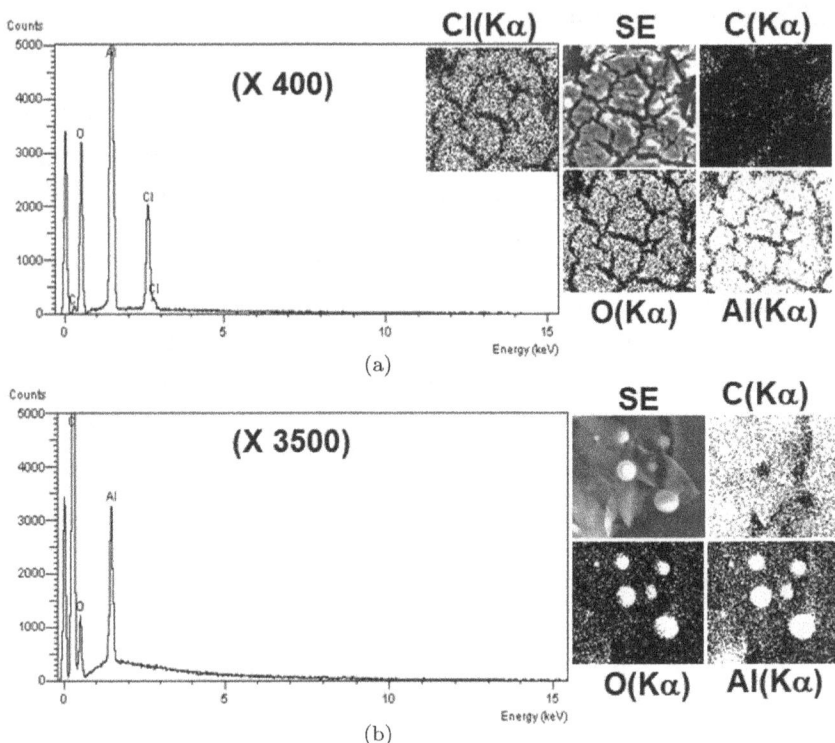

Fig. 5.1. ED X-ray spectra and SEM images of salt residues on a graphite platform containing 100 μg AlCl₃ heated to (a) 500°C, and (b) 1700°C [Castro *et al.* (2004b).]

However, it seems that Ca, Sr and Ba chlorides do not undergo hydrolysis. Interferences from chlorides also appear to be directly related to their vaporization/decomposition rate and temperature. Formation of intercalation compounds of graphite with several metals in the presence of chloride on the surface and sub-surface layers of the graphite tube has also been proved [Bulska and Ortner (2000)].

There is agreement in that the loss of some analytes lies in the formation of volatile chlorides, but there are doubts whether the reactions occur in the gaseous phase or in the condensed phase, as also about the nature of the losses (volatilization during pyrolysis and/or atomization, expulsion, or others). For example, TlCl is considered to be formed in the condensed phase during the pyrolysis stage and lost before the atomization stage [Qiao *et al.* (1993); Mahmood and

Jackson (1996)]. In the determination of Mn or Cd, the interfering chloride forms on the graphite surface and not in the gaseous phase [Wang and Holcombe (1992)]. If $MgCl_2$ is the interfering matrix at pyrolysis temperatures above 700°C, the analyte is lost during the pyrolysis stage, but at lower pyrolysis temperatures, the analyte is lost as $MnCl_2$ during the atomization stage [Byrne *et al.* (1992)]. In contrast, when NaCl is the interfering matrix, Mn is only lost during the atomization stage [Byrne *et al.* (1993)]. This is due to the fact that the appearance time of the volatile chloride coincides with that of Mn, suggesting a mechanism in the gaseous phase. Wall, platform and probe atomization show the same interference from NaCl, but wall atomization shows greater interference than the other atomization modes when $MgCl_2$ is the interfering substance [Carroll *et al.* (1992)]. Chemical effects in the vapor phase are larger for $MgCl_2$ than for NaCl, while losses through occlusion and expulsion due to co-vaporization predominate for NaCl. The interference follows the order: $NiCl_2$ > $MgCl_2$ > NaCl [Chaudhry *et al.* (1993)]. The presence of palladium nitrate hinders thermal vaporization for sodium chloride even together with protons (Fig. 5.2) [Castro *et al.* (2003)].

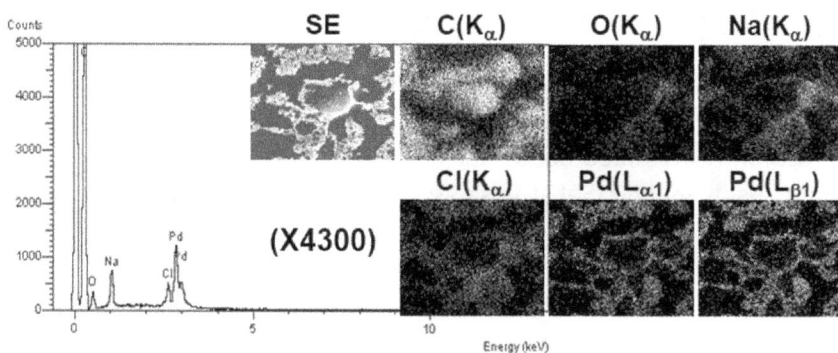

Fig. 5.2. Secondary electron (SE) image and X-rays maps from the same area for carbon, oxygen, sodium, chlorine and palladium. The EDS X-ray spectrum was derived from a graphite platform used to heat a Cd solution (10 ng/mL) together with sodium chloride (4000 mg/L), Pd nitrate (4000 mg/L) and protons (4000 mg/L) (as nitric acid) according to the temperature program up to a corrected temperature value of the platform surface of 850°C (4300X) [Castro *et al.* (2003)].

The interference mechanism for nickel chloride in the determination of zinc and cobalt [Akman and Döner (1995)] and for cobalt chloride in the determination of zinc [Döner and Akman (1994)] depends on the pyrolysis temperature. In the presence of an excess of nickel chloride, the analytes Zn and Co chlorides form in the condensed phase as a consequence of the reaction between the analyte species and HCl(g) generated by the hydrolysis of the nickel chloride. The analyte (Zn and Co) chlorides formed are eliminated during the pyrolysis stage or at the beginning of the atomization stage. At lower pyrolysis temperatures, where the nickel chloride is not fully hydrolyzed, the sensitivity decreases, this being attributable to: (i) expulsion of the analyte (Zn and Co) species together with the decomposition products of the nickel chloride which expand rapidly; and/or (ii) a reaction in the gaseous phase between the analyte (Zn, Co) atoms and chlorine in the atomization stage [Akman and Döner (1995)].

5.2.2. Effect of Sulphates

Although the effect of sulphur (as sulphate) on electrothermal atomization of analytes, such as selenium, has not been definitively clarified, the interference produced by sulphates might originate in one of the following:

- expulsion of the analyte (Se) atoms together with the matrix that is violently volatized before the atomization stage,
- losses of analyte (Se) atoms as a consequence of the formation and volatilization of SeO_2 in the pyrolysis stage owing to the presence of the oxidizing matrix SO_3 [Welz et al. (1992)],
- spectral interference due to the formation of metal sulphides [Aller (1996)]. In the presence of sulphates, some metal sulphides (Fig. 5.3) [Aller (1996); Castro et al. (2004a,b)], elemental sulphur [Castro et al. (2004a,b); Lemme et al. (2004)] and sulphur oxides [Raseleka and Human (2004)], seems to be present in condensed phase. Formation of gas phase CS emitting in the range 250–270 nm has also been proposed [Katskov et al. (2004)].

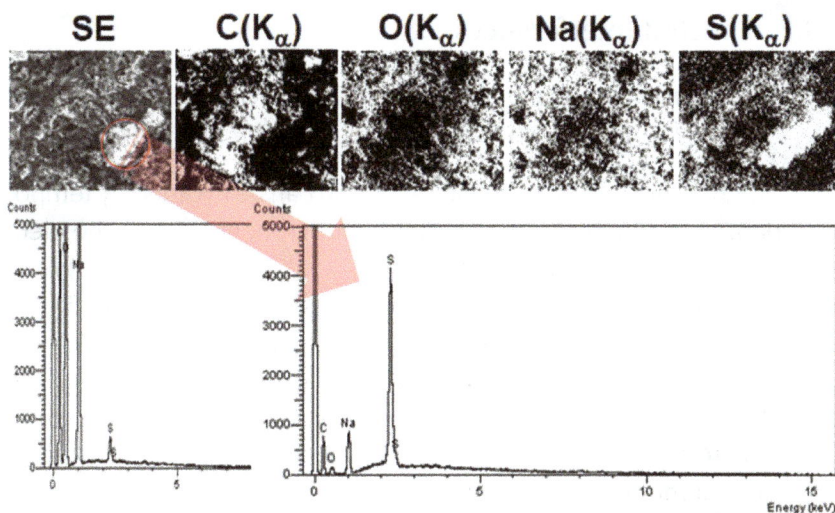

Fig. 5.3. ED X-ray spectra and element mappings from different areas on a platform containing 100 μg of sodium sulphate heated to 1200°C [Castro *et al.* (2004a)].

The origin of the sulphate interference observed when determining Se in the presence of Pd as a chemical modifier can be attributed to the formation of thermally stable $PdSO_4$ [Fischer and Rademeyer (1999)]. However, at temperatures below 550°C, no changes were observed when Na_2SO_4 and Pd were heated both together on the graphite platform, while sodium selenite was already reduced during drying, much more in the presence of pre-reduced palladium [Volynsky *et al.* (2001)]. Sulphur also interferes at other wavelengths, and for elements such as Pb [Kurfürst and Pauwels (1994)] and Mn [Akman and Tekgül (1999)]. The interference produced by a sulphate matrix in the determination of Pb appears to be due to the formation of PbS, which is lost through volatilization, as it is more volatile than Pb oxide [Imai *et al.* (1992)].

5.3. Correction of Interferences in ETAAS

At this point, comments will be made on the methods currently in use to eliminate or compensate for spectral and non-spectral interferences.

5.3.1. Spectral Interferences

5.3.1.1. *Temperature programme modification and employment of chemical modifiers*

Some interferences (whether spectral or not) can be avoided by temporal separation of the analyte and matrix components during the electrothermal heating. A new idea used for analytical purposes derives from the study of the differences in the diffusion rate of atomic and molecular vapors and their interaction with the graphite surface [Katskov *et al.* (1995)]. A general procedure, based on transferring the analyte during the pyrolysis stage from a solid sampling platform to the graphite tube wall, has been proposed to avoid spectral and/or non-spectral interferences [Maia *et al.* (2002)]. Similarly, they may be corrected by increasing the pyrolysis temperature, since this causes dissociation of the absorbing (interfering) molecules. In other cases, it is necessary only to modify the pyrolysis time, normally decreasing it, because in this way there is no time for such molecules to be generated. The addition of chemical modifiers such as a mixture of platinum and nickel [Bauslaugh *et al.* (1984)] and other chemicals, such as hydrogen [Frech and Cedergren (1976b)], allows the work to be carried out at increased pyrolysis temperatures, avoiding some molecular spectral interferences.

5.3.1.2. *Background correction*

The parameters affecting the background spectrum can vary temporarily or spatially, so that it is necessary to take *in situ* measurements of both the background and analytical signals if accurate analytical results are to be obtained. Background correction is especially needful when the sample components can scatter and/or absorb radiation at the analytical wavelength, particularly for low analyte concentrations, or the radiation sources contribute to non-specific emissions to the analytical signal.

The methods used for background correction in ETAAS can provide considerable improvement in the spectral signal/noise (S/N) ratio, what really represents a particular case of the noise reduction through modulation of the analytical wavelength. The methods used to improve the S/N ratio normally involve either increasing the signal or decreasing the spectral noise. In the first case, the noise has to be independent of the signal. To reduce the noise and improve the S/N ratio, the following general methods are employed: modulation procedures, and methods using a pulsed source and detector (electronic integration filtering).

Modulation can be achieved in three main ways: amplitude modulation, wavelength modulation, and modulation of the analysis sequence of sample and blank. Modulation does not improve the S/N ratio if the noise is white noise. The aim of modulation is to separate the frequency components that relate to the analytical signal from those components corresponding to noise. In the **amplitude modulation**, the analyte signal is connected and disconnected by means of a mechanical chopper, or by modulating (pulsing) electrically the source intensity. **Wavelength modulation** is based on rapid and repeated alteration of the analytical wavelength within a given wavelengths range. In the method involving **modulation of the analysis sequence of the sample and blank**, it is possible to discriminate a specific phenomenon as being from the sample in terms of a wavelength modulation, as long as the phenomenon is present in both the sample and blank and the latter's matrix has been exactly matched to that of the sample.

In those methods using **pulsed source/detector systems**, a pulsed source is employed with a detector having no delay or with a delay time between the two such that there is detection of the signal only 50% of the time.

The simplest methods for background correction are those that make use of a continuous source and those with alternative discrete lines. However, these methods are normally used to measure continuum backgrounds absorbance showing values below unity, as: (i) two different radiation sources cannot be always perfectly aligned; and (ii) interfering substances do not always show constant absorption coefficients across the spectral bandwidth selected for analysis.

Much closer to an ideal technique for background correction would be those utilizing just one line-emission source, because measurements can be taken even for backgrounds with absorbance values higher than 2. These alternative techniques for background correction also make use of modulation of: (i) the distribution of wavelengths of the line-emission source (the self-inversion lamp method [Smith and Hieftje (1983)]); and (ii) the polarization of the radiation from the primary line source; or (iii) the absorption spectrum of the analyte in the absorption cell (the Zeeman effect method [Koizumi *et al.* (1977)]) in two situations: when the absorbance from the analyte and background are determined simultaneously, and when only the background absorption is considered to be measured.

In the continuum emission source technique, a method for modulation of the wavelength is used that combines a reflector plate with a continuum emission source and a high-resolution *echelle* spectrometer. This method shows better S/N ratios when compared with those obtained with instruments having line emission sources for wavelengths above 230 nm. However, the opposite is also true for wavelengths below 230 nm. Nonetheless, a high-resolution continuum source atomic absorption spectrometer based on a xenon short-arc lamp, a transversely heated graphite furnace module with longitudinal Zeeman option, a double échelle monochromator and a linear array CCD detector has been developed for the correction of structured molecular background in the determination of selenium and arsenic [Becker-Ross *et al.* (2000)].

Zeeman ETAAS can use either modulation of the polarization or modulation of the magnetic field. A Zeeman effect ETAAS technique utilizing modulation of high-frequency polarized light has been introduced [Sholupov and Ganeyev (1995)]. Radiation emitted by the furnace can be removed of by means of high-frequency modulation of the hollow-cathode lamp in Zeeman AAS signals [Loos-Vollebregt and Ochten (1990)]. Heating the graphite tube also eliminates the oscillations problem in the spectral background [Loos-Vollebregt and Wrouwe (1997)]. Nonetheless, the greater cost involved in using a method based on the Zeeman-effect and the lower sensitivity yielded by the line-inversion method figure are among the main drawbacks of these methods.

5.3.1.2.1. Alternative wavelength (secondary line)

In the two-wavelength method, total absorbance (atomic plus background absorption) is measured at the analyte resonance line emitted by the hollow-cathode lamp (HCL), while non-specific absorbance (background absorption) is measured at a nearby line (the alternative line) (from the same HCL or a continuum source) (Table 5.6) at which the analyte atoms supposedly do not absorb. Alternatively, it is also possible to use a different HCL of an element not present in the sample to select the alternative wavelength.

The wavelength utilized for measuring the background absorbance must be close to the analyte line. One of the problems is precisely that of finding an appropriate nearby line. The maximum distance permissible is around 10 nm for the majority of elements. To ensure that there is adequate correction, the spectral background between these two lines should be constant. Another factor to be kept in mind is the lack of consistency over time of the spectral background, i.e.

Table 5.6. Secondary Lines used for Background Compensation

Element	Analyte Line λ (nm) (from HCL)	Non-absorbent Line λ (nm) (From Continuum Source)
Aluminium	309.2	307.0
Antimonium	217.6	217.9
Cadmium	228.8	227.6
Zinc	213.9	212.5
Cobalt	240.7	238.8
Copper	324.7	323.4
Chromium	357.9	352.0
Tin	286.3	283.9
Iron	248.3	247.2
Indium	304.0	305.7
Magnesium	285.2	281.7
Molibdenum	313.3	323.4
Nickel	232.0	231.6
Platinum	265.9	270.2
Lead	283.3	282.0
Vanadium	318.4	312.5

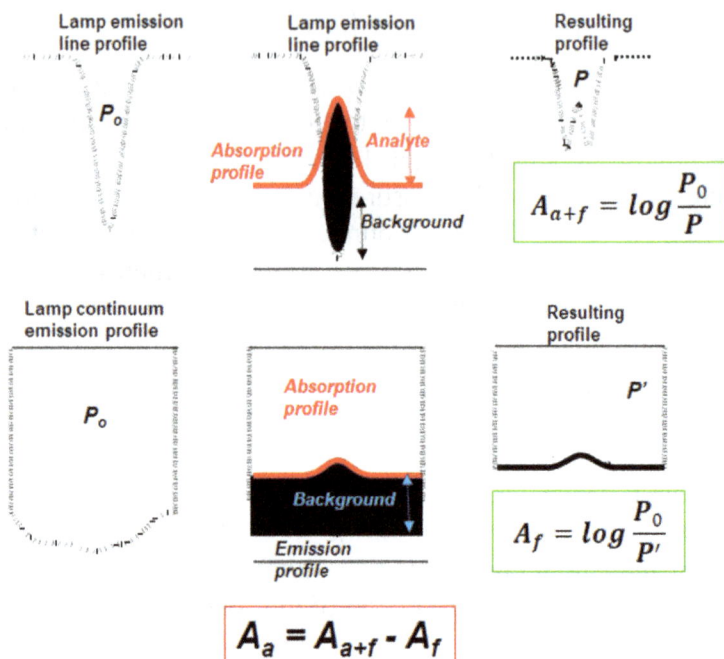

Lamp emission line profile

P_o

Lamp emission line profile

Absorption profile   Analyte

Background

Resulting profile

P

$$A_{a+f} = log \frac{P_0}{P}$$

Lamp continuum emission profile

P_o

Absorption profile

Background

Emission profile

Resulting profile

P'

$$A_f = log \frac{P_0}{P'}$$

$$A_a = A_{a+f} - A_f$$

Fig. 5.4. Background correction method using a continuum emission source (emission spectrum).

that for successive measurements the spectral background signal will not necessarily remain constant.

5.3.1.2.2. Continuum radiation

This method employs a HCL in combination with an arc deuterium lamp as a secondary continuum radiation source (Fig. 5.4) [Marshall *et al.* (1985)]. The rays proceeding from the HCL and the deuterium lamp are made to pass through the electrothermal atomizer with coincident trajectories. The absorption of continuum radiation coming from the deuterium lamp occurs fundamentally, thanks to molecules, i.e. the fraction of the continuous radiation that decreases as a consequence of absorption by the analyte atoms is quantitatively negligible, but not null (some error is assumed). Alternatively, the radiation from the analyte HCL is absorbed by both the analyte atoms and molecules. Electronic subtraction of the absorbance (background) measured with the deuterium lamp from the absorbance (atomic plus background) measured with the HCL, provides the corrected AAS signal for the analyte.

The background correction system using a deuterium lamp permits correction of spectral backgrounds with high absorbance levels. There is no real reason why a well-designed deuterium lamp system should not be able to correct spectral backgrounds with the same absorbance levels as the alternative Zeeman and Smith-Hieftje techniques. The reason for not correcting absorbance levels higher than 2 (where 99% of the light is absorbed) is more the increase in noise than any fundamental failure in the correction system.

Many spectral interferences do not disappear when the deuterium lamp background correction system is utilized, but can be eliminated by making use of one of the methods described below (Zeeman and Smith-Hieftje).

5.3.1.2.3. Zeeman effect

The Zeeman effect (Figs. 5.5 and 5.6) [Koizumi *et al.* (1977); Stephens and Murphy (1978); Loos-Vollebregt and Galan (1978, 1988); Fernández *et al.* (1981); Knowles and Frary (1988)] originates in

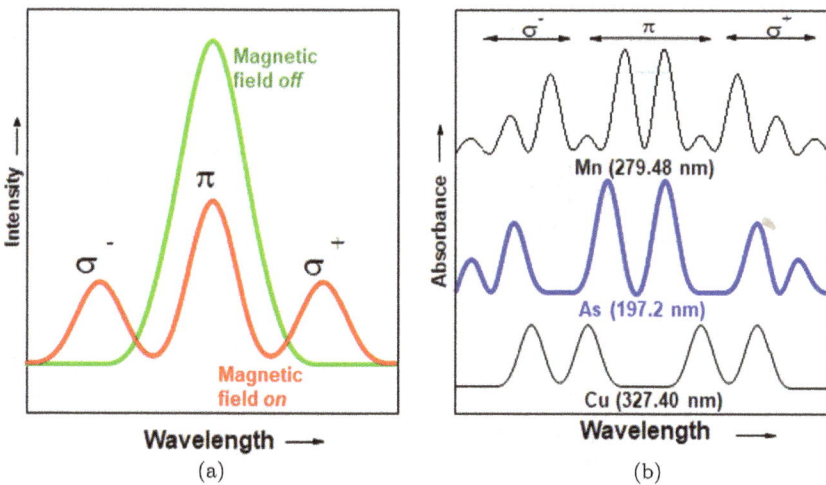

Fig. 5.5. Normal (a), and anomalous (b) Zeeman effect.

Type of transition	Element (wavelength)	Spliting patterns
$^1S_0$-$^1P_1$	Be 234.8 nm Mg 285.2 nm Ca 422.7 nm Sr 460.7 nm Ba 553.5 nm Zn 213.9 nm Cd 228.8 nm Hg 184.9 nm	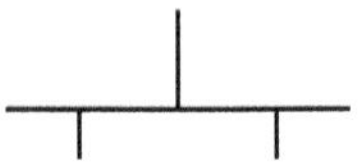
$^1S_0$-$^3P_1$; $^3P_0$-$^3P_1$; $^3P_2$-$^3P_2$	Zn 307.5 nm Cd 326.0 nm Hg 253.7 nm Sn 286.3 nm Pb 283.3 nm Si 251.6 nm	
$^6S_{5/2}$-$^6P_{7/2}$	Se 196.0 nm Te 214.3 nm	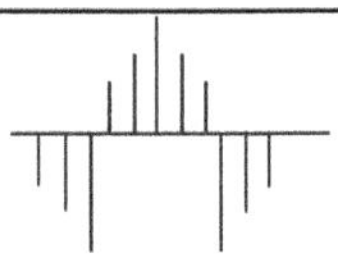
$^4S_{3/2}$-$^4P_{1/2}$	As 197.2 nm Sb 231.1 nm Bi 306.8 nm	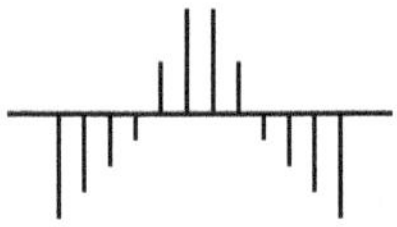
$^3F_4$-$^3G_5$	Al 309.2 nm Ti 365.3 nm Cr 357.9 nm Mo 313.5 nm Fe 248.3 nm Ni 232.0 nm Mn 279.5 nm Co 240.7 nm	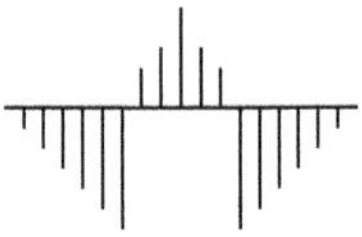
$^2S_{1/2}$-$^2P_{3/2}$; $^2S_{1/2}$-$^2P_{1/2}$	Li 670.7 nm Na 589.0 nm K 766.5 nm Cu 324.8 nm Ag 328.1 nm Au 242.8 nm	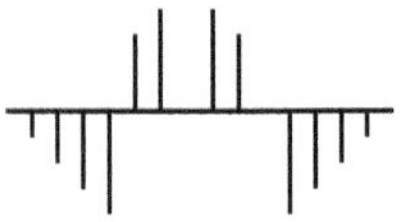
$^4S_{3/2}$-$^4P_{3/2}$	As 193.8 nm Sb 217.6 nm	

Fig. 5.6. Typical examples of the line splitting with the normal and anomalous Zeeman effects for some elements [Koizumi and Yasuda (1976a,b)].

the splitting of the spectral lines in the presence of a magnetic field. In practice, the Zeeman background correction method is based on the siting of an electromagnet around the radiation source (lamp) (**direct Zeeman**) [Koizumi and Yasuda (1976a)] or the atomizer (**inverse Zeeman**) [Koizumi and Yasuda (1976b)]. This magnet may operate with direct current or alternating current at a frequency which is normally 50 Hz or 100 Hz, or even ≥200 Hz [Gleisner *et al.* (2003)].

With the Zeeman method utilizing a modulated magnetic field, the total absorbance is measured with the magnetic field turned off, while the signal due to the background is measured with this field switched on. Thus, when the electromagnet is disconnected (is in *off*), radiation from the HCL is absorbed by both the analyte atoms and the molecular species vaporized in the atomizer (spectral background). In contrast, when the electromagnet is connected (is in *on*), the profile of the atomic emission line from the HCL (direct Zeeman) (Fig. 5.7a) or the atomic absorption line of the analyte (inverse Zeeman) (Fig. 5.7b) of the analyte is divided into components σ and π, which are polarized so as to be perpendicular and parallel, respectively, to the direction of the magnetic field. The two σ components (Fig. 5.5) are displaced away from the position of the emission line of the HCL and absorption of the π component is blocked by using a polarizer. Hence, when the electromagnetic field is *on* only background absorption is measured (Fig. 5.7). Subtraction of the average obtained with the field connected from the average yielded when the field is disconnected gives the corrected atomic absorption signal of the analyte.

For the analyte atoms in the presence of a magnetic field (inverse Zeeman), the atomic absorption lines are split, thanks to the Zeeman effect, and the cloud of atoms (in the vapor phase) behaves like a dichroic (and bi-refringent) medium. With a transverse orientation of the magnetic field, the cloud of the analyte atoms acts as a linear dichroic medium. In this case, the π component of the spectral profile of the analyte will absorb only linearly polarized light which is oriented parallel to the axis of the transverse magnetic field; in contrast, the (displaced) σ components of the spectral profile will absorb linearly polarized light that is oriented perpendicularly to the axis of the field.

Magnetic field	Field direction	Polarizer	Emission spectrum	Absorption spectrum	Signal discrimination Magnetic field	
					On (BG)	Off (BG+A)
Direct	⊥	↷ π/σ				
Direct	=	No				
Alternating	⊥	→ σ				
Alternating	=	No				

(a)

Magnetic field	Field direction	Polarizer	Emission spectrum	Absorption spectrum	Signal discrimination Magnetic field	
					On (BG)	Off (BG+A)
Direct	⊥	↷ π/σ				
Direct	=	No				
Alternating	⊥	→ σ				
Alternating	=	No				

(b)

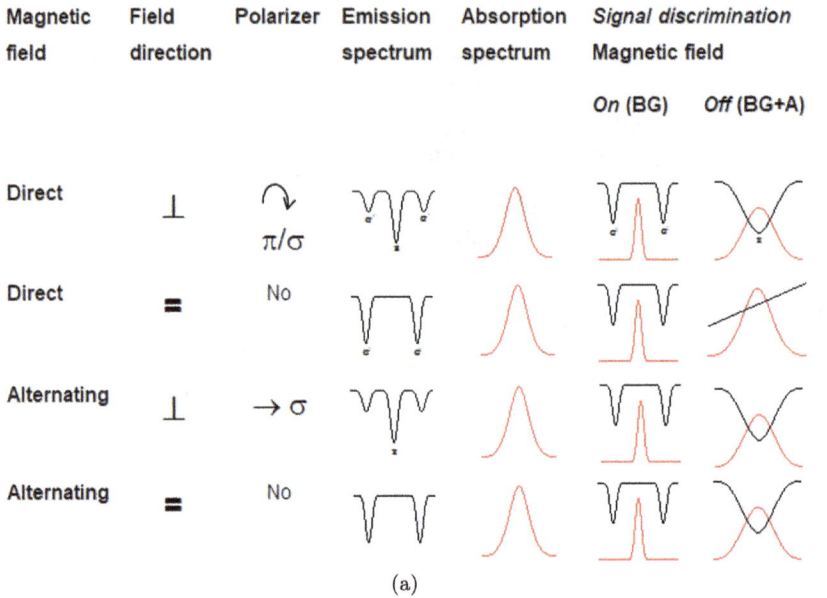

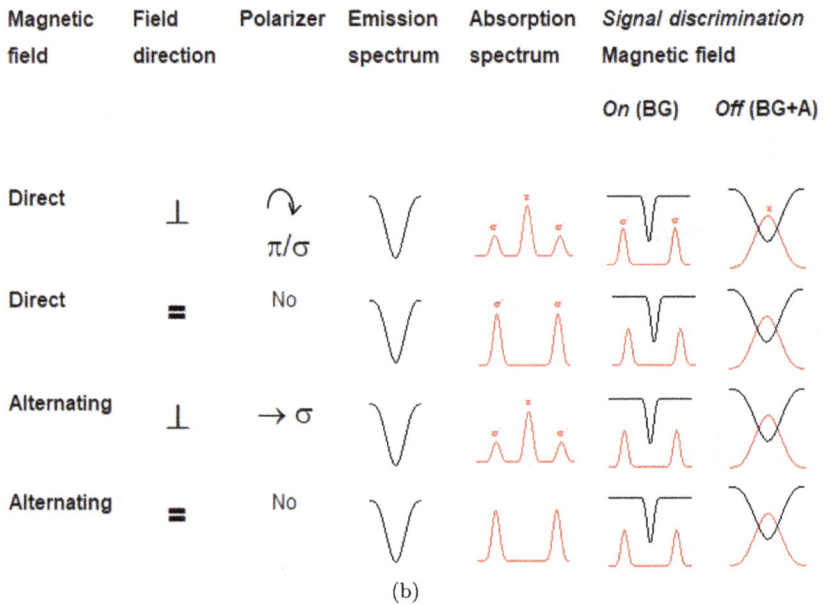

Fig. 5.7. Possibilities for the spectral background correction by means of (a) direct and (b) inverse Zeeman effect AAS. (BG: Background absorbance; A: analyte absorbance.)

Zeeman effect AAS can be divided into two groups, depending on the direction of the magnetic field relative to the beam from the radiation source.

Transverse Zeeman AAS. In this case, the magnetic field applied is perpendicular to the optical axis. Transverse Zeeman effect AAS needs a polarizer to separate the radiation from the HCL into two components. This Zeeman AAS requires an electromagnet which can operate with direct current (*d.c.*) and alternating current (*a.c.*).

In transverse *d.c.* Zeeman AAS, a rotary polarizer is placed between the excitation source and the sample (atomizer), and this divides the beam from the HCL into two linearly polarized components. One of these components oscillates in parallel to the magnetic field, and the other perpendicularly to it. Both components have the same wavelength. When the parallel component is transmitted, it will be absorbed by the central un-shifted π component from the analyte (analyte and background will both absorb light). In the next quarter-cycle, only the perpendicular source is transmitted. This perpendicular light will not be absorbed by the central π component from the analyte (even though it is at exactly the same wavelength) because its polarization is different. In an ideal case, the perpendicularly polarized sidebands far enough away one from the other so as not to absorb any perpendicular radiation, and hence in this instance only the spectral background would be absorbed. The difference between the signal from the perpendicularly polarized component and the parallel component corrects the background absorption.

In transverse *a.c.* Zeeman AAS, an *a.c.* electromagnet is used together with a static linear polarizer between the atomizer and the monochromator, this transmitting solely that component from the source which is polarized perpendicularly to the electromagnet's field for all measurements. Analyte and background absorbance is measured for a strength value of the magnetic field equal to zero. In reality, the σ component from the analyte in the atomizer is not displaced. For the maximum strength of the field, the background absorption is measured (in the ideal case) without there being absorption by the analyte. The σ components are shifted and the π component does not absorb the component from the source that is polarized perpendicularly to the magnetic field.

Longitudinal Zeeman AAS. In this case, the magnetic field applied (a longitudinal a.c. magnetic field around the absorption line) (inverse zeeman) is parallel to the optical axis. There is no need for a polarizer in this development of the spectral background correction system based on the Zeeman's effect, because the π component has totally disappeared and the σ components are circularly polarized. With this system, it is possible to measure the real background absorption present in the original resonance line of the analyte when the magnetic field is *off*.

The principal advantage of the inverse longitudinal a.c. Zeeman's effect background correction method lies in the fact that the background correction is carried out at the same exactly wavelength utilized for measuring the atomic absorption, permitting accurate correction even for structured backgrounds. Moreover, only one excitation source is used, so there is no need to adjust and align the radiation beams.

A few particular applications of the Zeeman effect use double *echelle* monochromators (high resolution) to eliminate some interferences such as the background over-compensation found in determining Cd ($\lambda = 228.8$ nm) in the presence of the mixture ($Mg(NO_3)_2 + NH_4H_2PO_4$) utilized as a chemical modifier [Heitmann *et al.* (1996)]. Zeeman-ETAAS has also been found to be useful for the determination of isotopic ratios. The isotope-shift Zeeman effect at the boron 208.9 nm line has been used for the determination of $^{10}B/^{11}B$ ratio. The shifted σ^- component of the ^{10}B absorption line coincides with the π component of the ^{11}B absorption line. The intensity of the σ^- component of the ^{10}B isotope increases with the $^{10}B/^{11}B$ ratio, and the net absorbance decreases linearly with increasing ^{10}B isotopic fraction for a given total concentration of analyte [Thangavel *et al.* (2006)].

5.3.1.2.4. Self-reversal line. Smith–Hieftje effect or the pulsed lamp method

A fourth procedure for background correction is based on the broadening and self-inversion of the spectral lines [Smith and Hieftje (1983); Siemer (1983)]. When a HCL operates at current intensities about 10 times greater than normal (i.e. up to 600 mA), the emission line broadens considerably and the intensity in the line center decreases (self-inversion) owing to the auto-absorption process. When the line profile broadens and is self-reversed, sensitivity in the atomic absorption measurements can be reduced whether or not because absorption of the radiation from the HCL due to molecular species is not always null during the high-current pulse. Thus, if the HCL operates in pulsed mode, alternating high and low current intensities with a frequency of 100 Hz, it is possible to take advantageous measurements representing, in alternation, absorption due to the analyte plus the background and absorption only due to the background (Fig. 5.8). With the Smith–Hieftje method, the total absorbance is measured during a low-current pulse from the lamp and the background during a high-current pulse. Subtraction of the signal obtained at high current intensities yields a measurement of net atomic absorption. Interferences caused by structured background and absorption line overlapping of the analyte from matrix components are many times compensated accurately [Oppermann *et al.* (2003)].

5.3.1.3. *Errors in background correction*

Timing errors

The above background correction methods can fall into errors on some occasions, since it is sometimes necessary to use an alternative wavelength so as to avoid spectral interferences, even when the Zeeman correction method is in use. Although the problem of a structured background has been studied in some detail, the spectral background changing rapidly over time has received scant attention hitherto. Many applications of the graphite tube involve signals that change quickly and in these cases the speed of the technique used for background correction has a critical effect on the reliability of measurements.

The majority of modern atomic absorption spectrometers incorporate *simultaneous* background correction methods, although in reality they carry out two measurements slightly separated in time (not truly simultaneous). One measurement relates to the total signal (atomic plus background signal), while the other measurement corresponds to the background signal (theoretically alone). The background signal is

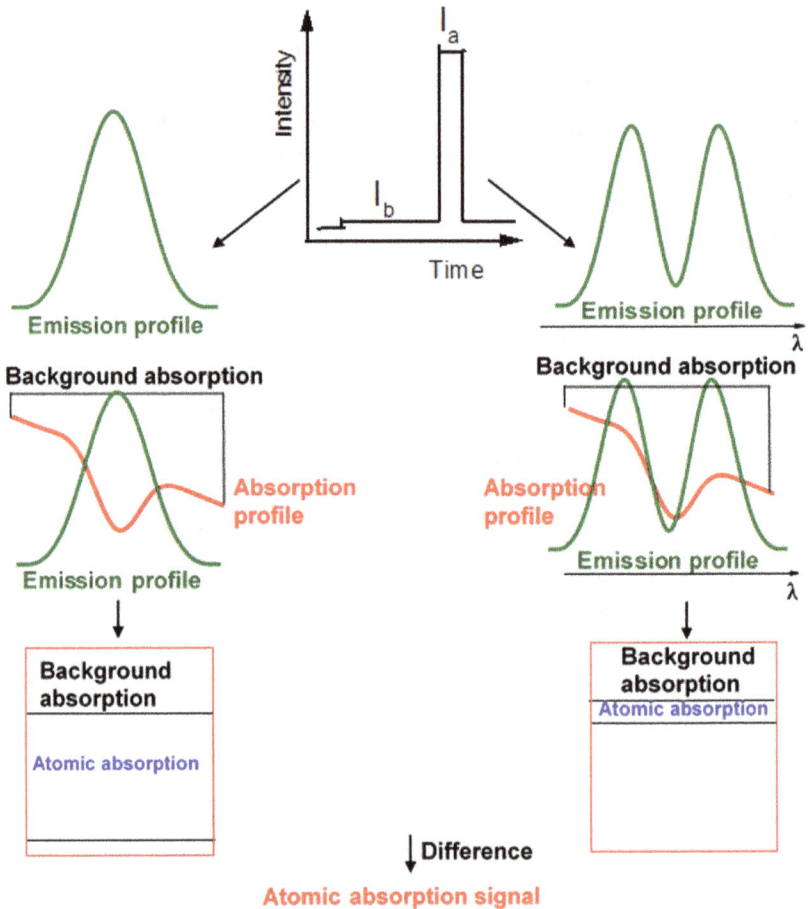

Fig. 5.8. Spectral background correction with the Smith–Hieftje method.

subtracted electronically from the total absorbance to yield the corrected atomic absorbance (without background).

The separation in time between the measurement of the total absorbance and the spectral background can vary from less than 1 ms up to 10 ms [Liddell *et al.* (1987)] (Fig. 5.9), depending on the correction method.

During each current cycle, it is usual to perform 2 to 4 measurements of the signals from the HCL and the deuterium lamp in the sample beam and 1 to 2 in the reference beam (for those systems

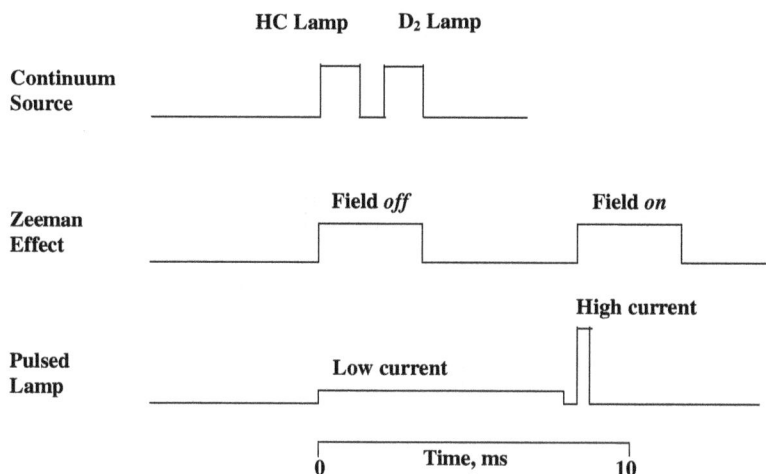

Fig. 5.9. Measurement timings for the three commonest background correction methods [Liddell *et al.* (1987)].

with a double beam). Hence, at a frequency of 60 Hz, there would be 240 measurements per second on the sample beam for background and total absorbance (plus 120 measurements of the reference beam). More conventional (slower) spectrophotometers perform 60 measurements per second (one for each current cycle).

Between each of the two lamp pulses, there is a period of time in which any emission by the graphite tube is measured and corrected. As the emission is measured four times in the sample beam, in each current cycle it is easy to correct changes or fluctuations in the emission.

In the majority of the double-beam spectrophotometers, the sample and the reference light beams take the same time for each measurement. However, the purpose of the reference beam is simply to compensate for the unstable noise from the lamp and long-term drift. The effect of these fluctuations in the intensity of the lamp can be eliminated by measuring the reference beam for a short period of time in each current cycle and thereafter applying electronic filtering. On the other hand, it is not possible to apply electronic filtering to the sample beam, as a rapid response is needed to follow the changing signals quickly. In consequence, the only way of reducing noise in

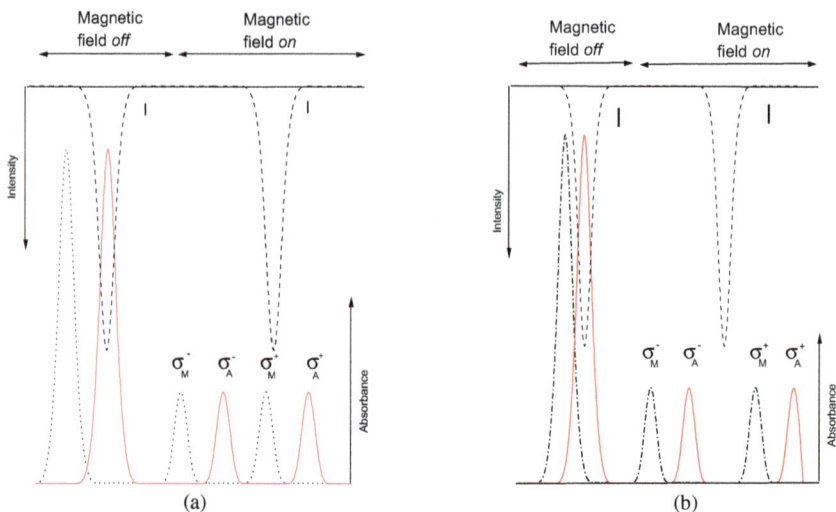

Fig. 5.10. Spectral background correction by excess (a), and by defect (b), using the Zeeman correction system causing interference [Zhong *et al.* (1994)].

the sample beam is: (i) to utilize a more intense light source; or (ii) to measure over a longer time-span. By reducing the time spent on the reference beam, it is possible to employ double the time used by most spectrophotometers on the sample beam. This leads to a considerable reduction in noise and an improvement in the detection limit [Liddell *et al.* (1987)].

Timing errors in background correction can be reduced to below 10^{-5} A, by means of Zeeman effect ETAAS utilizing high-frequency modulation polarized light [Sholupov and Ganeyev (1995)]. Similarly, it is possible to reduce the detection limit and increase the linear response range, at least by up to four orders of magnitude [Harnly and Holcombe (1985)].

Over- and under-correction

Other errors deriving from the use of a background correction method are over-correction and under-correction (Fig. 5.10). The measurement of the spectral background may be in error if the background arises from molecular spectra with many electronic excitation lines.

Particularly, the continuum source background correction method triggers some errors when the structure of the background changes rapidly with wavelength; i.e. when there is a structured background. On the other hand, some molecules (OH, PO, NO, NO_2 and SO_2) show a Zeeman's effect [Massmann (1982); Zong *et al.* (1994)] and hence, background measurements on two different polarization planes, whether measuring with or without a magnetic field, can be different. Background measurement by Zeeman techniques may cause serious systematic errors if the spectral background is caused by sharp rotational lines of bi- or tri-atomic molecules, but less from polyatomic molecules [Massmann (1982)].

The error arising in the measurement of the background signal constituted only by a peak presents the shape of the peak derivative. The shape of the error profile while the background peak lasts is independent of the delay between the measurement of the total and background absorbance values, but the magnitude of the error depends very strongly on the separation time between the two signals. One way to reduce this error is to use spectral background measurements taken before and after the total absorbance measurements to calculate the background magnitude corresponding to the time at which the total absorbance is measured [Harnly and Holcombe (1985)]. If the background signal changes linearly over time, this interpolation method is highly effective in reducing the error. However, when the background signal has a peak or a deep, there can still be a serious error or inaccuracy. In this case, errors in the spectral background correction may be decreased by utilizing three or four values of the background to establish a quadratic equation permitting estimation of the inaccuracy level. These errors are proportional to the third and fourth derivative of the function in question for a fit of three and four points, respectively [Holcombe and Harnly (1986)]. Hence, the less time elapsing between measurements of the background and the total absorbance, the more accurate the correction will be.

Non-alignment of beams

The spatial distribution of the beam proceeding from the lamp changes substantially under the different excitation conditions normally utilized in the background correction systems for each analysis (the use of two pulses of light: the sample beam, SB, and reference beam, RB). In general, the non-coincidence of the two light beams may give rise to certain forms of interference as a consequence of failures in optical components other than those of the lamp [Siemer (1984)]. Thus, some imperfections in the quality or alignment of the mirrors, lenses, beam splitters, polarizers, and so forth, can cause a failure to coincide of the RB and SB beams, especially using mobile components. The Zeeman correction is not immune to this problem.

Owing to the fact that with ETAAS instruments only one (fixed) set of optical components is used, small changes in the position of the emission zone within the lamp necessarily cause some lack of coincidence between the beams within the atomizer. This lack of coincidence is also observed with a single-pulse instrument, which indicates that the position of the maximum emission of light within the lamp changes significantly during pulses. However, in general the error will not be detected during routine analysis of samples, unless the error is big enough in relation to the signal from the analyte atoms to produce a negative net absorbance response.

Roll-over effect

In ETAAS, the absorbance-time profiles for high analyte concentrations differ from the normal and corrected AAS signals when the roll-over phenomenon takes place (Fig. 5.11).

For high analyte concentrations triggering roll-over, the atomic absorption signal shows a deformed peak; i.e. the absorbance-time profile increases with the atom concentration, but absorbance diminishes before attainment of the maximum atom population in the tube. From this moment on, the atom concentration in the tube decreases, but the signal strengthens and returns to the initial maximum before decreasing to zero again.

The roll-over phenomenon has been extensively studied in Zeeman AAS [Loos-Vollebregt and Galan (1980, 1982, 1984)] and using pulsed HCLs [Galan and Loos-Vollebregt (1984); Aller (1993)]. The Smith–Hieftje technique for background correction in AAS was at first

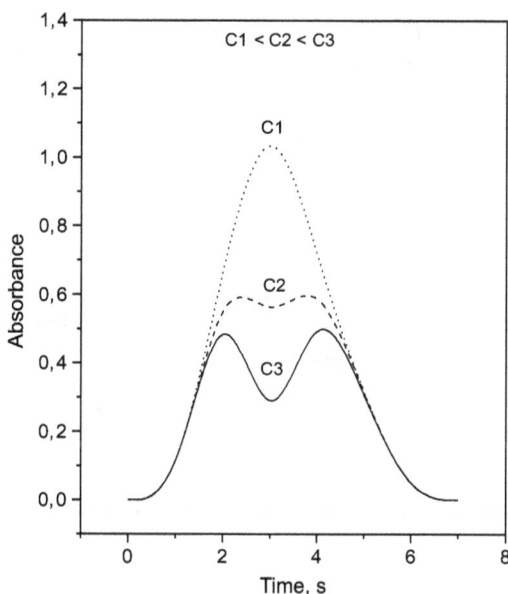

Fig. 5.11. Absorbance profiles showing the roll-over effect for increasing concentrations (C) of analyte ($C_1 < C_2 < C_3$).

accepted as a roll-over effect-free system. However, this effect has also been observed for several elements with this background correction system, both in flame AAS and in ETAAS [Galan and Loos-Vollebregt (1984); Aller (1993)]. Galan and Loos-Vollebregt (1984) and Loos-Vollebregt and Galan (1986) established that the maximum values on either side of the dip remain constant and that the roll-over phenomenon arises in the HCL, having been related to self-absorption of the emission lines in the HCLs for Zeeman AAS. Nonetheless, the role of other parameters (line overlapping, Zeeman splitting, strength of the magnetic field applied, non-linear absorption coefficients, wavelength for excitation and spectral profiles) has been investigated as a possible source generating the roll-over effect and its influence on the calibration line.

The roll-over effect has also been attributed to the presence of stray light, arising primarily through a scattering process. However, if near the resonance line there are weak spectral lines, these may also behave similarly to stray light when their relative intensity increases.

Nonetheless, the shape of the absorbance profile in this case is different from what is noted with stray light. The relatively low maxima observed on the Smith–Hieftje calibration line are the result of a considerable loss of sensitivity or of a high level of stray light during the lamp's low current pulse. All the same, the influence of the weak wings of the line (the Lorentz component) on the shape of the absorbance profile for high analyte concentrations is the principal cause of saturation for the calibration line using the Smith–Hieftje technique.

On the basis of a theoretical analysis, Galan and Loos-Vollebregt (1984) revealed the inter-relationship between the roll-over effect, analytical sensitivity and wavelength proximity in background correction systems. However, some instrumental parameters, such as the atomizer type and the thermal programme can also influence this inter-relationship and consequently the roll-over phenomenon. Thus, stray light originates in the HCL, but can be modified (increased or decreased) in the atomizer in different ways, depending on whether there is wall or platform atomization. The level of stray light depends on the wavelength region utilized for the analytical determination. Likewise, the presence of certain matrix components, and of chemical modifiers may also affect sensitivity and hence the roll-over phenomenon [Aller (1993)]. Under thsese situations, a sample dilution is necessary [Koizumi *et al.* (1982)].

5.3.2. Non-Spectral Interferences

A great deal of non-spectral interferences (ionic and physico-chemical, such as analyte losses and formation of compounds either in the vapor phase or in the condensed phase) may be corrected by the use of chemical modifiers (see Chapter 6). The probe atomization shows no (or minor) interferences from metal chlorides and sulphates, not needing the use of a chemical modifier [Zakharov *et al.* (2005)].

5.3.2.1. *Ionization*

Such interferences can be corrected (or at least compensated for) by the use of an ionization buffer, i.e. by adding a surplus of an easily ionizable element (normally an alkaline metal), both to standards and samples, modifying the analyte atom population present in the atom-izer, in accordance with the following scheme of ionizations

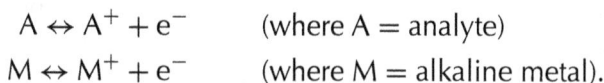

$$A \leftrightarrow A^+ + e^- \quad \text{(where A = analyte)}$$
$$M \leftrightarrow M^+ + e^- \quad \text{(where M = alkaline metal).}$$

The presence of the second reaction displaces the ionization bal-ance of the analyte (first reaction) towards the left, increasing the atom population of the analyte (A).

5.3.2.2. Physico-chemical interferences

5.3.2.2.1. Condensed phase: Matrix effect

The use of chemical modifiers (see Chapter 6) and the hot sample injection [Sotera *et al.* (1983)] can be very efficacious procedures in minimizing matrix effects. The use of platform atomization in com-bination with chemical modifiers is able to eliminate matrix inter-ference, the effect of which may be 10^3–10^4 times greater than that accepted with wall atomization. Moreover, the use of measurements of integrated absorbances may also present some advantages in certain cases.

Low pyrolysis temperatures, fast heating rates, and low atomiza-tion temperatures tend to shift the analyte absorption pulse towards the isothermal region of the wall temperature curve, consequently result-ing in higher peak-height sensitivity and greater freedom from matrix interferences [Chakrabarti *et al.* (1983)]. Differences in volatility of two analyte species allows the possibility of separation without using a previous isolation stage [Arpadjan and Krivan (1986)].

The memory effect can be eliminated or decreased by utilizing higher atomization temperatures or longer atomization and cleaning times.

5.3.2.2.2. Vapor phase

Interferences of this sort have been eliminated by the use of more isothermal atomizers, thanks to the atomized sample being introduced into a hot inert gas. Non-stop-flow working conditions have also been proposed as a means to reduce interferences in the gas phase [Belarra *et al.* (2000)].

Interference triggered by chlorides can also be avoided by using chemical modifiers [Welz *et al.* (1988); Hulanicki *et al.* (1990)] with addition of protons or ammonia in order to form volatile compounds, such as HCl or NH₄Cl [Castro *et al.* (2003); Feo *et al.* (2003)].

The trapping of the analyte atoms as a consequence of condensation can be decreased with the use of a small flow (mini-flow) of the purge gas. This mini-flow reduces diffusion of atoms towards the cold ends of the tube (since the gas flow direction is counter to that of the atomic diffusion), although it does reduce sensitivity [L'vov and Frech (1993)].

5.3.3. Other Problems

To reduce the magnitude of errors due to the non-coincidence of the light beams, the following has been suggested [Siemer (1984)]: (i) not focusing the light beams too centrally in the atomizer (this central position is the most usual practice), although it results in increased noise; (ii) restricting measurements to the middle portion of the light beams, even though this can reduce the light intensity with which the spectrometer must function; and (iii) utilizing additional optical components to produce beams that are particularly homogeneous (use of a diffuse reflectance integration sphere or of optical fibers to carry the light).

References

Akman, S. and Döner, G. (1995). Nickel chloride interferences on zinc and cobalt in graphite furnace atomic absorption spectrometry using a dual cavity platform. *Spectrochim. Acta, Part B,* **50**: 975–984.

Akman, S. and Tekgül, H.I. (1999). The interference effect of more than one anion and cation in graphite furnace atomic absorption spectrometry. Part 2. Effect of sodium, magnesium, sulphate and chloride mixtures on the atomization of manganese. *Spectrochim. Acta, Part B,* **54**: 505–514.

Aller, A.J. (1993). Roll-over in graphite furmace atomic absorption spectrometry with a gold pulsed hollow-cathode lamp. *Anal. Chim. Acta,* **284**: 361–366.

Aller, A.J. (1996). Remarks on the interference of sulfate in the determination of selenium by electrothermal atomic absorption spectrometry. *Anal. Sci.*, **12**: 977–980.

Aller, A.J. and García-Olalla, C. (1992). Spectral interferences on the determination of selenium by electrothermal atomic absorption spectrometry. *J. Anal. At. Spectrom.*, **7**: 753–760.

Arpadjan, S. and Krivan, V. (1986). Preatomization separation of chromium(III) from chromium(VI) in the grapghite furnace. *Anal. Chem.*, **58**: 2611–2614.

Bauslaugh, J.; Radziuk, B.; Saeed, K. and Thomassen, Y. (1984). Reduction of effects of structured non-specific absorption in the determination of arsenic and selenium by electrothermal atomic absorption spectrometry. *Anal. Chim. Acta*, **165**: 149–157.

Becker-Ross, H.; Florek, S. and Heitmann, U. (2000). Observation, identification and correction of structured molecular background by means of continuum source AAS-determination of selenium and arsenic in human urine. *J. Anal. At. Spectrom.*, **15**: 137–141.

Bektas, S. and Akman, S. (1990). Effects of alkali and alkaline earth metal chlorides on the atomization of lead and chromium in graphite furnace atomic absorption spectrometry. *Anal. Sci.*, **6**: 547–554.

Belarra, M.A.; Crespo, C.; Resano, M. and Castillo, J.R. (2000). Direct determination of copper and lead in sewage sludge by solid sampling-graphite furnace atomic absorption spectrometry — study of the interference reduction in the gaseous phase working in non-stop flow conditions. *Spectrochim. Acta, Part B*, **55**: 865–874.

Brumbaugh, W.G. and Koirtyohann, S.R. (1988). Effects of surface on the atomization of lead by graphite furnace. *Anal. Chem.*, **60**: 1051–1055.

Bulska, E. and Ortner, H.M. (2000). Intercalation compounds of graphite in atomic absorption spectrometry. *Spectrochim. Acta, Part B*, **55**: 491–499.

Byrne, J.P.; Chakrabarti, C.L.; Gregoire, D.C.; Lamoureux, M. and Ly, T. (1992). Mechanisms of chloride interferences in atomic absorption spectrometry using a graphite furnace atomizer investigated by electrothermal vaporization inductively coupled plasma mass spectrometry. Part 1. Effect of magnesium chloride matrix and ascorbic acid chemical modifier on manganese. *J. Anal. At. Spectrom.*, **7**: 371–382.

Byrne, J.P.; Lamoureux, M.; Chakrabarti, C.L.; Ly, T. and Gregoire, D.C. (1993). Mechanisms of chloride interferences in atomic absorption spectrometry using a graphite furnace atomizer investigated by electrothermal vaporization inductively coupled plasma mass spectrometry. Part 2. Effect of sodium chloride matrix and ascorbic acid chemical modifier on manganese. *J. Anal. At. Spectrom.*, **8**: 599–709.

Carnrick, G.R.; Barnett, W. and Slavin, W. (1986). Spectral interferences using the Zeeman effect for furnace atomic absorption spectroscopy. *Spectrochim. Acta, Part B*, **41**: 991–997.

Carroll, J.; Miller-Ihli, N.J.; Harmly, J.M.; O'Haver, T.C. and Littlejohn, D. (1992). Comparison of sodium chloride and manganese chloride interferences in contiuum source atomic absorption spectrometry with wall, platform and probe electrothermal atomization. *J. Anal. At. Spectrom.*, **7**: 533–538.

Castro, M.A. and Aller, A.J. (2003). Mechanistic study of the aluminum interference in the determination of arsenic by electrothermal atomic absorption spectrometry. *Spectrochim. Acta, Part B*, **58**: 901–918.

Castro, M.A. and Aller, A.J. (2013). *Interferentes y Modificadores en un Atomizador Electrotérmico.* Editorial Académica Española. (EAE), Saarbrücken, Deutschland. ISBN: 978-3-659-07115-7.

Castro, M.A.; Faulds, K.; Smith, W.E.; Aller, A.J. and Littlejohn, D. (2004a). Identification of condensed-phse species on the thermal transformation of alkaline and alkaline earth metal sulphates on a graphite platform. *Spectrochim. Acta, Part B,* **59**: 827–839.

Castro, M.A.; Faulds, K.; Smith, W.E.; Aller, A.J. and Littlejohn, D. (2004b). Characterization of condensed phase species produced during the thermal treatment of metal chlorides on a graphite platform using surface analysis techniques. *Spectrochim. Acta, Part B*, **59**: 1935–1942.

Castro, M.A.; Feo, J.C. and Aller, A.J. (2003). Thermal stability and spatial distribution of sodium chloride alone and in the presence of several metal salts on a graphite platform. *J. Anal. At. Spectrom.*, **18**: 260–267.

Chakrabarti, C.L.; Wu, S. and Bertels, P.C. (1983). Isothermal atomization from a platform in graphite furnace atomic absorption spectrometry. *Spectrochim. Acta, Part B*, **38**: 1041–1060.

Chaudhry, M.M.; Mouillere, D.; Ottaway, B.J.; Littlejohn, D. and Whitley, J.E. (1993). Investigation of chloride salt decomposition and pre-atomization interferences in electrothermal atomic absorption spectrometry. *J. Anal. At. Spectrom.*, **7**: 701–706.

Daminelli, G.; Katskov, D.A.; Mafolo, R.M. and Kántor, T. (1999b). Atomic and molecular spectra of vapours evolved in a graphite furnace. Part 2. Magnesium chloride. *Spectrochim. Acta, Part B*, **54**: 683–697.

Daminelli, G.; Katskov, D.A.; Mafolo, R.M. and Tittarelli, P. (1999a). Atomic and molecular spectra of vapours evolved in a graphite furnace. Part 1. Alkali halides. *Spectrochim. Acta, Part B*, **54**: 669–682.

Daminelli, G.; Katskov, D.A.; Marais, P.J.J.G. and Tittarelli, P. (1998). Characterization of the vapor-phase molecular and atomic absorption from sea water matrices in electrothermal atomic absorption spectrometry. *Spectrochimica Acta, Part B*, **53**: 945–964.

Doidge, P.S. (1991). Baseline shifts in inverse Zeeman-effect graphite furnace atomic absorption spectrometry ascribed to the Zeeman effect of the CN molecule. *Spectrochim. Acta, Part B*, **46**: 1779–1787.

Doidge, P.S. (1992). Autoionizing lines of aluminium — Their role in spectral interferences and atom number measurement of Al in a graphite-furnace. *Spectrochim. Acta, Part B*, **47**: 569–571.

Döner, G. and Akman, S. (1994). Effect of cobalt chloride on the atomization of zinc in electrothermal atomic absorption spectrometry. *J. Anal. At. Spectrom.*, **9**: 333–336.

Epstein, M.S.; Turk, G.C. and Yu, L.J. (1994). A spectral interference in the deter-mination of arsenic in high-purity lead and lead-base alloys using electrother-mal atomic absorption spectrometry and Zeeman-effect background correction. *Spectrochim. Acta, Part B,* **49**: 1681–1688.

Erspamer, J.P. and Niemczyk, T.M. (1982). Vaporization of some chloride matrices in graphite furnace atomic absorption spectrometry. *Anal. Chem.,* **54**: 538–540.

Fassel, V.A.; Rasmuson, J.O. and Cowley, T.G. (1968). Spectral line interferences in atomic absorption spectroscopy. *Spectrochim. Acta, Part B,* **23**: 579–586.

Feo, J.C.; Castro, M.A.; Lumbreras, J.M.; de Celis, B. and Aller, A.J. (2003). Nickel as a chemical modifier for sensitivity enhancement and fast atomization processes in electrothermal atomic absorption spectrometric determination of cadmium in biological and environmental samples. *Anal. Sci.,* **19**: 1631–1636.

Fernández, F.J.; Bohler, W.; Beaty, M.M. and Barnett, W.B. (1981). Correction for high background levels using the Zeeman effect. *At. Spectrosc.,* **2**: 73–80.

Fernández, P.; Marchante-Gayón, J.M. and Sanz-Medel, A. (2000). The atomisation system influence on the extent of interferences on the determination of volatile elements (cadmium and lead) in biological materials by ETAAS. *Quim. Anal. (Barcelona),* **19**: 225–232.

Fischer, I.L. and Rademeyer, C.I. (1999). Kinetics of selenium atomization in electrothermal atomization atomic absorption spectrometry (ETA-AAS). Part 3. Chemical interference of sulphate using palladium modifiers. *Spectrochim. Acta, Part B,* **54**: 975–983.

Frech, W. (1997). Non-spectral interference effects in platform-equipped graphite atomisers. *Spectrochim. Acta, Part B,* **52**: 1333–1340.

Frech, W. and Cedergren, A. (1976a). Investigations of reactions involved in flame-less atomic absorption procedures. Part I. Application of high-temperature equi-librium calculations to a multicomponent system with special reference to the interference from chlorine in the flameless atomic absorption method for lead in steel. *Anal. Chim. Acta,* **82**: 83–92.

Frech, W. and Cedergren, A. (1976b). Investigations of reactions involved in flame-less atomic absorption procedures. Part II. An experimental study of the rôle of hydrogen in eliminating the interference from chlorine in the determination of lead in steel. *Anal. Chim. Acta,* **82**: 93–102.

Frech, W. and L'vov, B.V. (1993). Matrix vapours and physical interference effects in graphite furnace atomic absorption spectrometry. *Spectrochim. Acta, Part B,* **48**: 1371–1379.

Frech, W.; L'vov, B.V. and Romanova, N.P. (1992). Condensation of matrix vapours in the gaseous phase in graphite furnace atomic absorption spectrometry. *Spec-trochim. Acta, Part B,* **47**: 1461–1469.

Frigge, C. and Jackwerth, E. (1992). Spectral interferences in the determination of traces of Pd in the presence of lead by atomic-absorption spectroscopy. *Spec-trochim. Acta, Part B,* **47**: 787–791.

Galan, L. de and Loos-Vollebregt, M.T.C. de (1984). Roll-over of analytical curves in atomic absorption spectrometry arising from background correction with pulsed hollow-cathode lamps. *Spectrochim. Acta, Part B,* **39**: 1011–1019.

García-Olalla, C. and Aller, A.J. (1992). Peak profile characteristics and atomization mechanisms for selenium in the presence of mercury for graphite furnace atomic absoption spectrometry. *Fresenius' J. Anal. Chem.*, **342**: 70–75.

Gleisner, H.; Eichardt, K and Welz, B. (2003). Optimization of analytical performance of a graphite furnace atomic absorption spectrometer with Zeeman-effect background correction using variable magnetic field strength. *Spectrochim. Acta, Part B*, **58**: 1663–1678.

Grotti, M. and Frache, R. (1997). Investigation of the formation of solid phase compounds between tellurium and interfering elements in graphite furnace atomic absorption spectrometry. *Spectrochim. Acta, Part B*, **52**: 1247–1258.

Harnly, J.M. and Holcombe, J.A. (1985). Background correction errors originating from nonsimultaneous sampling for graphite furnace atomic absorption spectrometry. *Anal. Chem.*, **57**: 1983–1986.

Heitmann. U.; Schütz, M.; Becker, Roß, H. and Florek, S. (1996). Measurements on the Zeeman-splitting of analytical lines by means of a continuum source graphite furnace atomic absorption spectrometer with a linear charge coupled device array. *Spectrochim. Acta, Part B*, **51**: 1095–1105.

Holcombe, J.A. and Harnly, J.M. (1986). Minimization of background correction errors using nonlinear estimates of the changing background in carbon furnace atomic absorption spectrometry. *Anal. Chem.*, **58**: 2606–2611.

Hulanicki, A.; Bulska, E. and Dittrich, K. (1990). Elimination of interferences in the electrothermal atomisation of manganese in atomic absorption spectrometry. *J. Anal. At. Spectrom.*, **5**: 209–213.

Imai, S.; Saito, K. and Hayashi, Y. (1991). Chemical reactions in an electrothermal graphite furnace in the presence of lead, magnesium, chloride and chromium nitrate. *Anal. Sci.*, **7**: 893–896.

Imai, S.; Tanaka, T.; Saito, K. and Hayashi, Y. (1992). Interference mechanism of sulfate matrices in graphite-furnace atomic-absorption spectrometry for lead. *Anal. Sci.*, **8**: 885–887.

Kantor, T. and Bezúr, L. (1986). Volatilization studies of cadmium compounds by the combined quartz furnace and flame atomic absorption method: Effects of magnesium chloride and ascorbic acid additives. *J. Anal. At. Spectrom.*, **1**: 9–17.

Katskov, D.A.; Lemme, M. and Tittarelli, P. (2004). Atomic and molecular spectra of vapors evolved in graphite furnace, Part 6: Sulfur. *Spectrochim. Acta, Part B*, **59**: 101–114.

Katskov, D.A.; Mofolo, R. M. and Tittarelli, P. (2000). Atomic and molecular spectra of vapors evolved in a graphite furnace, Part 3: Alkaline earth fluorides. *Spectrochim. Acta, Part B*, **55**: 1577–1590.

Katskov, D.A.; Mofolo, R. M. and Tittarelli, P. (2001). Atomic and molecular spectra of vapors evolved in a graphite furnace, Part 4: Alkaline earth chlorides. *Spectrochim. Acta, Part B*, **56**: 57–67.

Katskov, D.A.; Schwarzer, R.; Marais, P.J.J. and McCrindle, R.I. (1995). Diffusion of molecular vapors through heated graphite. *Spectrochim. Acta, Part B*, **50**: 763–780.

Knowles, M. and Frary, B.D. (1988). Zeeman AAS applied to element determinations in complex matrices. *International Laboratory*, **April**: 52–64.

Koizumi, H.; Sawakabu, H. and Koga, M. (1982). Correction for double valued calibration curves in Zeeman effect atomic absorption spectrometry. *Anal. Chem.*, **54**: 1029–1032.

Koizumi, H. and Yasuda, K. (1976a). An application of the Zeeman effect to atomic absorption spectrometry: a new method for background correction. *Spectrochim. Acta, Part B*, **31**: 237–255.

Koizumi, H. and Yasuda, K. (1976b). A novel method for atomic absorption spectroscopy based on the analyte-Zeeman effect. *Spectrochim. Acta, Part B*, **31**: 523–535.

Koizumi, H.; Yasuda, K. and Katayama, M. (1977). Atomic absorption spectrophotometry based on the polarization characteristics of the Zeeman effect. *Anal. Chem.*, **49**: 1106–1112.

Koshino, Y. and Narukawa, A. (1993). Investigation and elimination of sodium nitrate-borate interference of manganese in electrothermal atomic absorption spectrometry. *The Analyst*, **118**: 1027–1030.

Krasowski, J.A. and Copeland, T.R. (1979). Matrix interferences in furnace atomic absorption spectrometry. *Anal. Chem.*, **51**: 1843–1849.

Kurfürst, U. and Pauwels, J. (1994). Spectral interference on the lead 283.3 nm line in Zeeman-effect atomic absorption spectrometry. *J. Anal. At. Spectrom.*, **9**: 531–534.

Lamoureux, M.M.; Chakrabarti, C. L.; Hutton, J.C.; Gilmutdinov, A.Kh.; Zakharov, Y.A. and Gregoire, D.C. (1995). Mechanism of aluminium spike formation and dissipation in electrothermal atomic absorption spectrometry. *Spectrochim. Acta, Part B*, **50**: 1847–1867.

Lemme, M.; Katskov, D. A. and Tittarelli, P. (2004). Atomic and molecular spectra of vapors evolved in graphite furnace, Part 7: Alkaline metal sulfates and sulfides. *Spectrochim. Acta, Part B*, **59**: 115–124.

Liddell, R.P.; Athanasopoulos, N.; Grey, R.G. and Routh, M.W. (1987). The effect of background correction speed on the accuracy of AA measurements. *International Laboratory*, **April**: 82–87.

Loos-Vollebregt, M.T.C. de and Galan, L. de (1978). Theory of Zeeman atomic absorption spectrometry. *Spectrochim. Acta, Part B*, **33**: 495–511.

Loos-Vollebregt, M.T.C. de and Galan, L. de (1980). The shape of analytical curves in Zeeman atomic absorption spectrometry. II. Theoretical analysis and experimental evidence for absorption maximum in the analytical curve. *Appl. Spectrosc.*, **34**: 464–472.

Loos-Vollebregt, M.T.C. de and Galan, L. de (1982). Correction for background absorption and stray radiation in a.c. modulated Zeeman atomic absorption spectrometry. *Spectrochim. Acta, Part B*, **37**: 659–672.

Loos-Vollebregt, M.T.C. de and Galan, L. de (1984). The shape of analytical curves in Zeeman atomic absorption spectrometry. II. Extended dynamic range. *Appl. Spectrosc.*, **38**: 141–148.

Loos-Vollebregt, M.T.C. de and Galan, L. de (1986). Stray light in Zeeman and pulsed hollow cathode lamp atomic absorption spectrometry. *Spectrochim. Acta, Part B*, **41**: 597–610.

Loos-Vollebregt, M.T.C. de and Galan, L. de (1988). Longitudinal a.c. Zeeman AAS with a transverse heated graphite furnace. *Spectrochim. Acta, Part B*, **43**: 1147–1156.

Loos-Vollebregt, M.T.C. de and Ochten, P.J. van (1990). Influence of emission from the furnace wall on Zeeman atomic absorption spectrometric signals. *J. Anal. At. Spectrom.*, **5**: 183–187.

Loos-Vollebregt, M.T.C. de and Vrouwe, E.X. (1997). Spectral phenomena in graphite furnace AAS. *Spectrochim. Acta, Part B*, **52**: 1341–1349.

L'vov, B.V. and Frech, W. (1993). Matrix vapours and physical interference effects in graphite furnace atomic absorption spectrometry. I. End-heated tubes. *Spectrochim. Acta, Part B*, **48**: 425–433.

L'vov, B.V.; Romanova, N.P. and Polzik, L.K. (1991). Formation of soot during the decomposition of gaseous carbides in graphite furnace atomic absorption spectrometry. *Spectrochim. Acta, Part B*, **46**: 1001–1008.

Mahmood, T.M. and Jackson, K.W. (1996). Wall-to-platform migration in electrothermal atomic absorption spectrometry. Part 1. Investigation of the mechanism of chloride interference on thallium. *Spectrochim. Acta, Part B*, **51**: 1155–1162.

Maia, S.M.; Welz, B.; Ganzarolli, E. and Curtius, A.J. (2002). Feasibility of eliminating interferences in graphite furnace atomic absorption spectrometry using analyte transfer to the permanently modified graphite tube surface. *Spectrochim. Acta, Part B*, **57**: 473–484.

Majidi, V.; Ratliff, J. and Owens, M. (1991). Investigation of transient molecular absorption in a graphite furnace by laser-induced plasmas. *Appl. Spectrosc.*, **45**: 473–476.

Marshall, J.; Carroll, J.; Littlejohn, D.; Ottaway, J.M.; O'Haver, T.C. and Harnly, J.M. (1985). Microcomputer controlled background correction for ETA-AES and ETA-continuum source AAS. *Anal. Proceed.*, **22**: 67–69.

Massmann, H. (1982). The origin of systematic errors in background measurements in Zeeman atomic-absorption spectrometry. *Talanta*, **29**: 1051–1055.

Ohlsson, K.E.A. and Frech, W. (1989). Photographic observation of molecular spectra in inverse Zeeman-effect graphite furnace atomic absorption spectrometry. *J. Anal. At. Spectrom.*, **4**: 379–385.

Ohlsson, K.E.A. and Frech, W. (1991). Quantitative *in situ* spectroscopic measurements and thermodynamic equilibria modelling of CN and C_2 in a graphite furnace. *Spectrochim. Acta, Part B*, **46**: 559–581.

Ohta, K.; Aoki, W. and Mizuno, T. (1992). An investigation of the interference mechanism of the atomization of copper in a molybdenum tube electrothermal atomizer. *Chem. Anal. (Warsaw)*, **37**: 51–62.

Oppermann, U.; Schram, J. and Felkel, D. (2003). Improved background compensation in atomic absorption spectrometry using the high speed self reversal method. *Spectrochim. Acta, Part B*, **58**: 1567–1572.

Özcan, M. and Akman, S. (2000). Investigation of the effect of some inorganic salts on the determination of tin in graphite furnace atomic absorption spectrometry. *Spectrochim. Acta, Part B*, **55**: 509–515.

Panichev, N.A.; Ma, Q.; Sturgeon, R.E.; Chakrabarti, C.L. and Pavski, V. (1999). Condensation of analyte vapor species in graphite furnace atomic absorption spectrometry. *Spectrochim. Acta, Part B*, **54**: 719–731.

Pritchard, M.W. and Reeves, R.D. (1976). Non-atomic absorption from matrix salts volatilized from graphite atomizers in atomic absorption spectrometry. *Anal. Chim. Acta*, **82**: 103–111.

Qiao, H.; Mahmood, T.M. and Jackson, K.W. (1993). Mechanism of the action of palladium in reducing chloride interference in electrothermal atomic absorption spectrometry. *Spectrochim. Acta, Part B*, **48**: 1495–1503.

Raseleka, R.M. and Human, H.G.C. (2004). Identification of molecules in graphite furnace by laser ionization time-of-flight mass spectrometry: sulphur and chlorine containing compounds. *J. Anal. At. Spectrom.*, **19**: 899–905.

Sadler, D.A.; Boulo, P.R.; Soraghan, J.S. and Littlejohn, D. (1998a). Tutorial guide to the use of wavelet transforms to determine peak shape parameters for interference detection in graphite-furnace atomic absorption spectrometry. *Spectrochim. Acta, Part B*, **53**: 821–835.

Sadler, D.A.; Littlejohn, D.; Boulo, P.R. and Soraghan, J.S. (1998b). Application of wavelet transforms to determine peak shape parameters for interference detection in graphite-furnace atomic absorption spectrometry. *Spectrochim. Acta, Part B*, **53**: 1015–1030.

Shekiro, J.M.; Skogerboe, R.K. and Taylor, H.E. (1988). Mechanistic characterization of chloride interferences in electrothermal atomization systems. *Anal. Chem.*, **60**: 2578–2582.

Shepard, M.R.; Jones, B.T. and Butcher, D.J. (1998). High-resolution, time-resolved spectra of indium and aluminium atoms, fluorides, chlorides, and oxides in a graphite tube furnace. *Appl. Spectrosc.*, **52**: 430–437.

Sholupov, S.E. and Ganeyev, A.A. (1995). Zeeman atomic absorption spectrometry using high frequency modulated light polarization. *Spectrochim. Acta, Part B*, **50**: 1227–1236.

Siemer, D.D. (1983). An alternate approach to background correction in atomic absorption spectrometry. *Appl. Spectros.*, **37**: 552–557.

Siemer, D.D. (1984). Consequences of light beam misalignment in background corrected atomic absorption spectrometers. *Anal. Chem.*, **56**: 1517–1519.

Slavin, W.; Carnrick, G.R. and Manning, D.C. (1984). Chloride interferences in graphite furnace atomic absorption spectrometry. *Anal. Chem.*, **56**: 163–168.

Smith, S.B. and Hieftje, G.M. (1983). A new background-correction method for atomic absorption spectrometry. *Appl. Spectroc.*, **37**: 419–424.

Sotera, J.J.; Cristiano, L.C.; Conley, M.K. and Kahn, H.L. (1983). Reduction of matrix interferences in furnace atomic absorption spectrometry. *Anal. Chem.*, **55**: 204–208.

Stephens, R. and Murphy, G.F. (1978). Applications of the Zeeman effect to analytical atomic spectroscopy — VII. Line Interferences. *Talanta*, **25**: 441–445.

Sturgeon, R.E. and Berman, S.S. (1983). Determination of the efficiency of the graphite furnace for atomic absorption spectrometry. *Anal. Chem.*, **55**: 190–200.

Thangavel, S.; Rao, S.V.; Dash, K. and Arumachalam, J. (2006). Determination of boron isotope ratios by Zeeman effect background correction-graphite furnace atomic absorption spectrometry. *Spectrochim. Acta, Part B*, **61**: 314–318.

Todolí, J.-L. and Mermet, J.-M. (1999). Acid interferences in atomic spectrometry: analyte signal effects and subsequent reduction. *Spectrochim. Acta, Part B*, **54**: 895–929.

Tsunoda, K.-I.; Haraguchi, H. and Fuwa, K. (1980). Studies on the occurrence of atoms and molecules of aluminium, gallium, indium and their monohalides in an electrothermal carbon furnace. *Spectrochim. Acta, Part B*, **35**: 715–729.

Tsunoda, K.; Haraguchi, H. and Fuwa, K. (1985). Halide interferences in an electrothermal graphite furnace atomic absorption spectrometry with Group IIIB elements as studied by atomic and molecular absorption signal profiles. *Spectrochim. Acta, Part B*, **40**: 1651–1661.

Voloshin, A.V.; Gil'mutdinov, A.Kh. and Zakharov, Yu.A. (2003). Influence of the matrix on the atomic absorption of a transversely heated graphite atomizer. *J. Appl. Spectros.*, **70**: 942–947.

Voloshin, A.V.; Gil'mutdinov, A.Kh. and Zakharov, Yu.A. (2004). Spatiotemporal dynamics of vapors of chloride matrices in a transversely heated graphite furnace for atomic absorption spectrometry. *J. Anal. Chem.*, **59**: 134–140.

Volynsky, A.B.; Stakheev, A.Yu.; Telegina, N.S.; Senin, V.G.; Kustov, L.M. and Wennrich, R. (2001). Low-temperature transformations of sodium sulphate and sodium selenite in the presence of pre-reduced palladium modifier in graphite furnaces for electrothermal atomic absorption spectrometry. *Spectrochim. Acta, Part B*, **56**: 1387–1396.

Wang, P. and Holcombe, J.A. (1992). Pressure-regulated electrothermal atomizer for atomic absorption spectrometry. *Spectrochim. Acta, Part B*, **47**: 1277–1286.

Welz, B.; Akman, S. and Schlemmer, G. (1985). Investigations of interferences in graphite furnace atomic-absorption spectrometry using a dual cavity platform. Part 1. Influence of nickel chloride on the determination of antimony. *Analyst*, **110**: 459–465.

Welz, B.; Bozsai, G.; Sperling, M. and Radziuk, B. (1992). Palladium nitrate-magnesium nitrate modifier for electrothermal atomic absorption spectrometry. *J. Anal. At. Spectrom.*, **7**: 505–509.

Welz, B.; Schlemmer, G. and Madakavi, J.R. (1988). Investigation and elimination of chloride interference on thallium in graphite furnace atomic absorption spectrometry. *Anal. Chem.*, **60**: 2567–2572.

Wibetoe, G. and Langmyhr, F.J. (1984). Spectral interferences and background overcompensation in Zeeman-corrected atomic absorption spectrometry. Part 1. The effect of iron on 30 elements and 49 element lines. *Anal. Chim. Acta*, **165**: 87–96.

Wibetoe, G. and Langmyhr, F.J. (1985). Spectral interferences and background overcompensation in inverse Zeeman-corrected atomic absorption spectrometry. Part 2. The effect of cobalt, manganese and nickel on 30 elements and 53 element lines. *Anal. Chim. Acta*, **176**: 33–40.

Wibetoe, G. and Langmyhr, F.J. (1986). Spectral interferences and background over-compensation in inverse Zeeman-corrected atomic absorption spectrometry. Part 3. A study of eighteen cases of spectral interferences. *Anal. Chim. Acta,* **186**: 155–162.

Wibetoe, G. and Langmyhr, F.J. (1987). Interferences in inverse Zeeman-corrected atomic absorption spectrometry caused by Zeeman spliting of molecules. *Anal. Chim. Acta,* **198**: 81–86.

Yudelevich, I.G.; Katskov, D.A.; Papina, T.S. and Dittrich, K. (1989). Interference caused by indium in the atomization of Ag, Bi, Cd, Sn, and Tl in ETA-AAS. *Talanta,* **36**: 657–664.

Zakharov, Yu.A.; Gil'mutdinov, A.Kh. and Bokorina, O.B. (2005). Electrothermal atomization of a substance with fractional condensation of the element being determined on a probe. *J. Appl. Spectrosc.,* **72**: 132–137.

Zong, Y.Y.; Parsons, P.J. and Slavin, W. (1994). Background overcorrection problems for lead in the presence of phosphate with various metals in Zeeman graphite furnace atomic absorption spectrometry. *Spectrochim. Acta, Part B,* **49**: 1667–1680.

Chapter 6

Chemical Modifiers

6.1. Introduction

In ETAAS, the drying, pyrolysis and atomization stages are combined in order to optimize the analytical procedure. The purpose of the pyrolysis stage is to unify the behavior of the different analyte species in the atomizer. To achieve this purpose, various substances, termed chemical modifiers, are in common use [Tsalev and Slaveykova (1992)].

Chemical modifiers can be added to the sample and/or to the standards so as to eliminate interferences and especially to avoid the thermal volatilization, and consequently loss of analytes during the pyrolysis stage. The last aim of the chemical modification process is to achieve the greatest possible sensitivity. Through the chemical modification process analysis is facilitated, since there is *in situ* alteration of the thermochemical behavior of the analyte, the matrix components and/or the surface of the atomizer during the thermal cycle to which the sample is subjected, which can also affect gaseous components. The mechanism for this modification is complex and on occasions, very varied.

Investigations carried out into the chemical modification process can be categorized into two groups: *optimization* and *fundamental studies*. Practical work on optimization is directed towards discovering the most appropriate amount and composition for the chemical modifier and the optimum experimental conditions (especially the temperature program) for achieving complete analyte atomization. In contrast, the principal interest of fundamental studies is concentrated on getting to know the physical and chemical transformations undergone by both the analyte and the modifier in the atomizer during the heating cycle. In other words, they attempt to identify the intermediate and final species, their oxidation states, chemicals and crystallographic forms, together with their time and spatial distribution on the graphite surface and in the gas phase.

Any study of the chemical modification process implies an in-depth acquaintance with the behavior of the chemical modifiers within the graphite tube. Nevertheless, so as to achieve the maximum efficiency in the modification process, it is advisable also to be aware of the behavior of the analytes. Among the most important thermochemical processes undergone by analytes in a graphite furnace, reduction of their oxides and transformation into carbides are very common. Hence, attempts have been made to regroup analytes on the basis of these two factors, although attention has also been paid to other characteristics, such as: electronic configuration, bond energies, melting points, Pearson's ionic classification, acidic or base character of oxides, volatility, and the like. Consideration of their volatility would allow analytes to be classified into the three following general groups.

(i) Very volatile elements (P, Ga, In, Ge, Sn, As, Sb, Se, Te), which exhibit very high electronegativity and oxygen affinity, have non-metallic properties and their oxides show acidic character. They can be stabilized at very high temperatures by modifiers such as oxides and noble metals.

(ii) Elements showing very pronounced metallic properties (Zn, Cd, Ag, Au, Tl, Pb, Bi). They can be stabilized at low temperatures by modifiers in the oxides group, and at high temperatures by modifiers of the noble metal type.

(iii) Analytes of medium volatilization (Cu, Mn), evidencing an intermediate behavior.

6.2. Types of Chemical Modifiers

As happens with analytes, chemical modifiers may also be grouped as a function of different criteria (Fig. 6.1) [Volynsky (2003)], mainly: chemical nature, constitution, and physical state.

6.2.1. Chemical Nature

This criterion allows the chemical modifiers to be split into the following groups.

- *Inorganic salts*: These are the most popular and preferred chemical modifiers.
  (a) Cations: Ni(II), Pd(II), Ag(I), Mg(II), NH_4^+, etc.
  (b) Anions: NO_3^-, PO_4^{3-}, HPO_4^{2-}, $H_2PO_4^-$, MoO_4^{2-}, VO_3^-, WO_4^{2-}, HSO_4^-, F^- (NH_4F, NaF), Cl^- (NH_4Cl, NaCl), Br^- (NH_4Br).
- *Organic compounds*: CHF_3, CCl_2F_2, among others.
- *Organometals*: Cu(II), Mg(II) cyclohexanobutyrate, La(III) acetylacetonate, Ni(II) sulphonate, Pd(II) acetylacetonate, Mg(II) acetylacetonate.
- *Acids*: HCl, HNO_3, H_3PO_4.

(A)

Mg									
Ca	Sc	Ti	V	Cr	Mn	*Fe*	*Co*	*Ni*	*Cu*
Sr	Y	*Zr*	Nb	*Mo*		*Ru*	*Rh*	*Pd*	Ag
Ba	La	Hf	Ta	*W*			*Ir*	*Pt*	Au
	Ce								

(B)

Mg									
Ca	Sc	*Ti*	*V*	*Cr*	*Mn*	*Fe*	*Co*	*Ni*	*Cu*
Sr	Y	Zr	*Nb*	*Mo*		Ru	Rh	Pd	Ag
Ba	La	Hf	*Ta*	W		Ir	Pt	Au	
	Ce								

(C)

Mg									Al
Ca	Sc	Ti	V	Cr	Mn		*Ni*	*Cu*	
Sr	Y	Zr		Mo		*Rh*	*Pd*	*Ag*	
Ba	La			W		*Ir*	*Pt*	*Au*	
	Ce								

Fig. 6.1. Chemical modifiers grouped on the basis of three criteria: (A) Discriminant analysis using 3 fundamental parameters: ionic radius, ionization potential and electronegativity. (B) Discriminant analysis using 6 fundamental parameters: ionic and atomic radius, ionization potential, electronegativity, and melting points of the element and the oxide. (C) Chemical modifiers divided into three groups: (a) oxides/carbides (bold-face type), (b) oxides → carbides (italic), (c) metals (bold and italic).

- *Organic acids, alcohols*: (Tensioactive substances, quelating agents, solvents, antifoam agents, etc.): ethylenediaminotetraacetic acid (EDTA), thiourea (this is the most frequently used chemical modifier with metal atomizers, mainly Mo atomizer [Nóbrega *et al.* (2004)]), cysteine, ascorbic acid, oxalic acid, tartaric acid, methanol, ethanol, Triton X-100, and the like.

- *Bases*: NaOH, KOH, NH_4OH, tetra-alkyl-ammonium hydroxide (TAAH), $Ca(OH)_2$, $Ba(OH)_2$.

- *Oxidants*:   Air, O_2, H_2O_2, NO_3^-, MnO_4^-, $Cr_2O_7^{2-}$.
- *Reducing agents*: Ascorbic acid, H_2, CO, NH_2OH, HCl, $(NH_2)_2$, H_2SO_4.

6.2.2. Constitution

The efficiency of a chemical modifier depends in large measure, not only on the chemical identity of its "active ingredient(s)," but also on an adequate mixing of its components so as to ensure maximum and many side-effects. Furthermore, it should not be forgotten that a part is played by the reaction medium, by other *chemicals*, by *humectant or dispersant additives, catalyzers*, and such like.

In reality, when it comes to the analysis of complex samples (urine, blood, sea water, slurries, re-extracts, solid samples, etc.) it is preferable to use *compounds* or *mixed* chemical modifiers rather than a single component. *Mixes* refer to two-modifier versions, while *compounds* or better *complexes* involve several (more than two) modifiers together.

6.2.3. Physical State

If the physical state is considered as a criterion, the following types of chemical modifiers may be distinguished.

- *Gases*. A difference should be made between inert gases (Ar, N_2, He) and active gases or gaseous chemical modifiers [O_2, H_2, CO, air, CH_4, Cl_2, CHF_3] [Ohta and Mizuno (1989)]. The most appropriate inert purge gas is argon. Nitrogen cannot really be considered as an inert gas, since it reacts with carbon at high temperatures, forming cyanogen $(CN)_2$.
- *Liquids (solutions)*. These are very useful for *in situ* utilization or premixing with liquid samples. These are the most frequently used chemical modifiers.
- *Solids*: Solid chemical modifiers include granulated metallic Ni used to stabilize Se up to temperatures near 2000°C; solid graphite utilized in the shape of ribbons (graphite-cloth ribbon) and graphite powder also utilized to determine Se; refractory oxides (WO_3,

Al_2O_3); tungsten surfaces coated with metal oxides (Y_2O_3, ZrO_2); and graphite surfaces pretreated with oxygen, metal carbides or noble metals (Pd, Rh, Ru, Pt, Ir) [Silva *et al.* (2002)]. The application of a thin layer of these metals to the internal surface of pyrolytic graphite coated graphite tubes can be achieved by injecting metal solutions or sputtering from a solid electrode (acting as a cathode) situated in the graphite tube center [Rademeyer *et al.* (1995)]. These graphite surfaces normally show lower efficiency in thermal stabilization than modifiers in solution [Ortner *et al.* (2002)]. However, the efficiency of this thermochemical modifier based on Ir appears to be **permanent** during the whole lifetime of the graphite tube [Rademeyer *et al.* (1995)]. Moreover, by electrodeposition of noble metals (Pd, Ir and Rh) on the surface of the graphite tube, metals penetrate into the pyrolytic graphite structure, and pre-atomization losses of volatile elements (As, Se) are reduced with a decreased formation of oxides and carbides [Bulska *et al.* (2001)]. Additionally, a pretreated tube offers a long life in analytical use [Bulska and Jedral (1995)]. In general, the permanent modifier concept shows other advantages (simpler and faster heating programs, decreased detection limits, increased tube lifetime, improvement in hydride trapping) against the solution-injection procedure. Obviously, some potential drawbacks are also present (multiple peaks, overstabilization of some analytes, lower maximum applicable temperatures).

6.3. Action Mechanisms of the Chemical Modifiers

The action mechanism of the chemical modifiers may be the result of some of the following processes: (a) formation of strong chemical bonds between the analyte and modifier; (b) trapping of these species within the metal (analyte); and (c) catalytic effect of many reduction processes. Physical processes may also play a key role in the modifying action of metals such as Pd. That is, not only chemical reactions in the solid phase, but also occlusion and sorption of the analyte within the modifier are processes of great importance [Dočekalová *et al.* (1991); Johannessen *et al.* (1993)].

The following general conclusions about the use of the chemical modifiers can be drawn.

- The chemical modifiers are always used in concentrations considerably higher than those of the analytes.
- The thermal stabilization effect depends upon the modifier concentration.
- The majority of the chemical modifiers are compounds that form refractory chemical species (oxides, carbides, alloys, and similar) within the graphite tube.
- Normally, chemical modifiers mixtures provide a more effective thermal stabilization than their individual components do when applied separately in the same concentration.
- In many cases, the chemical modifiers trigger changes in the form and position of the atomic absorption signals, usually increasing the appearance temperature of the analyte.
- The rate constant for the analyte vaporization process in the presence of chemical modifiers is almost always considerably lower than in the case of the analyte alone.
- Quite often, those chemical modifiers showing similar properties and applied in identical concentrations (in weight terms), yield the outcome that modifiers with lower atomic mass are the more effective in the thermal stabilization of the volatile analytes.

The next sections will cover the general mechanisms of action of the chemical modifiers, grouping them into the following two categories: inorganic chemical modifiers and organic chemical modifiers.

6.3.1. Inorganic Chemical Modifiers

The effect of the inorganic chemical modifiers on the analyte behavior is usually due to the formation of thermally stable *homogeneous alloys*, among which it is possible to distinguish *solid solutions* (substitutional and interstitial, with the consequent formation of *isomorphic mixtures*) and *compounds* (*intermetallic* and *non-stoichiometric*, whether *electronic* or *interstitial*), although there may also be *catalytic processes* [Aller 2003], and occasionally *physical mechanisms*. The formation of these compounds tends to overestimate the role of reducing agents, $C(s)$ and $CO(g)$, since these are supposed to act both on the analyte and on the modifier.

Solubility in the solid state or, what comes to the same thing, the formation of mixed crystals, requires the presence of a crystalline lattice between particles of differing components. For two substances to be able to form mixed crystals, they need to be *isomorphic*, i.e. they need to have identical crystalline structures, nodal distances and attractive forces (creating and maintaining the lattice) of an identical nature. Nonetheless, there are substances with different crystallographic forms, capable of forming mixed crystals at a given fixed composition. Similarly, if the substances forming mixed crystals are metals, they may give rise to intermetallic compounds. Intermetallic compounds and mixed crystals are considered to be chemical compounds (stoichiometric or non-stoichiometric).

When the content of the solution constituents varies, giving rise to pure solid phases, their activity coefficient is unitary. However, isomorphic substitution in the crystalline lattice by an extraneous constituent is an important factor through which the activity of the solid phase may decrease.

Solid solutions

For a solid solution to be formed, the metals must be soluble in each other and form what are termed mixed crystals, those which exist in the solid (homogeneous) phase. Solid solutions may be of three types: substitutional, interstitial and omission.

In *substitutional solid solutions*, an element or ion (cation or anion) is replaced by other. Both ions not necessarily need to show the same charge state, requiring in this case a third ion to compensate the charge, and the process is named coupled substitution.

Interstitial solid solutions are produced when ions are localized into the interstitial framework of the crystal.

Omission solid solution is possible when an ion with a large charge replaces two or more ions, thus compensating for their charge.

Application of the solid solution model to the analyte vaporization process in the presence of a chemical modifier is possible if the following basic premises are accepted.

(i) The chemical modifier forms a thin solid layer (in comparison with the crystal lattice parameter) on the graphite surface.

(ii) The analyte species are randomly distributed into the solid crystal lattice, which is formed by the modifier species. In these cases, the analyte-modifier solid solution is extremely diluted and the predominant factor for its stabilization is the entropy of the mixture.

(iii) The analyte elimination process from the gaseous phase is rapid, mainly through convection.

(iv) The analyte vaporization takes place under close isothermal conditions.

(v) The analyte losses are controlled by quasi-static processes.

Under these considerations, the chemical modifier concentration has a decisive influence on the analyte loss rate and on the thermal stabilization. When solid solutions are formed, the relationship between the analyte loss rate and its concentration is the same as that existing between the analyte's vapor pressure and its concentration in a mixture.

The regular solution model is widely used in the study of solid and liquid solutions. The underlying suppositions in this model are that: (i) the species forming the solution in a common crystalline lattice are completely randomly distributed; (ii) the entropy of the mixture is entirely configurational; and (iii) only the interaction between the pairs of particles forming the solution are considered. Equation (6.1) represents a definition of the configurational entropy of a mixture, ΔS_{MIX}

$$\Delta S_{MIX} = -R(X_A \, Ln \, X_A + X_M \, Ln \, X_M), \qquad (6.1)$$

where R is the universal constant for gases ($R = 8.314$ J mol^{-1} K^{-1}), while X_A and X_M are the molar fractions of the analyte and the modifier, respectively [Mandjukov et al. (1995)].

Unlimited mutual solubility in solid and liquid phases indicates that solutions are stabilized by the energy and entropy of the mixture. In such cases, negative variations from Raoult's law can be expected, as well as greater decreases in the partial vapor pressure of the analyte and stronger effects of thermal stabilization, even in the presence of only a low concentration of modifier. In contrast, when the phase diagram for the analyte-modifier system shows an instance of limited solubility, positive variations from Raoult's law are to be expected. It should be kept in mind that approximation to a solid solution does not exclude the formation of intermetallic compounds.

The chemical modifier concentrations normally utilized correspond to values for molar fractions that are close to unity. In these cases, the analyte-modifier solid solution is stabilized by the predominant effect of entropy of the mixture so that the formation of intermetallic compounds is not very likely. Nonetheless, when the molar fractions of the chemical modifier have a value close to those in which intermetallic compounds exist, or may exist, the phase diagrams can be used directly.

The solid solution model is of relevance only for those systems forming refractory species in the graphite furnace. Thermal stabilization of volatile analytes through the use of chemical modifiers may be seen as due to a kinetic effect. The analyte thermal stabilization can be explained by a drop in the analyte losses under certain pyrolysis conditions (temperature and time) [Mandjukov et al. (1995)].

According to the solid solution model there should be a strong influence of the chemical modifier concentration over the analyte vaporization rate. From the kinetic point of view, vaporization processes may be described as first order reactions. However, the values obtained for the rate constants and activation energies depend on the modifier concentration. In such cases, the Arrhenius plot is not normally applicable to investigation of the analyte vaporization in the presence of the modifier, and the influence of the chemical modifier concentration should be taken into account.

For large amounts of analyte, the atomic absorption signal returns very slowly to the baseline. This has been attributed to a secondary adsorption-desorption process at the cooler ends of the tube [Frech *et al.* (1992)], but this depends largely on the excess of modifier. Hence, application of the regular solution theory to the modification process allows us to conclude that the thermal stabilization process cannot be adequately described on the basis of a simple chemical reaction [Mandjukov *et al.* (1992)]. The limiting stage of the process leading to atomization is in many cases the diffusion of the analyte through the drops formed at its surface.

Isomorphic mixtures (criteria for isomorphism)

Isomorphism refers to the possibility of replacing an atom of one element with an atom of some other element in the solid phase without modification of the crystalline structure. The criteria for isomorphism between two elements are generally the following (Fig. 6.2) [Aller (2003)].

- *Goldschmidt criterion.* This criterion applies essentially to ionic crystals. Isomorphic mixtures can be formed over a wide range of concentrations, at temperatures well below melting point when the difference between the radius of the two ions, (R_a, R_b), does not exceed 15%.

$$(R_a - R_b)/R_a < \pm 0.15 \quad \text{(for } R_a < R_b\text{)}. \tag{6.2}$$

Li

Na Mg   *Al* **Si** <u>P</u>

Ca Sc *Ti* **V** *Cr* **Mn Fe Co Ni** <u>*Cu*</u> Zn Ga <u>*Ge*</u> <u>*As*</u> <u>Se</u>

Zr **Nb** *Mo*   **Ru Rh** Pd <u>*Ag*</u> **Cd** In *Sn Sb* <u>*Te*</u>

Ta *W* **Re Os Ir** Pt <u>*Au*</u> **Hg** Tl Pb **Bi**

Fig. 6.2. Possibilities for isomorphic replacement of the universal chemical modifier (Pd) with analytes in accordance with several criteria: (i) Vlasov (bold-face); (ii) Hume-Rothery (italics), and (iii) Darken–Gurry (underlined).

- *Hume-Rothery criterion.* This criterion is appropriate for covalent and metallic crystals, with Eq. (6.2) just quoted for the Goldschmidt criterion requiring to be fulfilled, but with R_a and R_b being the atomic radii. The Hume–Rothery criterion also establishes additional rules to achieve maximum stability: (i) The crystal structures of the two elements must be identical; (ii) valence of the solute and solvent atoms should be the same, but in general the solute valence should be higher than that of the solvent; and (iii) electronegativity of the solute and solvent should be similar to form a substitutional solid solution, because by contrary they form an intermetallic compound.
- *Fersman criterion.* This criterion sets a further important requirement, for there to be similarity between the polarization properties of the two components; in other words, similar electronegativity ($\hat{E}$),

$$\hat{E}_a - \hat{E}_b < \pm 0.4. \qquad (6.3)$$

- *Vlasov criterion* (Table 6.1). This criterion establishes that isomorphism between two elements is possible when either of the following two requirements are met.

Criterion (a)

$$(Z_a - Z_b)/Z_a < \pm 0.15 \quad \text{(for } Z_a > Z_b) \qquad (6.4)$$

or

$$(nZ_a - Z_b)/Z_a < 0.15 \quad \text{(for } Z_a < Z_b). \qquad (6.4')$$

Criterion (b)

$$(A_a - A_b)/A_a < 0.15 \quad \text{(for } A_a > A_b) \qquad (6.5)$$

or

$$(nA_a - A_b)/A_a < \pm 0.15 \quad \text{(for } A_a < A_b) \qquad (6.5')$$

where Z is the atomic number for elements a or b; A is the atomic mass for elements a or b, n is a whole number (1, 2 or 4); a is always the matrix (in this case it would be the modifier) and b is always an atom involved in replacement (here it would be the analyte).

Table 6.1.   Isomorfic Posibilities Between Several Analytes and Modifiers According to Vlasov's Criterion. (x) Good or Perfect Isomorfism, and (o) Moderate Isomorfism

Analyte	Cd	Hg	Ir	La	Mg	Mn	Mo	Ni	P	Pb	Pd	Pt	Rh	V	W	Y	Zr
Ag	x	x	x		o	x	x	o	o	o	x	x	x	x	x		x
Al	o			o	x	o	o	o	o		o		o	x			x
As			x						o					o	o		
Au	x	o	o			o	x			x	x	x	o	o	x	x	x
Ba	x			x	o	x		x	o	x							
Bi									x					o	o		x
Cd	—	o	x		o	o	x	x	o	o	x	x	x	x			
Co	x	o		x	x	x		x	x	o	x		x	x	x		
Cr	x	o	o		x	o	x	x	o	o	x	o	x	x	o	x	x
Cu	x	x		x	o	o		x	x	o			x				
Fe	x	o	x	o	x	x		x	x	o	x	x	x	x	o		o
Hg	o	—	x			o	x			o	x	x	x	o	x		
In	x	o				o	x	x	x	o	x	x	x	x	x		
Mn	x	o	o	o	x	—		x	x	o	x	o	x	o		o	o
Mo	x	x	x		o	x	—			x	x	x	x	x	x	x	x
Nb		x	x		o		o			x	x	x	x	x	x	x	x
Ni	x	o	o	x	x	x		—	x		o	o	o	x			
Pb	x	o	x		o	o	x	o		—	x	x	x	o	x	o	x
Sb	x			x	o	x		o	o		x			x			
Se			x	x					x			x			o	x	x
Sn	x			x	o	x		x	o		x			x	o	o	
Ti	x	o	o		x	o	x			o	x	o	x	x	o	x	x
Tl	x	x	x		o	x	o			x	x	x	x	o	x	x	x
V	x	o	o		x	o	x			o	x	o	x	—	o	x	x
W		x	x				o			x		x	x	o	—	x	x
Y		x	x	o		o	x			o			x		x	—	x
Zn	o	o		x	o	x		o	x	o							
Zr		x	x	o		o	x			x	o	x	x	x	o	x	—

Vlasov gives greater priority to criterion (*a*), as also to those cases in which the value of *n* is the lowest (1 or 2).

- *Darken–Gurry criterion.* This criterion establishes high solubility between two elements if the solute atoms have electronegativity and ionic radii values similar to those of the solvent atoms, i.e., very close localization in the electronegativity *vs.* ionic radii plot (Darken–Gurry map).

Intermetallic compounds

Intermetallic compounds are chemical combinations in which the metal atoms are in a defined stoichiometry. Intermetallic compounds formed between the metal analyte and the chemical modifier usually raise the pyrolysis and atomization temperatures from several hundred to 1000°C. These chemical reactions, which take place within the graphite tube, may be explained as a function of the thermochemical properties of the analyte and the chemical modifiers used. In many cases it is possible to establish some correlation with the thermal processes in metallurgy. Thus, the thermodynamic activity of alloys has an influence over their vapor pressures and melting points and plays a very important role in the analyte atomization. When the chemical modifiers (Pd, Pt and Ni) form alloys (intermetallic compounds) with the analytes (Pb, Tl, Cd, Se), the activity of the analyte in the alloy becomes less than that of the pure analyte or, alternatively, the activity coefficient of the analyte in the alloy decreases to below unity. The vapor pressure of the analyte in the alloy also drops relative to that of the pure analyte. So, there are few losses from vaporization during the pyrolysis stage (and before atomization begins); in these cases, atomization is shifted upwards to higher temperatures.

Non-stoichiometric compounds

Stoichiometric imbalance in a crystal may arise as a consequence of the substitutional, interstitial, and subtractive processes. In the **substitutional process**, certain atoms in the crystalline structure of a compound are replaced by atoms from another component. This sort of substitution is restricted to intermetallic systems in which there is no ionic repulsion of the Coulomb type. In **interstitial incorporation**, the *supernumerary* atoms or ions of a compound take up places in interstitial positions (this is an uncompensated Frenkel defect). Such a substitution may also occur in predominantly covalent compounds (semiconductors of groups III-V), as long as the atomic sizes and electronegativities are similar. In **subtractive incorporation**, a deficit of a component takes place when some of its positions in the crystalline lattice remain unfilled (this is analogous to an uncompensated Schottky defect). This kind of substitution occurs when the cation and anion substructures are *ideally* filled, and furthermore, some interstitial positions are also filled by one of the components. Subtractive incorporation happens when the component in excess shows a complete lattice, while the component in deficit shows a lattice with vacant positions. Such a process brings with it an increase in the crystal density.

Any crystal in contact with the vapor of one of its constituents is potentially a non-stoichiometric compound. If thermodynamic equilibrium were reached, the solid phase composition would depend on the concentration (activity) of this constituent in the vapor phase. In this way, the stoichiometric compound would be merely a particular case for which the number of cations in the solid phase is equal to the number of anions in the same solid phase.

When the gas pressure, X_2, due to the anion in contact with the crystal, MX, rises above the value, P^0, appropriate for attaining the stoichiometric composition, $MX_{1.0}$, the X^- concentration within the crystal must also rise in order to maintain equilibrium. One of the two following cases are possible.

(i) The excess of anions, X^-, within the crystal might occupy interstitial positions, so that an equivalent number of cations, M^+, would be oxidized to M^{2+} in order to neutralize the charge

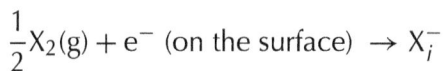

$$\frac{1}{2}X_2(g) + e^- \text{ (on the surface)} \rightarrow X_i^-$$

$$M_1^+ \rightarrow M_1^{2+} + e^-,$$

where (g) refers to the gaseous phase, subscript "*i*" indicates an ion in an interstitial position, and subscript "1" refers to an ion in the correct position for a normal lattice. The sole difference between non-stoichiometric compounds and solid solutions with impurities lies in the fact that in the first case the cation that alters the valence is of the same chemical element as the cation existing in the receptor compound.

(ii) Alternatively, the anions in excess may accumulate on the crystal surface, occupying normal positions in the lattice, and creating cationic vacancies which diffuse within the crystal. Simultaneously, an equivalent number of M^+ ions in the normal positions would be oxidized to M^{2+}.

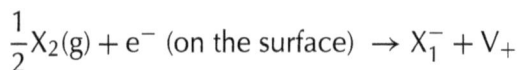

$$\frac{1}{2}X_2(g) + e^- \text{ (on the surface)} \rightarrow X_1^- + V_+$$

$$M_1^+ \rightarrow M_1^{2+} + e^-,$$

where, as previously, subscript "1" indicates that X^-, M^+ and M^{2+} are occupying normal positions in the lattice, and V_+ represents a vacant cationic position. This possibility for the incorporation of an excess of anions (or deficit of cations) is more common than the interstitial incorporation of an excess of anions, mentioned in the previous case (i)). Some examples may be found among certain metallic oxides ($Mn_{1-x}O$, $Fe_{1-x}O$, $Cu_{2-x}O$).

In contrast, when the pressure of the atomic gas, X_2, falls below the level P^0, sufficient for the existence of the stoichiometric compound, then once equilibrium is reached, two outcomes are also possible.

(i) In crystals with a tendency to present Schottky defects, the activity of the anion in the solid phase may be reduced through *vaporization* of a part of component X, leaving some anionic positions vacant and some electrons trapped in the proximity.

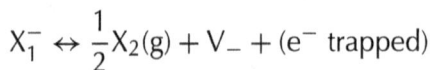

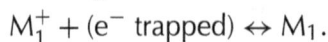

$$X_1^- \leftrightarrow \frac{1}{2}X_2(g) + V_- + (e^- \text{ trapped})$$

$$M_1^+ + (e^- \text{ trapped}) \leftrightarrow M_1.$$

This way of incorporating a deficit of anions is infrequent, although it may occur in certain alkaline halides and some oxides (TiO_{1-x}).

(ii) The other case is more common and occurring when the cationic and anionic structures are complete and the reduced cation, M, migrates to an interstitial position.

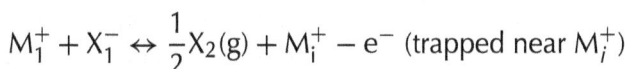

$$M_1^+ + X_1^- \leftrightarrow \frac{1}{2}X_2(g) + M_i^+ - e^- \text{ (trapped near } M_i^+)$$

Some examples of this sort can be found in zinc oxides ($Zn_{1+x}O$).

Deviations from ideal stoichiometry imply a change in the valence of some ions of the crystal, generally cations, having either a deficit or an excess of the metal. A stoichiometric deficit of the metal generally involves the presence of certain cations whose valence is higher than that of the cations in the stoichiometric compound. For this reason, it is normally limited to compounds of the transition metals (plus a few others like Pb), where the variability of the valence is permissible in energy terms. In contrast, an excess of the metal causes a reduction in the valence of some cations or in theory an increase in the valence of some anions, although this last possibility has never been observed. On the other hand, a reduction in the oxidation state of a cation undoubtedly can always occur, in principle even involving the formation of a neutral atom.

The necessary conditions for a compound to be stable over a certain composition range are the following.

- The energy needed to produce defects should not be very great.
- The energy difference between the two oxidation states should be reasonably small.
- If the crystalline lattice remain relatively undistorted, the size of the ions present in the two oxidation states should be similar.

These rules indicate when a deviation from stoichiometry would be probable, as also the sense in which it would take place.

The existence range of a stoichiometric phase shows a limiting value, which is governed by the interaction energy between pairs of similar defects. There is a critical composition whose structure permits a maximum number of defects. When this number is exceeded, the crystal splits into two phases.

The two greatest difficulties inherent in the concept of a compound that is far from the stoichiometric situation are the following: (i) a precise definition of what the crystalline structure is, since when a crystal contains a large number of defects randomly distributed, it becomes an arbitrary matter to say which ions are in normal positions in the lattice and which in interstitial positions (for example, certain chalco-genides of transition metals); and (ii) the enthalpy gained by ordering the defects in a regular distribution may more than compensate for the simultaneous decrease in entropy, particularly at lower temperatures ($\Delta G = \Delta H - T\Delta S$).

For major deviations from stoichiometry, four possibilities can be found.

- Homogeneous random distribution of the defects.
- Sub-micro-heterogeneity within a simple phase.
- Intermediate phases based on the ordering of defects in superstructures.
- Intermediate phases based on displacement structures.

Each of these possibilities is really a limiting situation and real cases frequently show characteristic of more than one type, depending on factors such as temperature, the value of the interaction energies, arrangement considerations, and so forth.

Catalytic processes

Catalytic processes are phenomena in which a relatively small quantity of an external substance (catalyzer) increases the chemical reaction rate without the catalyzer itself being consumed. The presence of a catalyzer permits chemical equilibrium to be reached via stages whose activation energies are lower than those through which the reaction would proceed in the absence of catalyzer. In ETAAS, it is not as important to improve selectivity of the reactions taking place in the atomizer, as to increase the reaction rate at a given temperature. This permits decomposition of highly volatile analyte compounds in the pyrolysis stage, allowing the use of lower atomization temperatures and reducing analyte losses.

Catalyzers may be homogeneous or heterogeneous. The most important heterogeneous catalyzers are metals (Fe, Ni, Pd, Pt, Ag), semiconductor oxides and sulphides (NiO, ZnO, MnO_2, Cr_2O_3, Bi_2O_3, MoO_3, WS_2), insulating oxides (Al_2O_3, SiO_2, MgO), and acids (H_3PO_4, H_2SO_4).

The most striking peculiarity of the catalytic process in ETAAS is the inverse relationship between the catalyzer and the analyte. Although the concentration is the same at the start and end of the process, the chemical reaction rate is proportional to the catalyzer concentration in accordance with the rate law. The catalyzer concentration is usually lower than that of the analytes. However, under typical ETAAS conditions, the chemical modifier amount is several times that of the analyte, even though there is also interaction between the chemical modifier and the matrix, significantly diminishing the amount of modifier remaining available for interaction with the analyte. Nonetheless, if the amount of chemical modifier is large (or if it shows weak interaction with the matrix components), the rate of the catalyzed processes, in which the analyte is involved, may be extremely high.

The platinum group metals were considered as acting as a chemical modifier through various processes [Volynsky (1996)].

- Reduction (or dissociation) of the analyte oxides and other compounds.
- Decomposition of volatile hydrides and organometallic compounds.
- Oxidation of both graphite and carbon monoxide.
- Other catalytic processes, such as: (i) gasification of graphite by water (catalyzer: La compounds); (ii) corrosion of the graphite tubes (catalyzer: Al, Be, Ca, Mg, Sr, Ba and Lanthanides); (iii) graphitization of the vitreous carbon (catalyzer: Fe, Mo and La compounds); (iv) uniform breaking of the co-extracted chelating agent (catalyzer: Cu); (v) atomization at low temperatures, Cd (catalyzer: Na_2EDTA); (vi) generation of hydrogen radicals in the Se hydride atomization (catalyzer: O_2); (vii) appearance of O_2 at 230°C ($2NO_2 \xrightarrow{CuO} 2NO + O_2$); and (viii) formation of refractory compounds, Pb-C (catalyzer: H_2).

Physical mechanisms

Some of the physical mechanisms through which metal chemical modifiers act, are related to adsorption and desorption of hydrogen or an increase in the atomization temperature [Hirokawa *et al.* (1992)]. Pd and Pt adsorb hydrogen and desorb it again at higher temperatures. This capability of Pd has been successfully used to trap hydrides [Hilligsøe *et al.* (1997)]. The hydrogen released can act as a reducer of the analyte together with other compounds, such as CO and C_nH_m, at high temperatures [Terui *et al.* (1991a)]. Physical trapping of the solid analyte species is also possible.

6.3.1.1. *Metals*

The behavior of metals as chemical modifiers may be grouped into three general categories.

(i) Chemical modifiers based on Mg, Ca, Sr, Ba, Sc, Y, La, Ce, Al, which form oxides and carbides. Their refractory oxides (with melting points above 2000°C) persist within the graphite furnace up to temperatures ranging between 1300°C and 2000°C. At higher temperatures, they will form carbides. These substances may react or be hydrolyzed through the action of water or diluted acids [Nowka *et al.* (2000); Sabé *et al.* (1999)].

The predominant effect of the elements of this group on thermal stabilization is the formation of oxide mixtures between the analyte and chemical modifier [Slavin *et al.* (1982)], dispersed inside the refractory matrix of the chemical modifier. One key factor is usually the maximum temperature available for the oxide in contact with the graphite ($MgO > CaO < SrO > BaO > Sc_2O_3 > Y_2O_3 > La_2O_3$).

From an experimental point of view, magnesium nitrate is the best modifier in this group for a number of moderately volatile analytes (V, Cr, Mn, Fe, Co, Ni, and Cu). All these analytes are isomorphous with Mg and can form oxides perfectly mixed with MgO. There is a close relationship between the maximum temperature for thermal pretreatment in the presence of magnesium nitrate as modifier and the appropriate temperature for reduction of the analyte oxide in the graphite tube.

Metal chemical modifiers, like Mg, which are not reduced to their elemental state, not forming alloys with the analyte during the pyrolysis and atomization stages, participate actively in the elimination of some interfering elements like Cl^-. These chemical modifiers suppress the vaporization of metal halides (chlorides) with higher vapor pressures (the effect of supportive distillation). Mg reacts easily with Cl at high temperatures (thanks to its considerable formation free energy at high temperatures), forming $MgCl_2$, which has a high vaporization temperature. Besides, unlike what happens with alkaline metals, $MgCl_2$ decomposes into MgO and/or Mg_2OCl_2 in the presence of H_2O at temperatures above 150°C in accordance with the following equation

$$2MgCl_2 + H_2O \leftrightarrow Mg_2OCl_2 + 2HCl$$

or alternatively

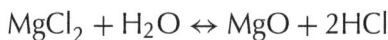

$$MgCl_2 + H_2O \leftrightarrow MgO + 2HCl$$

although in reality the two following reactions would take place

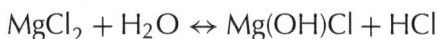

$$MgCl_2 + H_2O \leftrightarrow Mg(OH)Cl + HCl$$
$$Mg(OH)Cl \leftrightarrow MgO + HCl.$$

Thus, Cl is eliminated during the pyrolysis stage when $Mg(NO_3)_2$ is added as a chemical modifier [Ouishi *et al.* (1994)]. These effects are similar to those occurring with compounds with lower vapor pressures (such as sulphates and phosphates) (SO_4^{2-} or PO_4^{3-}) as modifiers, which replace compounds with higher vapor pressures (such as chlorides).

Lanthanum, used as a chemical modifier, forms some carbide with graphite, but is not reduced thermochemically to metal. Thus, the analyte is occluded in the La carbide and/or oxide matrix, and atomized after the matrix components. This behavior occurs essentially when the surface of the graphite tube is coated with La carbide.

Comparisons of various chemical modifiers often show a very clear relationship between the maximum temperature for pretreatment and the atomic mass of the chemical modifier (Sc > Y > La > Ce; Mg > Ca > Sr, when applied in the same concentrations). This tendency is not very strong, especially in the case of modifiers with similar atomic masses but different chemical properties. This trend can be affected by differences in the interaction energy between analyte species and the modifiers.

(ii) There are other chemical modifiers (Ti, Zr, Hf, V, Nb, Ta, Cr, Mo, W, Mn, Th) whose oxides are expected to be transformed into carbides at temperatures rather lower than those relating to the chemical modifiers mentioned above. Their carbides are resistant to the action of water and diluted acids, and are thus less corrosive for the graphite tube (except for Cr and Mn carbides). Due to the slow kinetics of the heterogeneous reactions, some oxygen will persist with these modifiers in the form of residual metal-oxygen, carbon-oxygen, or carbon-oxygen-metal bonds (oxicarbides of Zr, W, Ti, V, and so on) [Tsalev *et al.* (1990a,b); Nowka *et al.* (2000); Castro *et al.* (2002); Castro *et al.* (2005); Gong *et al.* (1998); Acar *et al.* (2000); Lima *et al.* (1999)]. However, for Zr and Th modifiers, the amount of oxygen is higher than residual [Castro *et al.* (2005, 2007); Muñoz and Aller (2006)], and the formation of metal carbides seems to be not conclusive even at high temperatures (Fig. 6.3). Nonetheless, the majority of these modifiers are utilized to produce carbide coatings on the graphite tube or

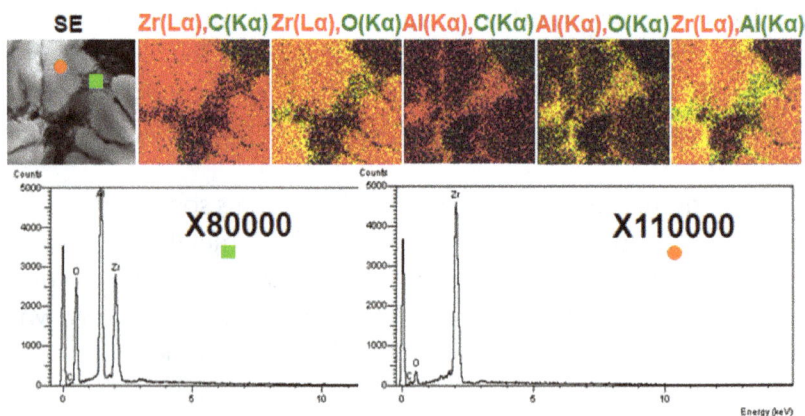

Fig. 6.3. ED X-ray spectra with the corresponding SEM mappings and colored superimposed two-element distribution mappings, taken from a non-pyrolytic platform containing 100 mg each of aluminium and zirconium salts and heated to 1800°C. The two ED X-ray spectra were taken from the two different marked locations. [Castro *et al.* (2005)].

platform [Brueggemeyer and Fricke (1986); Vieira *et al.* (2004); Flores *et al.* (2004)].

Vanadium (V), as NH_4VO_3 or in an aqueous solution as V_2O_5 (pH > 9) alone or mixed with Pd, has proved to have considerable efficiency in the thermal stabilization of many analytes [Tsalev *et al.* (1990a); García-Olalla and Aller (1991); Castro *et al.* (2002)]. The presence of mixtures of oxides dispersed within the refractory modifier oxide may persist up to 1000–1200°C. The possibility for an isomorphic replacement in structural units of these modifiers (MoO_4^{2-}, WO_4^{2-}, or VO_4^{3-}) by analyte isomorphic species, such as AsO_4^{3-}, PO_4^{3-}, SbO_4^{3-}, SeO_4^{2-} and the like, cannot be excluded.

Some analytes of high to medium volatility (Al, As, Cu, Co, Fe, Ge, P, Se and Si) may also be trapped within tetrahedral vacancies that exist in the globular structures of Mo(VI) or W(VI) heteropolyacids. Such structures ($PMo_{12}O_{40}^{3-}$) could be constituted of MoO_6 octohedrons, while the elements mentioned could be included as substitutes for phosphorus in the heteropolyanion.

The thermal stability of mixed oxides appears to be greater for those analyte-modifier pairs having the highest ionic character level (the greatest difference in electronegativity). In reality, for each modifier in this group and within the same group in the periodic system, the greater the acidity of the analyte oxide, the greater the maximum available temperature for the pyrolysis stage.

(iii) There is a further, third, group of chemical modifiers (Ni, Cu, Rh, Pd, Ag, Ir, Pt, Au) which are reduced to their elemental state in the thermal pretreatment stage (generally well below 1000°C, except for Cu) [Dèdina *et al.* (1987); Gong *et al.* (1998); Acar (2001); Sabé *et al.* (1999); Lima *et al.* (1999); Wojciechowski *et al.* (2001); Uggerud *et al.* (1999); Volynsky *et al.* (2000, 2001); Volynsky (2000); Tsalev *et al.* (2000, 2001); Volynsky and Krivan (1996); Volynsky and Wennrich (2001); Bulska *et al.* (2001); Muñoz and Aller (2006)]. The great efficiency of these modifiers might be due to the formation of solid solutions and/or any other compound involving the analyte trapped within the reduced modifier. Hence, the rapid reduction of the modifier is essential in this group, so as to transform the modifier (and perhaps

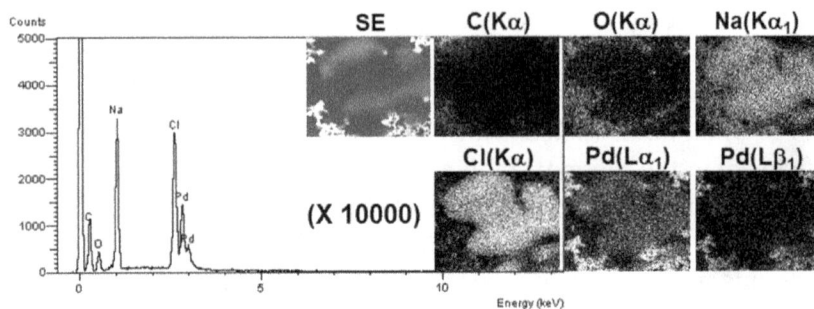

Fig. 6.4. Secondary electron (SE) image and X-rays maps from the same area for carbon, oxygen, sodium, chlorine and palladium. The EDS X-ray spectrum was derived from a graphite platform used to heat a Cd solution (10 ng mL^{-1}) together with sodium chloride (4000 mg L^{-1}), Pd nitrate (4000 mg L^{-1}) and protons (4000 mg L^{-1}) (as nitric acid) according to the temperature program, but up to a temperature value of the platform surface of 900°C (magnification, 10000×) [Feo *et al.* (2003)].

also the analyte) into a suitable form: reactive, elemental, and highly dispersed. This reduction also triggers separation of the metal from its counter-ion, many times chloride, and thus facilitating complete recovery of the analytes (such as Bi, Hg, Ga, In, and Tl). Many of these chemical modifiers have also been used under the permanent modifier concept [Slaveykova *et al.* (1997)].

It has been stated that Pd acts as a good chemical modifier in nitric solutions [Matsumoto (1993)], because in the presence of a great excess of Cl$^-$ ions, Pd is transformed into an unstable chloride, this being the cause of some losses of Pd during the pyrolysis stage. However, the addition of Mg nitrate counteracts these losses of Pd, solving the interference problem from Cl$^-$ ions. Nonetheless, the combined action of the chemical modifier constituted by the Pd-Mg mixture has been questioned [Xiao-Quan and Bei (1995)]. Despite the drawbacks mentioned above, some chemical modifiers from this group (Ru, Rh, Pd, Ir) have been introduced as chlorides, because it is said that they are reduced to the elemental state in the graphite atomizer between 400°C and 800°C. However, this is more difficult to be true for palladium chloride (Fig. 6.4) [Castro *et al.* (2003); Feo *et al.* (2003)].

Colloidal Pd has also shown effectivity in the presence of chlorides [Volynsky and Krivan (1996)]; Volynsky (1997)]; Volynsky *et al.*

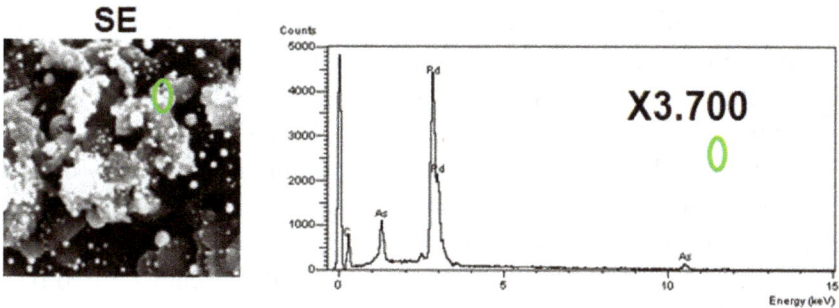

Fig. 6.5. ED X-ray spectra of pyrolytic graphite platform containing 100 μg of palladium chloride and 500 ng of As, after heating to 1200°C [Castro *et al.* (2007)].

(2001)]. Pd must be in its elemental state to be effective as a modifier. For this reason, the presence of a reducer such as ascorbic acid would favor the formation of analyte modifier alloys. When Pd is used together with ascorbic acid, the co-reduction of analytes, such as Se or As occurs during the drying and pyrolysis stage, and the alloying process ends during the pyrolysis stage and before atomization. The intensities of the atomic absorption signals grow, because Se or As alloyed with the modifier are atomized at higher temperatures and with faster kinetics [Terui *et al.* (1991a); Belarra *et al.* (1999)]. Nevertheless, in the majority of cases reduction of Pd happens rapidly during the pyrolysis stage without the need to add a reducing agent [Castro *et al.* (2007)] (Fig. 6.5). It is necessary to keep in mind that the behavior of the modifier changes according to the analysis conditions (fundamentally the pyrolysis temperature) and other elements present.

The action of Pd alone is similar to that of Pd-ascorbic acid and Pd-Mg mixtures in terms of the pyrolysis temperature attainable, the characteristic mass values, permissible interval for interferences, capability to improve atomization profiles, and relative standard deviations [Qiao and Jackson (1991)]. Nonetheless, the background absorption from modifiers based on Pd and the Pd-ascorbic acid mixture is much smaller than that from those based on a Pd-Mg mixture. The necessity to add ascorbic acid to Pd depends on the type of matrix, while the addition of $Mg(NO_3)_2$ is not essential [Xiao-Quan and Bei (1995); Bermejo-Barrera *et al.* (1998)], although this compound is sometimes utilized to facilitate the dry pyrolysis stage with biological samples (in general with organic matrices) [Kowalewska *et al.* (1999)]. These chemical modifiers are effective when added in relatively small quantities ($\cong \mu$g), producing considerable positive shifts in the atomic absorption peaks (maximum absorbance) of the analyte.

The effectiveness of metals used as chemical modifiers depends on the starting compound. Thus, Pd used as $(NH_4)_2PdCl_6$-$(NH_4)_3RhCl_6$ permits higher pyrolysis temperatures (920°C) than when a mixture of $PdCl_2$ and ascorbic acid is used in the ETAAS determination of Hg [De-Qiang *et al.* (1998)]. $PdCl_2$ decreases the temperature for reduction of PbO and Ga_2O_3 with graphite, whereas $NiCl_2$ catalyzes only the reduction of Ga_2O_3 [Volynsky *et al.* (1991)].

The stabilizing effect of Pd and other modifiers in its group (Pt, Ni) is more pronounced for relatively volatile analytes, such as Se, As, Hg, Te and Cd. These analytes are stabilized up to higher pyrolysis temperatures, thanks to the formation of intermetallic compounds or sinterization of an alloy with the analyte over a wide concentration range [Peng-yuan *et al.* (1992)] of differing stoichiometric characteristics. This is due to the higher melting points of these alloys as compared with the analytes and the pyrolysis temperatures, although just below the atomization temperatures. Nonetheless, it should be kept in mind that there may be eutectic phases (Ni-As, Se; Pd-Se, Te; Pt-As, Sb, Se; Fe-As) and/or intermetallic compounds (Pd-As, Se) with low melting points, since these compounds have the potential to vaporize the analyte during the pyrolysis stage [Hirokawa *et al.* (1992)], which varies around 1000°C. The presence of Pd(II) and Pd(0) gives rise to compounds like Pd_4Se and $Pd_{17}Se_{15}$, respectively (or compounds involving oxygen, $Se_xPd_yO_z$), according to studies carried out by XANES and EXAFS [Lamoureux *et al.* (1998)]. Other X-ray measurements of the pyrolyzed residues indicate the presence of many other similar compounds such as PdSn, Pd_3Sn, Pd_2Sn, Pd_3Sn_2, $PdSn_3$, Ge_9Pd_{23}, $GePd_2$, $MgGeO_3$, Pd_3Pb_2, Pd_3Pb [Gong *et al.* (1993); Li *et al.* (1996); Castro *et al.* (2003)] before the atomization stage, when a mixture of Pd and bovine serum is used as a modifier, but not if Pd is used alone [Gong *et al.* (1993)]. Pd alone stabilizes the behavior of Pb in the graphite atomizer owing to the formation of labile refractory Pd-Pb-C species [Dabeka (1992)].

The behavior of Ni as a chemical modifier in the atomization of Se [Hernández Carballo *et al.* (1999)], may be explained from the point of view of metallography using phase diagrams. In the presence of Ni, the atomic absorption signal for Se appears at temperatures of 1000–1300°C, corresponding to the liquid phase in the interval of existence of the eutectic Ni-NiSe. In this eutectic, a part of the mixture of Ni and Se begins to melt at around 1000°C. However, Se has a high vapour pressure and vaporizes, but Ni does not. In this way, the proportion of Ni grows and the melting point of the alloy climbs during the pyrolysis stage; at this point, the small quantities of residual Se present in the Ni alloy are atomized. At the pyrolysis temperature of around 1200°C, which is close to the melting point of Ni, Ni melts rapidly and becomes alloyed with Se. As a consequence, Se is retained in the graphite tube up to the atomization temperature without suffering any loss [Terui *et al.* (1991a)]. However, absorbance decreases at the temperature of 1500°C, higher than the melting point of Ni.

If the chemical modifier alloyed with the analyte shows melting points higher than the pyrolysis temperatures, but lower than the atomization temperature, the analyte atomization would be not affected by the presence of interfering species or chemical modifiers. If the alloys formed with chemical modifiers have activity coefficients greater than unity, no influence of these modifiers is observed. Thus, Mn forms intermetallic compounds with Pd but, thermodynamically speaking, Mn cannot be completely reduced before the atomization process. The melting points of Mn-Pd alloys are slightly higher than the pyrolysis temperature range, so that the positive effect of the modifier is virtually not observable. Moreover, Co does not form intermetallic compounds with Pd, and hence neither losses through volatilization nor positive effects of the modifier are observed.

Above the melting points of the alloys, the amount of the analyte (Cd) vaporized increases and produces an atomic absorption signal. However, the presence of certain interfering species, such as halide ions which form compounds with metals that have high vapor pressures, must be controlled. In this way, when Cl^- ions are present they cause losses through the analyte (Cd) vaporization at lower pyrolysis temperatures [Morishige *et al.* (1994); Ouishi *et al.* (1994)]. In these cases, other chemical modifiers must be used (for instance, NH_4NO_3 to vaporize and expel from the furnace Cl^- ions in the form of NH_4Cl; Mg^{2+} to form compounds with higher boiling points, such as $MgCl_2$; and SO_4^{2-} and PO_4^{3-} to prevent vaporization of compounds like metal halides with high vapor pressure ($PbCl_2$) [Hirokawa *et al.* (1992)].

The effective temperature of the atomic vapor of Se in the presence of Pd and Rh rises by some 400°C. Furthermore, the effective temperature of the atomic vapor of Se using platform atomization is around 200°C higher than using wall atomization. Similar conclusions were found with the use of other chemical modifiers (Ag, Au, Cd, Cu, Mg, Pd, Pt and Sb) for determining other analytes (Pb and Sn) [Ouishi *et al.* (1994)].

It is true that Pd shows some superiority over other modifiers for analytes in groups IIIB-VIB because Pd satisfies the requirements that should be met by any chemical modifier: (i) it does not reduce the lifetime of the tube; (ii) it does not produce undesirable background noise in proximity to the analytical wavelength; and (iii) it permits the use of higher pyrolysis temperatures [Matsumoto (1993)]. Nonetheless, some modifiers appear to be more effective than Pd, for instance $(NH_4)_3RhCl_6$ plus citric acid, in the Zeeman ETAAS determination of Se in the presence of phosphorus [Mei *et al.* (1998)], or nickel used for the determination of Cd in the presence of chlorides [Feo *et al.* (2003)]. However, Pd shifts the absorbance-time profile to a higher temperature (Fig. 6.6).

Volatile metals have also been used as chemical modifiers: Co [Narukawa *et al.* (1998)] to determine Bi in a W atomizer; Cd and Hg (together with Pd) to determine Se [García-Olalla and Aller (1991, 1992); Manzoori and Saleemi (1994)].

Iridium has been widely used as a *permanent* chemical modifier, by prior impregnation of the graphite tubes [Shuttler *et al.* (1992)] and the Zr or W platforms [Tsalev *et al.* (1995, 1996a,b); Slaveykova *et al.* (1997, 1999)], in the determination of hydride-forming elements.

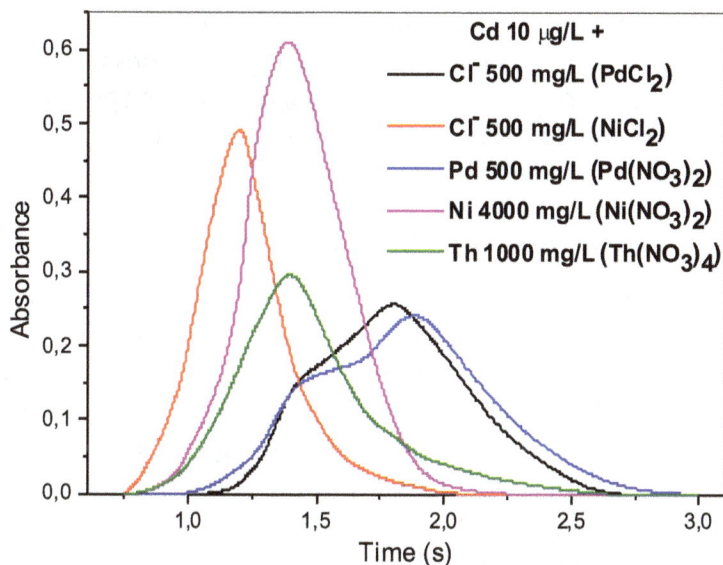

Fig. 6.6. Atomic absorption profiles for platform atomization of Cd together with different compounds [Feo *et al.* (2003)].

Other metals, such as W and Rh [Lima *et al.* (1998)] have also been evaluated as permanent chemical modifiers.

Some typical instances of metals that are not effective as chemical modifiers would be the following [Hirokawa *et al.* (1992)].

- Metals not forming a solid solution with the analyte in any range of concentrations (Fe-Pb).
- When alloys formed show melting points close to, but greater than, the melting point of the chemical modifier used (Ni-Cr, Mn, V; Pd-Cr, Mn,Ni).
- When the alloys and intermetallic compounds formed have melting points lower than the pyrolysis temperature; the melting point of the chemical modifier may be lower than that of the analyte, as with Sn, Zn, Sb and Bi which cannot be used as chemical modifiers. Nonetheless, these metals can be utilized to increase the appearance of the analyte signal in refractory materials by producing matrices with a low melting point. This behavior also constitutes one of the fundamental roles of a metal chemical modifier.

In such cases, it must be remembered that graphite, acting as the contact medium, shows the capability to form carbides with the analyte, matrix components and the modifier, with melting points, viscosity and vapor pressure differing from those of the metals.

6.3.1.2. *Inorganic (non-metallic) compounds*

Besides metals (and organic compounds), inorganic (non-metallic) compounds have also received some attention as chemical modifiers. The use of phosphates as chemical modifiers triggers the formation of metal pyrophosphates (with the analyte) and other intermediary species, $Sn_2P_2O_7$, SnP_2O_7, Pd_9P_2, $Pb_3(PO_4)_2$, $Pb_8P_2O_{13}$, PbO [Li et al. (1996)] during the thermal treatment. These pyrophosphates decompose at higher temperatures giving rise to the free analyte atomic vapor. However, the use of phosphates as chemical modifiers presents several drawbacks: high volatility, corrosive action on the graphite surface, and excessive background absorption. Nonetheless, this compound has been used widely, mainly mixed together with some metals [Acar (2001)].

Ammonium nitrate (and nitric acid) can be used as a chemical modifier in determining volatile elements [Şahin et al. (2005)]. This modifier reacts with NaCl to form $NaNO_3$ and NH_4Cl, and these are volatilized or decomposed at temperatures below 400°C [Chaudhry and Littlejohn (1992)]. The vaporization temperature of NaCl can be reduced from 800–1000°C to below 800°C using equimolar mixtures of diacid ammonium orthophosphate ($NH_2H_2PO_4$), ammonium nitrate, or nitric acid. NH_4NO_3 eliminates Cl^- ions at pyrolysis temperatures of about 200°C, but those remaining are not vaporized until about 1000°C. However, $NH_4H_2PO_4$ almost totally eliminates Cl^- ions at 500–600°C. The only chemical modifier, among those listed above that eliminates $SO_4^=$ ions, is $NH_4H_2PO_4$, which reduces the vaporization temperature from 1000°C to 700°C. The use of phosphate does not appear to be so effective as Pd, at least in the ETAAS determination of Pb using tungsten [Bruhn et al. (1998)] and graphite [Penninckx et al. (1992)] atomizers.

Fluoride has been utilized principally as a volatilizer for some analytes (Al, Mo) or silicon matrices, but there are also other applications in which it acts as a thermal stabilizer for Cd, thanks to the formation of CdF_2 with a high boiling point. Hydrofluoric acid has also been used to overcome the interference derived from the presence of NaCl [Cabon (2002); López-García *et al.* (1999)].

6.3.2. Organic Chemical Modifiers

Organic compounds decompose thermally during the pyrolysis stage and before the atomization process, generating hydrogen and hydrocarbons with differing molecular masses (the molecular mass range obtained with ascorbic acid and Triton X-1000 is between 12 daltons and 110 daltons) [Terui *et al.* (1991a)]. Ascorbic acid, alone or in combination with other metal modifiers, essentially Pd, has been widely used as a chemical modifier in the ETAAS analysis of complex matrices. The effectiveness of this chemical modifier (and of others, such as oxalic acid) is based on the formation of up to 1% v/v of active carbonaceous substances and reducing gases (H_2, CH_4, CO and CO_2). Thus, formation and evaporation of hydrochloric acid was followed at relatively low temperatures in the presence of metal chlorides [Kantor (1995)], varying with the following parameters: the chemical modifier utilized, pyrolysis temperature, surface characteristics of the pyrolytically graphite coated graphite tube [Byrne *et al.* (1993)] (Fig. 6.7), the uniformity in the analyte distribution on the atomizer surface, and the formation of reducing active centers on the graphite surface in the temperature range from 700 K to 1070 K, although a carbonaceous surface may also be produced. Nevertheless, oxalic acid suppresses the formation of carbon residues.

Pyrolysis or thermal destruction of organic compounds, like ascorbic acid, takes place in the following stages: (i) formation of carbon and intermediary gaseous carbon-based substances, such as CO and CO_2 (below 580 K); (ii) formation of active carbon species (between 600 K and 1100 K); (iii) formation of thermally stable amorphous carbon residues (between 1000 K and 1200 K), and later release of these residues through decomposition into active carbon species (between

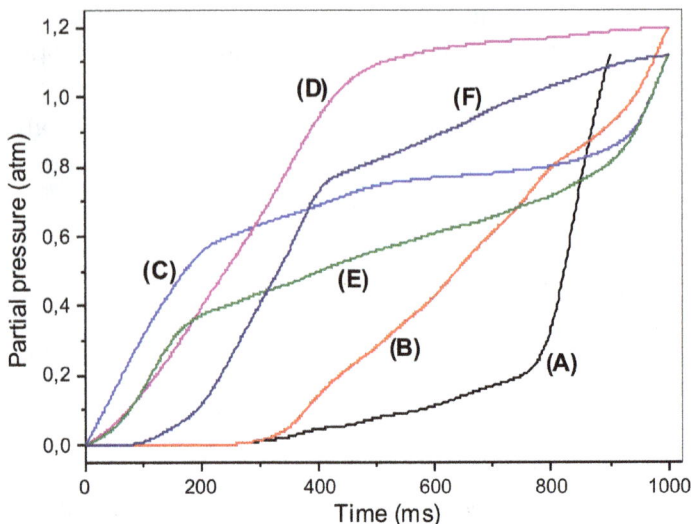

Fig. 6.7. Variation in the partial pressure of H_2 (A, C, E) and CO (B, D, F) during the atomization temperature ramp for 1% (v/v) samples (10 μL) in 1% (v/v) nitric acid alone (A, B), or together with 1% (w/v) ascorbic acid (C, D), and 1% (w/v) oxalic acid (E, F) [Byrne *et al.* (1993)].

1200 K and 2400 K); and (iv) changing of the thermally stable amorphous carbon species into not very well ordered pyrolytic graphite (above 2500 K) [Imai and Hayashi (1991); Imai *et al.* (1995)].

The addition of ascorbic acid up to 5% m/v has no effect on the vaporization of Cl^- and $SO_4^=$ ions. As a result of this drawback in the use of ascorbic acid, alternative organic compounds have been trialled. Thus, the use of hydroxylamine hydrochloride as a reducing agent has the advantage in that it can be mixed with Pd solutions, unlike ascorbic acid, which triggers precipitation of Pd.

These chemical modifiers (ascorbic and oxalic acids) produce alterations in the profiles of the analyte atomic absorption signals, as a result of the reactions occurring between the reducing gases (H_2, CO) formed during the atomization temperature ramp and the gaseous analyte molecules. Hence, the shift in the appearance temperature of the analyte (Pb) when pyrolytic graphite tubes are used depends on the amount of H_2 and CO produced during the pyrolysis of the organic chemical modifiers. This shift in the appearance temperature has been observed only when: (i) ascorbic acid and the analyte solutions are introduced into the tube simultaneously; and (ii) the pyrolysis temperature does not exceed 650°C. In this way, the greatest shift in the appearance temperature produced by ascorbic acid is in previously used pyrolytic tubes and for low pretreatment temperatures. These shifts in the absorption peak profiles cannot be explained from reactions involving the graphite surface, but interactions between the analyte and amorphous microporous carbon (with microporosities between 0.4 nm and 1 nm in size), being stronger than the analyte-analyte interactions [Iwamoto *et al.* (1997)]. Nevertheless, such shifts can be easily justified if changes in the composition of the gaseous phase in the furnace (essentially H_2 and the CO/CO_2 ratio) are noted to take place during the initial atomization stage [Gilchrist *et al.* (1990)].

Depending on the pyrolysis temperature, the analyte atomization processes might be described as a function of the following two mechanisms

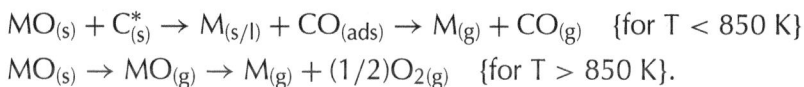

$$MO_{(s)} + C^*_{(s)} \rightarrow M_{(s/l)} + CO_{(ads)} \rightarrow M_{(g)} + CO_{(g)} \quad \{\text{for } T < 850 \text{ K}\}$$
$$MO_{(s)} \rightarrow MO_{(g)} \rightarrow M_{(g)} + (1/2)O_{2(g)} \quad \{\text{for } T > 850 \text{ K}\}.$$

The large shifts observed in the appearance temperatures of those elements which vaporize at between 580°C and 930°C arise through the formation of stable oxides on the graphite surface, as a result of a chemisorption process of O_2 molecules on the active centers over this temperature range.

Ascorbic acid also increases sensitivity in the determination of volatile analytes (Cd, Pb) when coated tubes are used. The graphite surface is activated by the use of coated tubes, increasing the elimination rate of O_2, decreasing the appearance temperatures, shifting the atomic absorption signals of the more volatile analytes, and increasing sensitivity as well. The graphite surface is degraded by O_2, but with time and at high temperatures, there is an annealing or graphitization of the surface if atomization takes place in the absence of O_2.

The carbonaceous residue of ascorbic acid provides a less variable atomization surface, thus reducing the differences in the appearance temperatures for any given element from one type of tube to another [Sturgeon and Berman (1985)].

The existence of small numbers of active centers on the surface is harmful for the heterogeneous gas–solid reactions ($O_2 + C_{(s)}$). On the other hand, a limited quantity of carbonaceous species released during the heating of the organic material (ascorbic acid) would damage the homogeneous reactions in the gas phase ($O_2 + CH_4$).

6.4. Advantages of the Use of Chemical Modifiers

The chemical modification process is accompanied by major changes in the chemical form, reactivity and structure of the atomizer surface. Hence, dissemination and wetting of the sample, intercalation of the analyte, modifier and some matrix components, the atomization mechanism and kinetics, and so forth, can all be strongly affected. Moreover, the gas chemical composition in the tube (partial pressure of the active gases) and presumably time variations of the total pressure may also be affected by: (i) the presence of various gases (air, O_2, CO, H_2, CO_2, CH_4, Cl_2 CHF_3) produced *in situ* (with cleaning or gathering effects relating to the gas phase) or deliberately added; and (ii) pre-oxidation of the graphite surface. Nonetheless, the addition of certain amounts of modifier gives different positive effects, fundamentally on the analyte and the matrix, although the graphite tube surface and the gas phase may also be affected [Volynsky (1998)].

6.4.1. Effects on the Analyte

The positive effect that the chemical modifier exercises over the analyte may be summed up as below.

(i) *Thermal stabilization* (of volatile or moderately volatile analytes), in other words *"preventing vaporization of the analyte during the drying and pyrolysis stages,"* thus allowing the use of higher pyrolysis temperatures. Stabilization of the analyte at high temperatures during the atomization stage takes place through migration and embedding of the analyte in the melted chemical modifier (palladium) [Chen and Jackson (1996); Qiao and Jackson (1991)]. The use of thermal programs with high pyrolysis temperatures gives the following advantages.

(a) Elimination of the volatile matrix components before the atomization stage.
(b) Hydrolyzation of some interfering species.
(c) Atomization delay until higher temperatures yields more isothermal conditions. Through the use of a slow atomization process, the effective vapor temperature or the gas phase temperature is not much affected by the chemical modifier. However, with a fast atomization, the effective vapor temperature is strongly affected by the presence of the modifier. Pd and Ni are widely used as elements indicating the temperature of the atomizer atmosphere. Other metals, like Sn, are not very suitable, as their dynamic range is too narrow, probably owing to the formation of an alloy with some matrix component [Terui *et al.* (1991b)].
(d) Better separation time between the analyte atomic absorption signal and the background absorption signals.
(e) Greater closeness of the pyrolysis and atomization temperatures, so that losses through the analyte vapor convection during atomization are reduced, unless an intermediate *cool-down step* is included in the temperature program.

(ii) *Isomorphic formation of different analyte species.* The analytical signals originating from different analyte species (several oxidation states or different bond types [for instance, Se(II), Se(IV), Se(VI), Se^0, $SeCH_3$, $SeCH_2CH_3$] always present in many complex samples) can be stabilized using a suitable chemical modifier. Nevertheless, certain precautions are necessary when it comes to any extrapolation, as chemical modifiers do not always (in fact almost never) present 100% effectiveness in the homogenization process, because it is usual for a given chemical modifier not to be able to stabilize all the chemical species of each analyte. Thus, chemical modifiers such as nitrates of Ni, Mg, Cu, Cu+Mg, Pd, Pd+Mg cannot stabilize to the same extent all the following chemical

Se species (selenite, selenate, seleno-methionine, and trimethyl-selenonium) [Johannessen *et al.* (1993); Laborda *et al.* (1993)].

(iii) *Increased volatility of the analyte during atomization,* fundamentally for non-volatile and carbide-forming analytes. In this case, the modifier acts predominantly as a volatilizer, facilitating atomization at low temperatures. These modifiers encourage atomization of the analyte in refractory matrices, so that they trigger the analyte atomization before the matrix vaporization. This also facilitates time separation of the analyte atomic signal relative to background signals, causing less interference in the gas phase. However, the chemical modifier may also increase the effective vapor temperature. Hence, employing Pd as a modifier for Pb, the Pb atomic vapor temperature is raised by some 200–300°C above that obtained when Ni is used as the chemical modifier [Hirano *et al.* (1992)]. Nevertheless, the use of lower atomization temperatures and shorter atomization times increases sensitivity and the lifetime of the tube.

Moreover, to encourage atomization of analytes such as Sb or Sn in refractory alloys like those with high contents of Ni, chemical modifiers, such as Zn, are used that can form alloys with the analyte and those matrix components whose vapor pressures are higher. Organic compounds forming complexes with the analyte and also with higher vapor pressures have been added as well.

The use of chemical modifiers may render more effective the atomization stage of the analyte as a consequence of its distribution among the fine carbon particles or on the furnace wall. Good contact with the atomizer surface is ensured by adding some easily-decomposed organic compound, such as tensioactives.

(iv) *Improving the cleaning stage.* The use of reactive gases during the cleaning stage permits better elimination, at lower temperatures and over shorter times, of the persistent residues deriving from the analytes and matrices that can form refractory carbides. In consequence, the lifetime of the graphite tube is increased, while memory effects and changes in sensitivity are substantially reduced.

(v) *In situ speciation.* Differences in volatility between two chemical species of the same analyte may serve to separate them during the preatomization stage. This has been true for Cr(III) and Cr(VI) [Arpadjan and Krivan (1986)].

6.4.2. Effects on the Matrix

The positive effects that the chemical modifiers may exercise on the matrix are as follows.

(i) *Increasing volatility of interfering substances,* that is *"vaporizing matrix components that interfere the analyte atomization."* Interfering species can be volatized and expelled from the atomizer during the pyrolysis stage. In this way, interferences in the condensed phase, the gas phase, and the background interferences are reduced or eliminated, while losses of analyte during atomization due to processes such as co-volatilization and convection also decrease. For this purpose, NH_4NO_3 constitutes a classical example. This compound reacts at high temperatures with Cl^- ion to form NH_4Cl, this being vaporized during the pyrolysis stage. Consequently, the analyte remains in the atomizer until the atomization stage.

(ii) *Chemical transformation of certain interfering substances* that is *preventing easy vaporization of the matrix components during atomization.* Some interfering species can be transformed into other less harmful species. Mg is usually employed as a modifier with samples containing high levels of Cl^- ions. Formation of $MgCl_2$ impedes vaporization of some analyte chlorides, whose vaporization points are close to 1000°C. For some volatile elements (Sb or As) in refractory materials (Ta and rare earths), carbon powder decreases the vaporization of the matrix components.

(iii) *Thermal stabilization of interfering species.* Matrices containing certain macrocomponents (PO_4^{3-}, Fe, etc.), giving rise to serious background absorption problems at the analytical wavelength, as a consequence of the formation of different absorbing species (P_2, PO, Fe, FeO, metal sulphides, chlorides of alkaline earth metals, Al

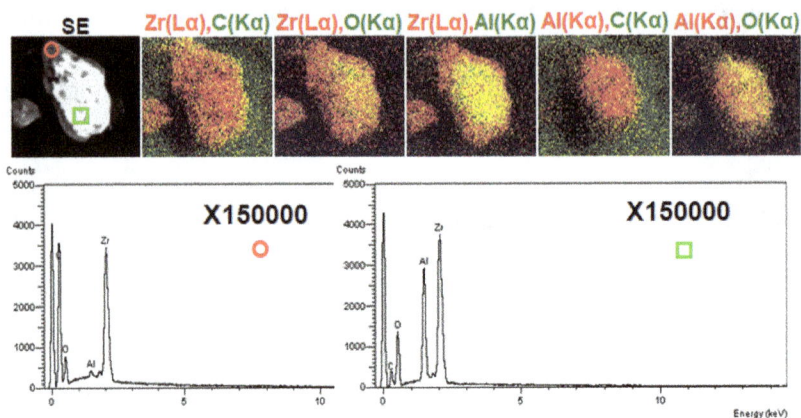

Fig. 6.8. ED X-ray spectra and colored superimposed two-element distribution mappings from a pyrolytic graphite platform containing 100 mg each of aluminium and ziconium salts heated to 1800°C [Castro *et al.* (2005)].

oxides) can be stabilized in the solid phase by the use of chemical modifiers [Castro *et al.* (2005)] (Fig. 6.8).

(iv) *Facilitating pyrolysis of the organic matrices.* Many modifiers [$Mg(NO_3)_2$, HNO_3, H_2SO_4, H_2O_2, air, O_2 and other oxidants] favor ashing of the organic matrices. In consequence, there is a reduction in background absorption, the composition of the solid residues, and variations in sensitivity.

(v) *Improving contact between the sample and the atomizer surface.* Detergents and organic additives reduce the surface tension of a liquid sample and of molten materials that may occur, giving better contact, greater covered surface or simply sample dilution. A similar effect can be produced by introducing the sample as an aerosol.

6.4.3. Other Analytical Advantages

The use of chemical modifiers may also improve the following analytical characteristic and techniques.

(i) *Analytical (metrological) characteristics.* Reliability (the prime aim), precision, and sensitivity/limit of detection (especially in real matrices that may contain depressing components or solvents and/or with low efficiency atomizers).

(ii) *Technical characteristics.* Simplified sample pretreatment; simple calibration (less complicated standards may be used); simpler and shorter temperature programs, possibility of multi-element determinations; longer lifetimes of the atomizer; improved stability and so less frequent recalibration; and less requirement for versatility of the equipment.

6.5. Problems (and Possible Solutions) Arising from the Use of Chemical Modifiers

The repetitive use of large amounts of chemical modifiers may show certain drawbacks, these being briefly noted below.

(i) *Blanks and standards.* Standards prepared with some elements (Al, Cd, Cr, Cu, Fe, Mn, Pb, Zn) and above all the chemical modifiers themselves (phosphates, ammonium nitrate and so on) must be purified by solvent extraction, ion exchange, electrolysis, and thermal treatment, in order to avoid incorporation of considerable exogenous potential interfering species.

(ii) *Contamination (memory effects).* Some chemical modifiers (i.e. Ag, Al, Au, Co, Cr, Mo, P, Pd, Sr, V) irreversibly contaminate graphite and the laboratory material, making it difficult or impossible to determine them immediately afterwards. In consequence, the use of *more universal* chemical modifiers which are rarely if ever determined through ETAAS would be the most practical solution.

(iii) *Spectral background.* The chemical modifier should not contribute to the direct overlapping of the analytical wavelength, nor in any way to the finely structured spectral background (not even temporarily, as a consequence of the rapidly changing background signals). The use of phosphate as a chemical modifier presents problems for the analysis of some analytes: Ag (328.1 nm), As (193.7 nm), Cd (326.1 nm), Co (240.7 nm, 242.5 nm, 252.1 nm), Fe (246.3 nm), Hg (253.7 nm), Pd (246.3 nm), Se (196.0 nm), Sn (286.3 nm) and Te (214.3 nm). Other chemical modifiers may also cause spectral interferences when

Table 6.2. Potential Spectral Interferences Between Some Analytes and Chemical Modifiers (or Matrix Components)

Modifier	Analyte (λ, nm)
Al {due to AlO (Zeeman)}	As (193.7 nm), Pb (217.0 nm)
Ba {due to BaO (Zeeman)}	Al (267.6 nm), Pb (283.3 nm), Sn (286.3 nm)
Co	Au (242.8 nm), Sb (231.1 nm), Se (204.0 nm)
Cu	Bi (223.1 nm), Sb (217.6 nm)
Fe	As (193.7 nm), Sb (217.6 nm), Sb (231.1 nm), Se (196.0 nm)
Ni	As (197.2 nm), P (213.6 nm), Sb (231.1 nm), Se (204.0 nm), Se (196.0 nm) for a Ni excess
Pd	Cu (324.7 nm), In (303.9 nm) and Tl (276.8 nm)
Rh	Ag (328.1 nm) (Zeeman)

a deuterium lamp is used as a background correction system (Table 6.2).

Such interferences may be overcome by some of the following ways: (a) decreasing the spectral bandwidth; (b) using different background correction systems (Zeeman, Smith–Hieftje); (c) using an alternative wavelengths; (d) utilizing *non-reversible* graphite tubes (decreasing background due to volatilization of PO species); (e) reducing the atomization temperature and/or atomization time; (f) selecting the atomization ramp mode; (g) taking the peak height as the analytical parameter, instead of the integrated area, as also making use of time resolution signals; and (h) using mixtures of chemical modifiers, which are often more effective and present a stable thermal behavior.

(iv) *Corrosive effect on graphite*. Certain chemical modifiers (including Ce, $Cr^{(VI)}$, $FeCl_3 + HNO_3$, La, Mn, H_2SO_4, $Cr_2O_7^=$, PO_4^{3-}, Sc, Sr, Y, NH_4^+ NO_3^-) are corrosive for graphite, substantially cutting back the useful lifetime of the tubes and platforms. The destruction of the pyrolytic graphite coating can also cause a drastic loss of sensitivity. The morphological changes that can take place on the graphite surface are the following: peeling, pitting, bubbling, defoliation and possibly warping and fractures.

The biggest problems are usually the outcome of the use of large amounts of La.

(v) *Toxicity.* The use of some chemical modifiers may give rise to toxic vapors (Ba, Hg, Mn, Ru, V), carcinogenic (Cr, Ni, tetra-alkyl-ammonium hydroxide) or radioactive (Th), so that they should be handled with all due care and precaution.

(vi) *Compatibility problems.* Some chemical modifiers are difficult to keep in solution, as they are easily hydrolyzed (Hf, Nb, Si, Ta, Ti, W, Zr; some noble metals). They should thus be used only in extreme conditions and in media that are mostly acids or bases (WO_4^{2-}, MoO_4^{2-}, VO_3^-, SiO_3^{2-}).

(vii) *Reduction in sensitivity.* A great excess of almost any of the chemical modifiers will produce a decrease in the analytical signal.

(viii) *Economical considerations.* High-purity reagents (Ag, Au, Ir, Pt, Pd, Rh, Ru and rare elements) are very costly. Hence, consumption of them should be reduced by using any of the following: recovery of the reagents, employment of mixtures of chemical modifiers *in situ* prepared or added separately into the tube.

6.6. Efficiency of the Chemical Modification Process

The effectiveness of any chemical modifier depends on several factors.

- *The modifier itself.* (a) Chemical nature; (b) chemical and physical states: magnesium nitrate acts as modifier but magnesium chloride is an interfering species. This includes the size and granulometry of the metallic modifier particles on the graphite surface (this being dependent on the reduction method); (c) chemical environment within the atomizer (other constituents, the medium, and so on): nitric acid reduces the effectiveness of some chemical modifiers (like Pd), but has no effect, or at most very little, on other modifiers (such as nickel); and (d) amount and mass ratio in respect of other species.
- *The analyte.* Chemical nature, type and number of species (oxidation states).

- *The matrix.* Composition and amount introduced into the atomizer.
- *The atomizer.* (a) Construction, efficiency, isothermal nature, atomization modes (wall, platform), and the like; (b) material (uncoated, pyrolytic graphite coated, pretreated, metallic); and (c) age (number of firing cycles) and history (types of samples previously analyzed).
- *The temperature programme.* (a) Pyrolysis ramp and temperature; and (b) atomization ramp and temperature.

To end and sum up, it is possible to conclude that any ideal or universal modifier should have the following characteristics.

(a) To increase thermal stability of analytes up to a minimum of 1000°C (at this temperature, many matrix components would have been previously eliminated, during the pyrolysis stage).
(b) To be a great purity substance.
(c) To be a substance not normally determined by ETAAS.
(d) To enlarge, or at least not to reduce, the lifetime of the atomizer (tube or platform).
(e) To produce no background absorption (structured or not).
(f) To show high efficiency for a large number of analytes.
(g) To work for different chemical forms of the analyte.
(h) To show low toxicity.
(i) To show considerable stability during storage.
(j) To be cheap.
(k) To show robustness in its action.
(l) To attain maximum efficiency using simple thermal programs.

References

Acar, O. (2001). Determination of cadmium and lead in biological samples by Zeeman ETAAS using various chemical modifiers. *Talanta*, **55**: 613–622.

Acar, O.; Türker, A.R. and Kiliç, Z. (2000). Determination of bismuth, indium and lead in spiked sea water by electrothermal atomic absorption spectrometry using tungsten containing chemical modifiers. *Spectrochim. Acta, Part B*, **55**: 1635–1641.

Aller, A.J. (2003). Espectroscopía de absorción atómica electrotérmica. Servicio de Publicaciones. Universidad de León, León, Spain.

Arpadjan, S. and Krivan, V. (1986). Preatomization separation of chromium (III) from chromium (VI) in the graphite furnace. *Anal. Chem.*, **58**: 2611–2614.

Belarra, M.A.; Resano, M.; Rodriguez, S.; Urchaga, J. and Castillo, J.A. (1999). The use of chemical modifiers in the determination of cadmium in sewage sludge and tin in PVC by solid sampling-graphite furnace atomic absorption spectrometry. *Spectrochim. Acta, Part B*, **54**: 787–795.

Bermejo-Barrera, P.; Moreda-Piñeiro, J.; Moreda-Piñeiro, A. and Bermejo-Barrera, A. (1998). Usefulness of the chemical modification and the multi-injection technique approaches in the electrothermal atomic absorption spectrometric determination of silver, arsenic, cadmium, chromium, mercury, nickel and lead in sea-water. *J. Anal. At. Spectrom.*, **13**: 777–786.

Brueggemeyer, T.W. and Fricke, F.L. (1986). Comparison of furnace atomization behaviour of aluminium from standard and thorium-treated L'vov platforms. *Anal. Chem.*, **58**: 1143–1148.

Bruhn, C.G.; Neira, J.Y.; Valenzuela, G.D. and Nóbrega, J.A. (1998). Chemical modifiers in a tungsten coil electrothermal atomizer. Part 1. Determination of lead in hair and blood. *J. Anal. At. Spectrom.*, **13**: 29–35.

Bulska, E. and Jedral, W. (1995). Application of palladium- and rhodium-plating of the graphite furnace in electrothermal atomic absorption spectrometry. *J. Anal. At. Spectrom.*, **10**: 49–53.

Bulska, E.; Thybusch, B. and Ortner, H.M. (2001). Surface and subsurface examination of graphite tubes after electrodeposition of noble metals for electrothermal atomic absorption spectrometry. *Spectrochim. Acta, Part B*, **56**: 363–373.

Byrne, J.P.; Chakrabarti, C.L.; Gilchrist, G.F.R.; Lamoureux, M.M. and Bertels, P. (1993). Chemical modification by ascorbic acid and oxalic in grafite furnace atomic absorption spectrometry. *Anal. Chem.*, **65**: 1267–1272.

Cabon, J.Y. (2002). Determination of Cd and Pb in seawater by graphite furnace atomic absorption spectrometry with the use of hydrofluoric acid as a chemical modifier. *Spectrochim. Acta, Part B*, **57**: 513–524.

Castro, M.A.; Feo, J.C. and Aller, A.J. (2003). Thermal stability and spatial distribution of sodium chloride alone and in the presence of several metal salts on a graphite platform. *J. Anal. At. Spectrom.*, **18**: 260–267.

Castro, M.A.; Aller, A.J.; McCabe, A.; Smith,W.E. and Littlejohn, D. (2005). Spectrometric and morphological characterization of condensed phase zirconium species produced during electrothermal heating on a graphite platform. *J. Anal. At. Spectrom.*, **20**: 385–394.

Castro, M.A.; Aller, A.J.; McCabe, A.; Smith,W.E. and Littlejohn, D. (2007). Spectrometric study of condensed phase species of thorium- and palladium-based modifiers in a complex matrix for electrothermal atomic absorption spectrometry. *J. Anal. At. Spectrom.*, **22**: 310–317.

Castro, M.A.; García-Olalla, C.; Robles, L.C. and Aller, A.J. (2002). Behaviour of thorium, zirconium, and vanadium as chemical modifiers in the determination of arsenic by electrothermal atomic absorption spectrometry. *Spectrochim. Acta, Part B*, **57**: 1–14.

Chaudhry, M.M. and Littlejohn, D. (1992). Ion chromatographic study of the effect of ammonium nitrate as a modifier in electrothermal atomic absorption spectrometry. *Analyst*, **117**: 713–715.

Chen, G. and Jackson, K.W. (1996). Low-temperature migration of lead, thallium, and selenium onto a palladium modifier during the analysis of solutions and slurries by electrothermal atomic absorption spectrometry. *Spectrochim. Acta, Part B*, **51**: 1505–1515.

Dabeka, R.W. (1992). Refractory behavior of lead in a graphite furnace when palladium is used as a matrix modifier. *Anal. Chem.*, **64**: 2419–2424.

Dèdina, J.; Frech, W.; Cedergren, A.; Lindberg, Y. and Lundberg, E. (1987). Determination of selenium by graphite furnace atomic absorption spectrometry. Part 2. Role of nickel for analyte stability. *J. Anal. At. Spectrom.*, **2**: 435–439.

De-Qiang, Z.; Zhe-ming, N. and Han-Wen, S. (1998). Stabilization of organic and inorganic mercury in the graphite furnace with $(NH_4)_2PdCl_6$-$(NH_4)_3RhCl_6$ as a mixed chemical modifier. *Spectrochim. Acta, Part B*, **53**: 1049–1055.

Dočekalová, H.; Dočekal, B.; Komárek, J., and Novotný, Y. (1991). Determination of selenium by electrothermal atomic absorption spectrometry. Part 1. Chemical Modifiers. *J. Anal. At. Spectrom.*, **6**: 661–668.

Feo, J.C.; Castro, M.A.; Lumbreras, J.M.; de Celis, B. and Aller, A.J. (2003). Nickel as a chemical modifier for sensitivity enhancement and fast atomization processes in electrothermal atomic absorption spectrometric determination of cadmium in biological and environmental samples. *Anal. Sci.*, **19**: 1631–1636.

Flores, A.V.; Pérez, C.A. and Arruda, A.Z. (2004). Evaluation of zirconium as a permanent chemical modifier using synchrotron radiation and imaging techniques for lithium determination in sediment slurry samples by ET AAS. *Talanta*, **62**: 619–626.

Frech, W.; Li, K.; Berglund, M. and Baxter, D.C. (1992). Effects of modifier mass and temperature graphite on analyte sensitivity in electrothermal atomic absorption spectrometry. *J. Anal. At. Spectrom.*, **7**: 141–145.

García-Olalla, C. and Aller, A.J. (1991). Determination of gold in ores by flame and graphite furnace atomic absorption spectrometry using a vanadium chemical modifier. *Anal. Chim. Acta*, **252**: 97–105.

García-Olalla, C. and Aller, A.J. (1992). Alternative mercury-palladium chemical modifier for the determination of selenium in coal fly ash by graphite furnace atomic absorption spectrometry. *Anal. Chim. Acta*, **259**: 295–303.

Gilchrist, G.F.R.; Chakrabarti, C.L.; Byrne, J.P. and Lamoureux, M. (1990). Gas-phase thermodynamic equilibrium model and chemical modification in graphite furnace atomic absorption spectrometry. *J. Anal. At. Spectrom.*, **5**: 175–181.

Gong, B.; Li, Hui; Ochiali, T.; Zheng, L.T. and Matsumoto, K. (1993). Enhancement effect of some matrix modifiers on the tin sensitivity in graphite furnace atomic absorption spectrometry and direct evidence for Pd_3Sn_2 formation during ashing in the presence of a palladium modifier. *Anal. Sci.*, **9**: 723–726.

Gong, B.; Liu, Y.; Li, J. and Lin, T. (1998). Comparison of chemical modifiers used for the determination of gold in ores by electrothermal atomic absorption spectrometry. *Anal. Chim. Acta*, **362**: 247–251.

Hernández Caraballo, E.A.; Alvarado, J.D. and Domínguez, J.R. (1999). Study of the electrothermal atomization of selenium (IV) in transversely-heated graphite atomizers in the presence of nickel compounds as chemical modifiers. *Spectrochim. Acta, Part B*, **54**: 1593–1606.

Hilligsøe, B.O.; Andersen, J.E.T. and Hansen, E.H. (1997). Investigations into the role of modifiers for entrapment of hydrides in flow injection hydride generation electrothermal atomic absorption spectrometry as exemplified by the determination of germanium. *J. Anal. At. Spectrom.*, **12**: 585–588.

Hirano, Y.; Nomura, Y.; Yasuda, K. and Hirokawa, K. (1992). Direct analysis of whole blood for trace elements by graphite furnace-atomic absorption spectrometry. *Anal. Sci.*, **8**: 427–431.

Hirokawa, K.; Yasuda, K. and Takada, K. (1992): Graphite furnace atomic absorption spectrometry and phase diagrams of alloys. *Anal. Sci.*, **8**: 411–417.

Imai, S. and Hayashi, Y. (1991). Effect of ascorbic acid on graphite furnace atomic absorption signals for lead. *Anal. Chem.*, **63**: 772–775.

Imai, S.; Nishiyama, Y.; Tanaka, T. and Hayashi, Y. (1995). Investigations of pyrolysed ascorbic acid in an electrothermal graphite furnace by inductively coupled argon plasma mass spectrometry and Raman spectrometry. *J. Anal. At. Spectrom.*, **10**: 439–442.

Iwamoto, E.; Itamoto, M.; Nishioka, K.; Imai, S.; Hayashi, Y. and Kumamaru, T. (1997). Effects of conditions for pyrolysis of ascorbic acid as chemical modifier on the vaporization mechanism of gold in electrothermal atomic absorption spectrometry. *J. Anal. At. Spectrom.*, **12**: 1293–1296.

Johannessen, J.K.; Gammelgaard, B.; Jøns, O. and Hansen, S.H. (1993). Comparison of chemical modifiers for simultaneous determination of different selenium compounds in serum and urine by Zeeman-effect electrothermal atomic absorption spectrometry. *J. Anal. At. Spectrom.*, **8**: 999–1004.

Kantor, A. (1995). On the mechanisms of organic acid modifiers used to eliminate magnesium chloride interferences in graphite furnace atomic absorption spectrometry. *Spectrochim. Acta, Part B*, **50**: 1599–1612.

Kowalewska, E.; Bulska, E. and Hulanicki, A. (1999). Organic palladium and palladium-magnesium chemical modifiers in direct determination of lead in fractions from distillation of crude oil by electrothermal atomic absorption analysis. *Spectrochim. Acta, Part B*, **54**: 835–843.

Laborda, F.; Viñuales, J.; Mir, J.M. and Castillo, J.R. (1993). Effect of nickel and palladium as chemical modifiers and influence of urine matrix on different chemical species of selenium in electrothermal atomic absorption spectrometry. *J. Anal. At. Spectrom.*, **8**: 737–743.

Lamourex, M.M.; Hutton, J.C. and Styris, D.L. (1998). Elucidation of mechanisms of palladium induced modification of selenium in electrothermal atomic absorption spectrometry by investigation of condensed species using synchrotron X-ray absorption spectroscopy. *Spectrochim. Acta, Part B*, **53**: 993–1002.

Li, H.; Gong, B. and Matsumoto, K. (1996). Tributyl phosphate as a sensitivity-enhancing solvent for organotin in carbon furnace atomic absorption spectrometry. *Anal. Chem.*, **6**: 2277–2280.

Lima, E.C.; Barbosa, F.Jr.; Krug, F.J. and Guaita, U. (1999). Tungsten-rhodium permanent chemical modifier for lead determination in digests of biological materials and sediments by electrothermal atomic absorption spectrometry. *J. Anal. At. Spectrom.*, **14**: 1601–1605.

Lima, E.C.; Krug, F.J. and Jackson, K.W. (1998). Evaluation of tungsten-rhodium coating on an integrated platform as a permanent chemical modifier for cadmium, lead, and selenium determination by electrothermal atomic absorption spectrometry. *Spectrochim. Acta, Part B*, **53**: 1791–1804.

López-García, I.; Sanchez-Merlos, M. and Hernández-Córdoba, M. (1999). Use of hydrofluoric acid to decrease the background signal caused by sodium chloride in electrothermal atomic absorption spectrometry. *Anal. Chim. Acta*, **396**: 279–284.

Mandjukov, P.B.; Tsakovski, S.L.; Simeonov, V.D. and Stratis, J.A. (1995). A solid solution theory for matrix modification in electrothermal atomic absorption spectrometry, *Spectrochim. Acta, Part B*, **50**: 1733–1746.

Mandjukov, P.B.; Vasileva, E.T. and Simeonov, V.D. (1992). Regular solution theory in model interpretation of the analyte losses during preatomization sample treatment in the presence of chemical modifiers in electrothermal atomization atomic absorption spectrometry. *Anal. Chem.*, **64**: 2596–2603.

Manzoori, J.L. and Saleemi, A. (1994). Determination of chromium in serum and lake water by electrothermal atomic absorption spectrometry using vanadium and molybdenum modifier. *J. Anal. At. Spectrom.*, **9**: 337–339.

Matsumoto, K. (1993). Palladium as a matrix modifier in graphite-furnace atomic absorption spectrometry of group IIIB-VIB elements. *Anal. Sci.*, **9**: 447–453.

Mei, L.; Zhe-ming, N. and Zhu, R. (1998). Determination of selenium in biological tissue samples rich in phosphorus using electrothermal atomization with Zeeman-effect background correction and $(NH_4)_3RhCl_6+$ citric acid as a mixed chemical modifier. *Spectrochim. Acta, Part B*, **53**: 1381–1389.

Morishige, Y.; Hirokawa, K. and Yasuda, K. (1994). The role of metallic matrix modifiers in graphite furnace atomic absorption spectrometry. *Fresenius' J. Anal. Chem.*, **350**: 410–412.

Muñoz, M.L and Aller, A.J. (2006). Appraisal of the chemical modification process for the determination of lead by ultrasonic slurry sampling-electrothermal atomic absorption spectrometry. *J. Anal. At. Spectrom.*, **21**: 329–337.

Narukawa, T.; Uzawa, A.; Yoshimura, W. and Okutani, T. (1998). Effect of cobalt as a chemical modifier for determination of bismuth by electrothermal atomic absorption spectrometry using a tungsten furnace. *Anal. Sci.*, **14**: 779–784.

Nóbrega, J.A.; Rust, J.; Calloway, C.P. and Jones, B.T. (2004). Use of modifiers with metal atomizers in electrothermal atomic absorption spectrometry: a short review. *Spectrochim. Acta, Part B*, **59**: 1337–1345.

Nowka, R.; Eichardt, K. and Welz, B. (2000). Investigation of chemical modifiers for the determination of boronb by electrothermal atomic absorption spectrometry. *Spectrochim. Acta, Part B*, **55**: 517–524.

Ohta, K. and Mizuno, T. (1989). Effect of hydrogen on atomic absorption of iron, cobalt and nickel. *Anal. Chim. Acta*, **217**: 377–382.

Ortner, H.M.; Bulska, E.; Rohr, U.; Schlemmer, G.; Weinbruch, S. and Welz, B. (2002). Modifiers and coatings in graphite furnace atomic absorption spectrometry-mechanisms of action (A tutorial review). *Spectrochim. Acta, Part B*, **57**: 1835–1853.

Ouishi, K.; Yasuda, K.; Morishige, Y. and Hirokawa, K. (1994). Role of metal matrix modifier in ashing and beginning of the atomization process in graphite furnace-atomic absorption spectrometry. *Fresenius' J. Anal. Chem.*, **348**: 195–200.

Peng-yuan, Y.; Zhe-ming, N.; Zhi-xia, Z.; Fu-chun, X. and An-bei, J. (1992). Study of palladium-analyte binary system in the graphite furnace by surface analytical techniques. *J. Anal. At. Spectrom.*, **7**: 515–519.

Penninckx, W.; Massart, D.L. and Smeyers-Verbeke, J. (1992). Effectiveness of palladium as a chemical modifier for the determination of lead in biological materials and foodstuffs by graphite furnace atomic absorption spectrometry. *Fresenius' J. Anal. Chem.*, **343**: 526–531.

Qiao, H. and Jackson, K.W. (1991). Mechanism of modification by palladium in graphite furnace atomic absorption spectrometry. *Spectrochim. Acta, Part B*, **46**: 1841–1859.

Rademeyer, C.J.; Radziuk, B.; Romanova, N.; Skaugset, N.P.; Skogstad, A. and Thomassen, Y. (1995). Permanent Iridium modifier for electrothermal atomic absorption spectrometry. *J. Anal. At. Spectrom.*, **10**: 739–745.

Sabé, R.; Rubio, R. and García-Beltrán, L. (1999). Study and comparison of several chemical modifiers for selenium determination in human serum by Zeeman electrothermal atomic absorption spectrometry. *Anal. Chim. Acta*, **398**: 279–287.

Şahin, F.; Volkan, M. and Ataman, O.Y. (2005). Effect of nitric acid for equal stabilization and sensitivity of different selenium species in electrothermal atomic absorption spectrometry. *Anal. Chim. Acta*, **547**: 126–131.

Shuttler, I.L.; Feuerstein, M. and Schlemmer, G. (1992). Long-term stability of a mixed palladium-iridium trapping reagent for *in situ* hydride trapping with a graphite electrothermal atomizer. *J. Anal. At. Spectrom.*, **7**: 1299–1301.

Silva, A.F. da; Welz, B. and Curtius, A.J. (2002). Noble metals as permanent chemical modifiers for the determination of mercury in environmental reference materials using solid sampling graphite furnace atomic absorption spectrometry and calibration against aqueous standards. *Spectrochim. Acta, Part B*, **57**: 2031–2045.

Slaveykova, V.I.; Lampugnani, L.; Tsalev, D.L.; Sabbatini, L. and Giglio, E.De (1999). Permanent iridium modifier deposited on tungsten and zirconium-treated platforms in electrothermal atomic absorption spectrometry: vaporization of bismuth, silver and tellurium. *Spectrochim. Acta, Part B*, **54**: 455–467.

Slaveykova, V.I.; Lampugnani, L.; Tsalev, D.L. and Sabbatini, L. (1997). Morphological and spectroscopic investigation of the behaviour of permanent iridium modifier deposited on pyrolytic graphite coated and zirconium treated platforms in ETAAS. *Spectrochim. Acta, Part B*, **52**: 2115–2126.

Slavin, W.; Carnrick, G.R. and Manning, D.C. (1982). Magnesium nitrate as a matrix modifier in the stabilized temperature platform furnace. *Anal. Chem.*, **54**: 621–624.

Sturgeon, R.E. and Berman, S.S. (1985). Absorption pulse shifting in grafite furnace atomic absorption spectrometry. *Anal. Chem.*, **57**: 1268–1275.

Terui, Y.; Yasuda, K. and Hirorawa, K. (1991a). Metallographical consideration on the mechanism of matrix modifier in graphite furnace atomic absorption spectrometry. *Anal. Sci.*, **7**: 397–402.

Terui, Y.; Yasuda, K. Hirokawa, K. (1991b). Measurement of effective vapor temperature in graphite furnace of atomic absorption spectrometry. *Anal. Sci.,* **7**: 599–604.

Tsalev, D.L. and y Slaveykova, V.I. (1992). Chemical modification in electrothermal atomic absorption spectrometry. Organization and clasification of data by multivariate methods. *J. Anal. At. Spectrom.* **7**: 147–153.

Tsalev, D.L.; D'Ulivo, A.; Lampugnani, L.; di Marco, M. and Zamboni, R. (1995). Thermally stabilized iridium on an integrated, carbide-coated platform as a permanent modifier for hydride-forming elements in electrothermal atomic absorption spectrometry. Part 1. Optimization studies. *J. Anal. At. Spectrom.,* **10**: 1003–1009.

Tsalev, D.L.; D'Ulivo, A.; Lampugnani, L.; di Marco, M. and Zamboni, R. (1996a). Thermally stabilized iridium on an integrated, carbide-coated platform as a permanent modifier for hydride-forming elements in electrothermal atomic absorption spectrometry. Part 2. Hydride generation and collection, and behaviour of some organoelement species. *J. Anal. At. Spectrom.,* **11**: 979–988.

Tsalev, D.L.; D'Ulivo, A.; Lampugnani, L.; di Marco, M. and Zamboni, R. (1996b). Thermally stabilized iridium on an integrated, carbide-coated platform as a permanent modifier for hydride-forming elements in electrothermal atomic absorption spectrometry. Part 3. Effect of L-cysteine. *J. Anal. At. Spectrom.,* **11**: 989–995.

Tsalev, D.L.; Dimitrov, T.A. and Mandjukov, P.B. (1990a). Study of vanadium (V) as a chemical modifier in electrothermal atomisation atomic absorption spectrometry. *J. Anal. At. Spectrom.,* **5**: 189–194.

Tsalev, D.L.; Lampugnani, L.; D'Ulivo, A.; Petrov, I.I.; Georgieva, R.; Marcucci, K. and Zamboni, R. (2001). Electrothermal atomic absorption spectrometric determination of selenium in biological fluids with rhodium modifier compared with hydride generation atomic spectrometric techniques. *Microchem. J.,* **70**: 103–113.

Tsalev, D.L.; Slaveykova, V.I. and Mandjukov, P.B. (1990b). Chemical modification in graphite-furnace atomic absorption absorption spectrometry. *Spectrochim. Acta Rev.,* **13**: 225–274.

Tsalev, D.L.; Slaveykova, V.I.; Lampugnani, L.; D'Ulivo, A. and Georgieva, R. (2000). Permanent modification in electrothermal atomic absorption spectrometry — advances, anticipations and reality. *Spectrochim. Acta, Part B,* **55**: 473–490.

Uggerud, H.Th. and Lund, W. (1999). Modifier effects from palladium and iridium in the determination of arsenic and antimony using electrothermal vaporisation inductively coupled plasma mass spectrometry. *Spectrochim. Acta, Part B,* **54**: 1625–1636.

Vieira, M.A.; Ribeiro, A.S. and Curtius, A.J. (2004). Slurry sampling of sediments and coals for the determination of Sn by HG-GF AAS with retention in the graphite tube treated with Th or W as permanent modifiers. *Anal. Bioanal. Chem.,* **380**: 570–577.

Volynsky, A.B. (1996). Catalytic processes in graphite furnaces for electrothermal atomic absorption spectrometry. *Spectrochim. Acta, Part B,* **51**: 1573–1589.

Volynsky, A.B. (1997). Collidal palladium — a promising chemical modifier for electrothermal atomic absorption spectrometry. *Spectrochim. Acta, Part B,* **52**: 1293–1304.

Volynsky, A.B. (1998). Investigation of the mechanisms of the action of chemical modifiers for electrothermal atomic absorption spectrometry: what for and how?. *Spectrochim. Acta, Part B*, **53**: 139–149.

Volynsky, A.B. (2000). Mechanism of action of platinum group modifiers in electrothermal atomic absorption spectrometry. *Spectrochim. Acta, Part B*, **55**: 103–150.

Volynsky, A.B. (2003). Chemical modifiers in modern electrothermal atomic absorption spectrometry. *J. Anal. Chem.*, **58**: 905–921.

Volynsky, A.B. and Krivan, V. (1996). Comparison of different forms of palladium used as chemical modifiers for the determination of selenium by electrothermal atomic absorption spectrometry. *J. Anal. At. Spectrom.*, **11**: 159–164.

Volynsky, A.B. and Wennrich, R. (2001). Comparative efficiency of Pd, Rh and Ru modifiers in electrothermal atomic absorption spectrometry for the simultaneous determination of As, Se and In in a sodium sulphate matrix. *J. Anal. At. Spectrom.*, **16**: 179–187.

Volynsky, A.B.; Akman, S.; Dogan, C.E. and Koklu, U. (2001). Application of colloidal palladium modifier for the determination of As, Sb and Pb in a spiked sea water sample by electrothermal atomic absorption spectrometry. *Spectrochim. Acta, Part B*, **56**: 2361–2369.

Volynsky, A.B.; Tikhomirov, S. and Elagin, A. (1991). Proposed mechanism for the action of palladium and nickel modifiers in electrothermal atomic absorption spectrometry. *Analyst*, **116**: 145–148.

Wojciechowski, M.; Piaścik, M. and Bulska, E. (2001). Noble metal modifiers for antimony determination by graphite furnace atomic absorption spectrometry in biological samples. *J. Anal. At. Spectrom.*, **16**: 99–101.

Xiao-Quan, S. and Bei, W. (1995). Is Palladium or Palladium-Ascorbic acid or Palladium-Magnesium nitrate a more universal chemical modifier for electrothermal atomic absorption spectrometry?. *J. Anal. At. Spectrom.*, **10**: 791–798.

Chapter 7

Electrothermal Atomization Mechanisms: *Theoretical and Practical Considerations*

7.1. Characteristics of the Electrothermal Atomic Absorption Signals

Electrothermal atomic absorption signals represent the absorbance obtained in a time interval, i.e. while the analyte atomic vapor is along the optical path (absorbance-time profiles). Figure 7.1 shows a typical absorbance-time profile. Two analytical parameters are usually distinguished: absorbance and integrated absorbance. Absorbance (dimensionless parameter) is measured at the maximum of the peak profile, while integrated absorbance (time units) is determined by the area under the absorbance-time curve between the appearance time (t_{ap}) and the end time (t_{end}).

From the peak profile, some general considerations about the electrothermal vaporization/atomization of the analyte can be derived.

The peak profile characteristics depend on various parameters affecting the vaporization rate and the atomization degree. Both processes are regulated by the composition of the matrix, the atomizer atmosphere, the atomizer surface, and the temperature program (Table 7.1) [Sturgeon and Berman (1985)]. On the other hand, the peak

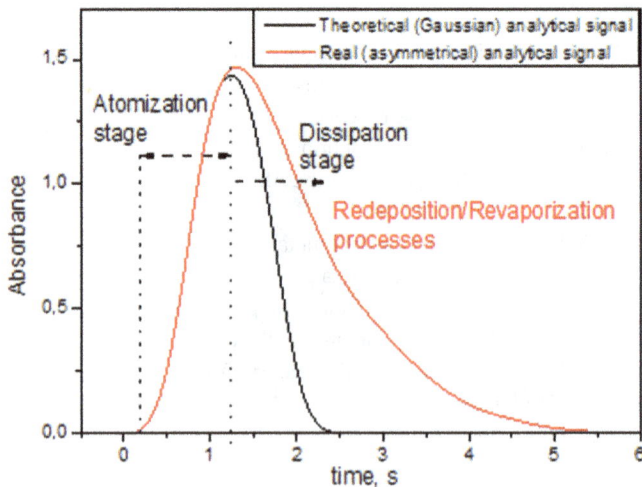

Fig. 7.1. Typical absorbance-time profile for an electrothermal atomic absorption signal.

Table 7.1. Appearance Temperature Shiftments (ΔT, K) for a Few Elements as a Function of the Graphite Tube Type and the Atmosphere Composition [Sturgeon and Berman (1985)]

Element	Tube	$(T_{app})_{Ar}$, K	$(\Delta T_{app})_{Ar/O_2}$, K	$(\Delta T_{app})_{H_2}$, K
Bi	Non-coated	1145	75	−125
	Coated	950	80	−45
	Vítreous	985	115	−50
Cd	Non-coated	950	50	−100
	Coated	770	25	−10
	Vítreous	810	50	0
Cr	Non-coated	1735	0	−35
	Coated	1730	45	0
	Vítreous	1795	85	0
Cu	Non-coated	1400	0	0
	Coated	1400	0	0
	Vítreous	1440	0	0

time and the end time depend directly and respectively on the growing and decaying kinetics of the gas phase atomic population. The gas phase atomic population is also affected by other parameters, such as the atomizer geometry, permeability of the graphite surface, and the heating rate. A minimum population of atoms in the graphite tube is necessary to derive an ETAAS signal, which is regulated by pressure, temperature and, in particular by the bond strength of the precursors of the gas phase analyte atoms.

The temperature distribution along and across the atomizer is also a main factor affecting the ETAAS signals. It was usually thought that temperature gradients generate smaller ETAAS signals. Nonetheless, under isothermal conditions, the ETAAS signals show higher peaks but smaller integrated absorbances. The main impact of the *isothermal atomizer* is really to provide a low memory effect [Güell and Holcombe (1991)]. As a result of the temperature gradients, heterogeneous spatial distributions of the analyte atoms and molecules are produced [Frech and Baxter (1996)]. However, when standards and samples generate the same relative spatial distribution of the analyte atoms, linearity and sensitivity can deteriorate, but precision is unaffected [Holcombe and Histen (1996)]. The spatial distribution of the analyte atoms and molecules do not shows a homogeneous pattern [Frech and Baxter (1996)]. However, the distribution density of the analyte atoms in the optical path at distances of about 1.2 mm out the tube shows values of 14%–17% with respect to the internal center of the tube [Hadgu *et al.* (1996, 1998)].

Peak shifts for some elements (Ag, As, Cd, Cu) have been associated with a migration process inside the graphite bulk, usually a few microns under the graphite surface. Migration process occurs as well at low as well as at high temperatures [Eloi *et al.* (1993); Jackson *et al.* (1995c)], and can start before the drying step, through imperfections of the graphite tube surface [Galbács *et al.* (1997)]. Peak shifts can suggest changes in the atomization mechanisms, but they are usually the result of slight differences in the heating rates [Güell *et al.* (1993)]. Peaks can be shifted towards shorter times with the gas flow [Huie and Curran (1990)] and with the amount of concomitant species [García-Olalla and Aller (1992)]. Reactivity of the atomizer substrate is the prime mediator in controlling the partial pressure of oxygen [Sturgeon *et al.* (1984)], which affects the peak shifts, mainly for those elements which undergo atomization *via* a thermal dissociation reaction [L'vov *et al.* (1981); Rayson and Holcombe (1982); L'vov and Ryabchuk (1982)].

Peak broadening has usually been attributed to a diffusion process of the analyte atoms inside the solid (or melted) chemical modifier. However, peak broadening can also be the result of a secondary adsorption of the analyte atoms on the cold edge of the platform where the chemical modifier may be placed [Frech *et al.* (1992a,b)]. This argument can also be used to explain the long tails of the ETAAS signals obtained at low temperatures, although L'vov *et al.* (1988) also attributed this effect to the formation of carbon shells on sample microparticles. In conclusion, many parameters may affect the atomization process. Nonetheless, several theoretical and experimental efforts have been carried out in order to understand their contribution to the electrothermal atomization process.

7.2. Electrothermal Atomization Mechanisms: *Theoretical Considerations*

In a general sense, the shape and height of the ETAAS signals (Fig. 7.1) can be affected by the following processes [Aller (2002)].

- Physical transformations (crystallization, melting, vaporization, etc.) and solid-phase chemical reactions between analyte and other matrix components during the heating stages.
- Analyte interactions (in the solid phase and/or in the gas phase) with the atomizer surface (adsorption, re-adsorption, re-combination, surface diffusion and diffusion inside the graphite bulk, and generation and decomposition of new compounds).
- Analyte de-sorption from the surface of solid particles (sublimation, evaporation, thermal dissociation {atomization} and de-proportion).
- Interaction between the gas phase atoms and the protective gas (generation, de-composition and transformation of different compounds, such as oxides, carbides, hydrides, oxi-carbides and inter-metallic compounds).
- Convection and/or diffusion of the gas phase atoms inside/outside the analysis zone.

Some of the above processes may show larger incidence than the others, whether they are thermodynamically more favorable or pose higher kinetic rates. This fact contributes to the temporal heterogeneous distribution of the analyte atoms into the atomizer. As a result, the concentration of the gas phase atomic population in the graphite tube during the analysis time is dependent on the kinetics of the global processes generating and transforming the analyte species. Hence, any theoretical model establishing a relationship between thermodynamic and/or kinetic parameters and the absorbance-time profile characteristics would be a useful tool for the elucidation of those processes experienced by the analyte in the atomizer [L'vov (1997a,b, 1998)].

7.2.1. Thermodynamic Aspects

In order to carry out thermodynamic studies about the processes taking part in the electrothermal atomization of any analyte, it would be necessary to know all the chemical reactions taking place in the atomizer, as well as the characteristic steps of the atomization process. Samples are usually introduced as nitrate or chloride solutions. Hence, we will consider a few general characteristics of the chemical transformations usually occurring in an ET atomizer during the decomposition process of these two salts.

7.2.1.1. *Vaporization and decomposition of metal nitrates and oxides*

The behavior of metal nitrates in the solid phase depends on the atomizer surface [Majidi *et al.* (1996)]. Graphite atomizers are the most frequently employed and consequently the following comments are focused on them. In general, metal nitrates, [$M(NO_3)_2$] precipitate on the graphite surface during the drying stage. Then, the nitrates are transformed into solid metal oxides [MO], gas nitrogen dioxide [NO_2] and oxygen during the pyrolysis stage [L'vov and Novichikhin (1995a,b)], according to the following reaction, which constitutes the gasification mechanism

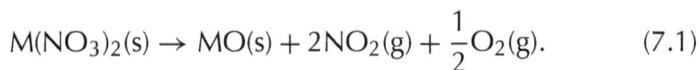

$$M(NO_3)_2(s) \rightarrow MO(s) + 2NO_2(g) + \frac{1}{2}O_2(g). \qquad (7.1)$$

The decomposition temperature can be thermodynamically predicted. Reaction (7.1) represents an *isothermal* decomposition of metal nitrates. However, with the graphite tubes usually employed in ETAAS, the following two mechanisms are possible for the formation of oxides due to the non-isothermality of the graphite surface

$$M(NO_3)_2(s) \rightarrow MO(g) + 2NO_2(g) + \frac{1}{2}O_2(g) \qquad (7.2)$$

or

$$M(NO_3)_2(s) \rightarrow MO(g) + 2NO(g) + (3/2)O_2(g). \qquad (7.3)$$

Both mechanisms are not thermodynamically favorable at between 400 K and 700 K. However, L'vov (1991) has suggested that in ETAAS, the reaction rates related to the mechanisms (7.2) and (7.3) are higher than that of reaction (7.1). This is due to the low diffusion rate for $NO_2(g)$ or $NO(g)$ through the solid reactants and products. L'vov has proposed that $MO(g)$ condensed on the graphite surface at 1 atm and, consequently, reduction of $MO(s)$ can take place without loss of analyte. This mechanism was also considered by McAllister (1994).

On the other hand, other authors [Jackson *et al.* (1995a,b)] have suggested that the best mechanism to explain the experimental data is a physical expulsion process: (i) liquid metal nitrates would be expelled from the tube during vacuum boiling; (ii) high vapor pressure of liquid metal nitrates; and (iii) fragmentation of small crystal or clusters produced during the crystalline reordering.

Once the metal oxide is formed, the atomization process is produced as a result of some of the following reactions.

- **Reduction of the metal oxide**, $MO(s)$, by graphite at high temperatures, an atom-formation mechanism early proposed by Campbell and Ottaway (1974),

$$MO(s) + C \xrightarrow{\Delta} M(s) + CO(g). \qquad (7.4)$$

- **Decomposition of the gas phase metal oxide**, $MO(g)$,

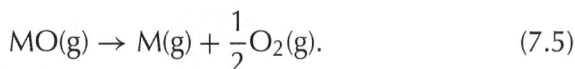

$$MO(g) \rightarrow M(g) + \frac{1}{2}O_2(g). \qquad (7.5)$$

- **Formation of the metal carbide**, at a temperature lower than the atomization/integration temperature,

$$MO(s) + 2C \Leftrightarrow MC(s) + CO(g). \qquad (7.6)$$

The metal carbide is subsequently dissociated. Nonetheless, those metal oxides having high vapor pressures at the atomization temperature are usually lost in large quantities before the atomization stage is achieved.

Vaporization of several oxides and salts has been largely studied by L'vov [L'vov *et al.* (2004); L'vov and Ugolkov (2003, 2004)] through the third-law method, which is based on the direct application of the basic equation of chemical thermodynamics

$$\Delta_r H_T^0 = T(\Delta_r S_T^0 - R \ln K_P), \tag{7.7}$$

where $\Delta_r S_T^0$ is the entropy change and K_P is the equilibrium constant for the following reaction, equivalent to reactions (7.2) and (7.3)

$$S(s) \rightarrow aA(g) + bB(g) \tag{7.8}$$

in terms of the partial pressures, P, of the gaseous products

$$K_P = P_A^a \, P_B^b. \tag{7.9}$$

The E parameter of the Arrhenius equation for reaction (7.8) is equal to

$$E = \frac{\Delta_r H_T^0}{v} \tag{7.10}$$

which can be reduced to

$$E = T \left(\frac{\Delta_r S_T^0}{v} - R \ln P_{eq} \right), \tag{7.10'}$$

where P_{eq} is the equivalent pressure of the gaseous product in reaction (7.8), v is the total number of moles of gaseous products, $(a + b)$, and $\Delta_r H_T^0$ is the change of the enthalpy in the reaction (7.8). In this case, a measurement at only one temperature is sufficient for the determination of E but knowledge of absolute values of the entropy of all components in the reaction is necessary for the calculation. Nonetheless, all these reactions are strongly affected by the free oxygen content in the electrothermal atomizer, which in turn depends on the atomizer type, matrix composition and modifier used [Volynsky et al. (1986)].

7.2.1.2. *Vaporization of metal chlorides*

Vaporization of metal chlorides is mainly regulated through their saturation pressure, P_{MCl}^S. Nonetheless, vaporization at the temperature that usually operate in the graphite tubes can show some particularities if adsorption/sorption processes occur. Besides the saturation pressure, other parameters such as the thermodynamic stability

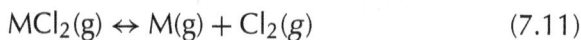

$$MCl_2(g) \leftrightarrow M(g) + Cl_2(g) \tag{7.11}$$

and the chemical stability

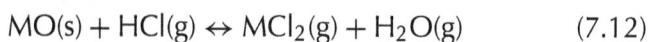

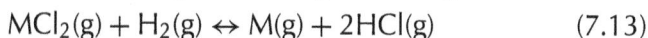

$$MO(s) + HCl(g) \leftrightarrow MCl_2(g) + H_2O(g) \tag{7.12}$$

$$MCl_2(g) + H_2(g) \leftrightarrow M(g) + 2HCl(g) \tag{7.13}$$

can also affect on the transformation of the metal chlorides in the graphite tube [Jóźwiak and Maniecki (2005)]. Evolution of these chemical reactions can be predicted using the conditional pressure and the temperature-dependent equilibrium constant, K_P, of the reactions. Thus, the higher value for the conditional pressure of the reaction (7.12), the higher the concentration of the gas phase species. However, even assuming that a metal chloride is vaporized, decomposition (dissociation) depends on the thermal and/or chemical stability in gas phase.

7.2.2. Kinetic Aspects

The atomization process can be schematically represented by

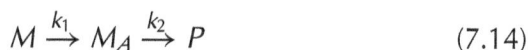

$$M \xrightarrow{k_1} M_A \xrightarrow{k_2} P \tag{7.14}$$

In this process, two main stages exist (Fig. 7.1): atomization (generation of the analyte atoms, M_A, characterized by the rate constant, k_1); and dissipation (loss of the analyte atoms, characterized by the rate constant, k_2). The main objective of any kinetic model is to derive those kinetic parameters related to all the analyte-involved atomization and dissipation processes. That is, to determine the rate constants (atomization, k_1, and dissipation, k_2) and the kinetic order, as well as the frequency factor and the activation energy. Two conditions have been assumed in all the theoretically developed models: atomization at constant temperature, and atomization at growing temperature. However, three general limitations are always present.

- Vaporization and atomization are different processes, but they usually occur simultaneously, with great difficulties to be studied as independent processes.

- The vaporization rate and the partial pressure of the analyte species cannot be directly measured.
- The atomization rate can be derived applying an *indirect* method to the atomic absorption signal profile, which result from the convolution of the supply/atomization and dissipation functions of the analyte atoms [Broek and Galan (1977); Torsi and Tessari (1973, 1975); Tessari and Torsi (1975); Paveri-Fontana *et al.* (1974)]. Changes in the number of the gas phase analyte atoms going in the graphite tube are always the balance between the supply and the dissipation processes of the analyte atoms, i.e.

$$\frac{\partial N(t)}{\partial t} = \acute{n}_1(t) - \acute{n}_2(t), \qquad (7.15)$$

where $\partial N(t)/\partial t$, represents the change rate in the number of the gas phase analyte atoms into the tube; $\acute{n}_1(t)$ and $\acute{n}_2(t)$ are the number of the analyte atoms inward and outward, respectively, in unit time.

The exothermal interaction of the matrix components and modifier with the substrate and the corresponding evolution and dissipation of chemical energy via the gas phase can affect the vaporization kinetics and atomization efficiency of analytes [Katskov *et al.* (2006)].

7.2.2.1. *Number of free atoms (atom density)*

The study of the absorbance-time profile allows us to visually characterize the kinetics of the generation and dissipation processes of the analyte atoms. General models describing generation and losses in the graphite tubes have been proposed [L'vov (1978); van den Broek and de Galan (1977); Falk and Schnürer (1989)]. Alternative equations were also derived for a graphite rod atomizer [Torsi and Tessari (1973, 1975); Paveri-Fontana *et al.* (1974)]. L'vov's treatment [L'vov (1978)] for the atomization under a growing temperature, provided that the integrated absorbance is proportional to the total number of atoms, N_0, injected into the analysis volume and to the residence time, τ_2, of the analyte atoms inside the graphite tube

$$\int_0^\infty N(t)dt = N_0\tau_2, \qquad (7.16)$$

where $N(t)$ is the number of atoms in the analysis volume at time t.

The maximum number, N_{max}, of the analyte atoms in the graphite tube during the atomization stage was written as follows

$$N_{max} = \eta(\tau_1/\tau_2)N_0, \tag{7.17}$$

where N_0 is the total number of the analyte atoms introduced into the atomizer; τ_1, the supply function; τ_2, the dissipation function or the residence time; and η, a correction factor which depends on the τ_1/τ_2 ratio and the vaporization mode. The vaporization mode is a characteristic of the heating rate of the atomizer and it depends on the physical and/or chemical processes experienced by the analyte during atomization. Of course, Eq. (7-17) represents a relatively simple model based on several assumptions.

- Complete sample atomization.
- Homogeneous cross-section distribution of the analyte atoms in the graphite tube.
- No sample injection hole.
- Constant longitudinal temperature distribution in the graphite tube.
- Constant longitudinal atom concentration gradient.
- Small diffusion losses of the analyte atoms during the atomization stage.
- Analysis volume filling the graphite tube.

The integrated value of N_{max} represents the integrated absorbance or the peak area, Q_a, and for an atomization constant temperature, N_{max} decreases if pressure grows, while Q_a stays constant. This is true if $\tau_2 \gg \tau_1$, with Lorentz's broadening prevailing against Doppler's broadening, and the absorbance is measured at the non-shifted peak profile maximum. Thus, a pressure range exists for which N_{max} is inversely proportional to the pressure, while Q_a and T are constants. For the opposite situation, if the pressure in the graphite tube decreases, any increased diffusion gives rise to a decrease in τ_2, with smaller values for N_{max} and Q_a [Hassell *et al.* (1988)].

For kinetic studies, it is usual to admit that the atom dissipation process is the result of only the diffusion process, as it is the case of the transversally heated graphite atomizers [Hadgu *et al.* (1996)]. The

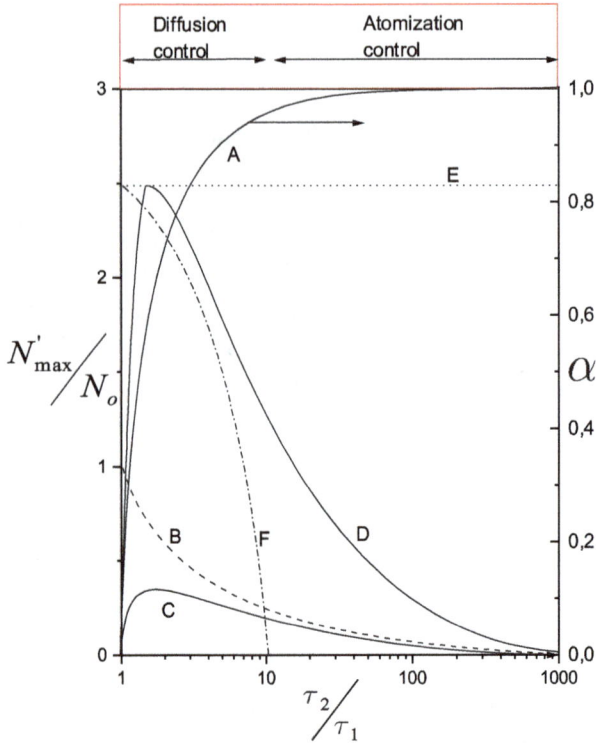

Fig. 7.2. (Line A) Fraction (α) of suitable atoms for excitation as a function of the supply and dissipation functions. (Lines B and C) Maximum relative number of atoms for a peak profile: (B) L'vov model [L'vov (1970)] and (C) Lonardo *et al.* model (1996) for N'_{max}. (Lines D, E and F) Maximum relative number of atoms for a peak profile: (D) Lonardo *et al.* model (1996) for N'_{max} without chemical interference, (E) as in (D) but $\Omega = 0$, and (F) as in (D) but with chemical interference, $\Omega = 0.11$ [Lonardo *et al.* (1996)].

maximum number of the analyte atoms in the graphite tube during the vaporization/atomization process, for a low pressure atomization and in the presence of chemical interference, also including the atom loss diffusion process, can be described by the following theoretical model [Lonardo *et al.* (1996)] (Fig. 7.2),

$$N'_{max} = \alpha_t N_{max} = \alpha_t \eta(\tau_1/\tau_2)N_0 = (1 - \{\tau_1/\tau_2\})\eta(\tau_1/\tau_2)N_0$$

$$(7.18)$$

where η and N_0 have the same meaning as in Eq. (7.17), while the parameter, α_t, is defined by the following expression

$$\alpha_t = \frac{\tau_t - \tau_1}{\tau_t} \approx \frac{\tau_2 - \tau_1}{\tau_2} \qquad (7.19)$$

and represents the average time fraction used by each atom in the gas phase, being $\tau_t = \tau_1 + \tau_2$, while $(\tau_t - \tau_1) \approx (\tau_2 - \tau_1)$. Variations in the working pressure alter the relative contributions from the supply and dissipation processes. For the flash atomization, $\tau_2 \gg \tau_1$ and hence $\tau_t \approx \tau_2$, which can be considered as the total time needed by each atom to be excited.

On the other hand, if N_0 is constant

$$N'_{max} = N_A(1 - \psi CN_\beta\tau_2) = N_{max}(1 - \Omega\tau_2), \qquad (7.20)$$

where N_A and N_β represent the number of the analyte atoms and the number of atoms from the matrix (interferent) existing in the analytical volume, and Ω $(=\psi CN_\beta)$ is a constant whose value is between 0 and 1, ψ is a value between 0 and 1, being constant for some atoms A (analyte) and B (matrix), while

$$C = \frac{\pi d^2 \sqrt{\frac{8kT}{\pi\mu}}}{V^2} \qquad (7.21)$$

is a constant for the temperature, T, the analytical volume, V, and a particular chemical species; the parameter, d, is the reduced diameter, k is the Boltzmann's constant, and μ is the reduced mass.

The time-dependent absorbance signal has also been expressed [Hsiech and Pardue (1993)] as a function of the rate constants for a first order process. The main advantage of this approach is the capability to reduce the effects from the atomization temperature, at least at between 2200°C and 2600°C for a few elements (Cr, Mn, K, Yb, and Fe). The assumption of first-order kinetics for the atom formation made in some of the earlier studies [Sturgeon *et al.* (1976); Smets (1980); Chung (1984); Akman *et al.* (1980)] can only be used to derive the activation energy at the earlier stages of the atomization process.

Attempts to consider high heating rates have also been continuously carried out [Torsi *et al.* (1995, 2000, 2005)]. An increase in the diffusion coefficient and the gas combination factor reduces the maximum atom population because of the faster rate of the atom loss. A high value for the frequency factor increases the rate of atom formation, but the contrary is true for the activation energy [Chakrabarti *et al.* (1983)].

However, the analyte atoms can generally be dissipated by a mechanism different to diffusion. In these cases, it is necessary to use new rate constants. Thus, Musil and Rubeska (1982) included the participation of an atom redeposition process, developing the following equation

$$N = \frac{N_0 k_1}{(k_1 - k_2)(k_1 - k_e - k_R)} \left[\frac{(r_1 + k_2)(r_2 + k_1)}{\sqrt{D}} e^{r_1 t} \right.$$
$$\left. + \frac{(r_1 + k_1)(r_2 + k_2)}{\sqrt{D}} e^{r_1 t} + (k_2 - k_1)e^{-k_1 t} \right], \qquad (7.22)$$

where N_0, k_1, k_2, k_R, k_e represent the total number of the analyte atoms, the rate constants corresponding to the atomization, revaporization, redeposition and dissipation, processes, respectively

$$D = (k_2 + k_e + k_R)^2 - 4k_2 k_e. \qquad (7.23)$$

In this model, if $k_2 = 0$ (absence of re-deposition), Eq. (7.22) is simplified

$$N = \frac{N_0 k_1}{(k_e - k_1)} (e^{-k_1 t} - e^{-k_e t}) \qquad (7.24)$$

which derives on Fuller's model [Fuller (1974, 1977)]. The suitability of the model proposed by Musil and Rubeska (1982) to the interpretation of the absorbance-time profiles obtained for several analytes was evaluated by Welz *et al.* (1988). Using atomizers made of polycrystalline electrographite, pyrolytic graphite and glassy carbon, it was possible to assign values to the rate constant for redeposition and revaporization.

L'vov (1970) also described two equations for the graphite tube atomization, for $t \leq \tau_1$

$$N_t = 2N_0\tau_2^2 \frac{\left[\left(\frac{t}{\tau_2}\right) - 1 + \exp\left(\frac{-t}{\tau_2}\right)\right]}{\tau_1^2}, \qquad (7.25)$$

and for $t \geq \tau_1$

$$N_t = 2N_0\tau_2^2 \frac{\left[\left(\frac{\tau_1}{\tau_2}\right) - 1 + \exp\left(\frac{-t}{\tau_2}\right)\right]\exp\left[\left(\frac{\tau_1-t}{\tau_2}\right)\right]}{\tau_1^2}, \qquad (7.26)$$

which can be used to determine the number of atoms, N_t, at any time, t as a function of the supply τ_1 and dissipation τ_2 functions and the total number of atoms injected, N_0.

Torsi and Tessari (1973, 1975), Tessari and Torsi (1975) and Paveri-Fontana *et al.* (1974) developed an alternative equation for a rod graphite atomizer

$$N_1(t) = \sigma q d\theta/dt = k\sigma q\theta = v\sigma q\theta \exp(-\Delta G/RT). \qquad (7.27)$$

However, if a linear temperature ramp exists with an initial temperature, T_0, at the time, $t = 0$, and with a heating rate, α

$$N_1(t) = v\sigma q\theta \exp(-\Delta G/R\{T_0 + \alpha t\}), \qquad (7.28)$$

where σ is the surface (cm^2), θ is the surface fraction covered by the analyte at the time t ($0 \leq \theta \leq 1$), q is the surface atom concentration (atoms/cm^2); and ΔG is the heat of vaporization; when $\theta = 1$, k is the rate constant for the vaporization process (which shows a normal Arrhenius-type dependence with temperature), and v is the frequency factor.

The absorbance obtained for non-isothermal conditions is *quasi* proportional to the gas phase atom density for low analyte densities. Under non-isothermal conditions, the number of atoms, N_p, in the analyte volume at the time of the peak maximum, t_p, was expressed as [Zhou *et al.* (1984)]

$$N_p = \frac{k_p}{k_R}N_0 \exp\left[-\frac{1}{\alpha}\frac{T_m^2(T_p - T_0)k_m}{T_p T_0}\right], \qquad (7.29)$$

where k_R is the rate constant of atom leaving the analyte volume, k_p and k_m are the rate constants of atom formation at the peak and temperature T_m, where $T_o < T_m < T_p$. If $T_m^2 \approx T_p T_0$ for the analytical conditions used, the peak absorbance, A_p, is given by

$$A_p = \beta N_0 \frac{k_p}{k_R} \exp\left[-\frac{1}{\alpha}(T_p - T_0 k_m)\right], \qquad (7.29')$$

where β relates the number of free atoms in the analytical volume to the absorbance and the exponential is the coverage at the peak position.

If the peak appears in the constant temperature region, the above equation should be

$$A_p = \beta N_0 \frac{k_p}{k_R} \exp[-\{(t - t_0)k_m + (t_p - t)k_p\}], \qquad (7.29'')$$

where the meaning for t is similar to that corresponding for temperatures, T, and $t - t_0 = (T_p - T_0)/\alpha$, α being the heating rate. However, if atomization proceeds at constant temperature, k_s and k_R, are constant and the peak absorbance is given by

$$A_p = \beta N_0 \frac{k_s}{k_R} \exp[-k_s(t_p - t_0)], \qquad (7.29''')$$

where k_s is the rate constant for atom formation under isothermal conditions. The above equation is similar to that derived by Fuller (1977) and Chakrabarti et al. (1981).

7.2.2.2. Activation energy and frequency factor

The Arrhenius equation

$$k_1 = v \exp(-E_a/RT) \qquad (7.30)$$

is largely used for calculation of the activation energy, E_a, and the frequency factor, v. The rate constant k_1 is temperature dependent, obtaining several values for k_1 at different temperatures. Smet's method [Smets (1980)] provides information about E_a from the slope of the plot of Ln k_1 versus $1/T$. The frequency factor is related to the interactions between the atomic vapor precursor species and the graphite surface. The frequency factor values grow with temperature if atomization is carried out from metal- and/or metal carbides-coated surfaces and chemical modifiers in solution are used. Smets (1980) established a relationship between the frequency factor values between 10^{10} s^{-1} and 10^{12} s^{-1} and those processes, such as sublimation, in which surface does not participate. Similarly, the frequency factor values between 10^{15} s^{-1} and 10^{18} s^{-1} are related to those processes for which surface is implicated in the reduction of the analyte oxides. Nonetheless, the frequency factor is not currently used for elucidation of the atomization mechanisms.

The determination of the activation energy for the atomization process can be carried out using different approaches, but the most commonly used are of Smets (1980) and Sturgeon *et al.* (1976). If the analyte is introduced in the graphite tube in very small amounts ($\leq$ng), it can be assumed that it is in the condensed phase and is distributed on the graphite surface forming a monolayer or submonolayer. The release of these species is considered to be the result of a de-sorption process, according to the following kinetic equation [Holcombe and Rayson (1983)]

$$-\frac{\partial \sigma}{\partial t} = v\sigma^n \exp\left(-\frac{E_a}{RT}\right), \tag{7.31}$$

where σ is the surface covered by the analyte at any time, n is the desorption order, v is the frequency factor, E_a is the activation energy of the de-sorption process, R is the ideal gas constant, and T is the absolute temperature. From Eq. (7.31) and assuming that absorbance, A, is proportional to $\frac{\partial N}{\partial t}$, since there is a direct relationship between σ and N (number of atoms leaving the graphite surface), it is possible to write a similar equation, after applying neperians

$$\text{Ln } A = n \text{ Ln } \sigma + \text{Ln } (v) - \frac{E_a}{RT} \tag{7.32}$$

known as Sturgeon's method [Sturgeon *et al.* (1976)], which assumes that the term σ does not change with temperature. Equations (7.30) and (7.31) show a similar pattern for a first order process [Frech *et al.* (1982)]. It has been stated that the use of nonlinear heating ramps shows minor effects on the activation energy values deduced from the Arrhenius plot [Histen and Holcombe (1998a)].

The Arrhenius plots usually curve, particularly owing to the assumptions made:

- existence of a stationary state in the earlier atomization stage,
- constant temperature inside the tube (or in any case, absence of important thermal gradients in the atomizer during the pretreatment stage),
- monolayer distribution of the analyte atoms on the graphite surface, covering a constant area and showing a nule bidimensional analyte mobility,
- first order atomization processes,
- analyte loss mechanisms only kinetically regulated and dependent upon the pretreatment conditions.

On the other hand, together with the above assumptions, the additional processes are not usually considered [Holcombe *et al.* (1982); Huie and Curran (1988); Gilmutdinov *et al.* (1995, 1996)]:

- atomic vapor dissipation by mechanisms other than diffusion,
- redeposition or recombination of the free analyte atoms,
- completely integrated atomic absorption signals,
- atom diffusion through the graphite walls,
- changes in the crystalline particle sizes generated during drying and pyrolysis stages,
- changes in the surface atomizer covering,
- interferences between the supply and dissipation functions, and
- spatially and temporally non-homogeneous distribution of atoms in the electrothermal atomizer.

The Arrhenius plot is affected by the degree of redeposition of the free analyte atoms, which is determined by temperature gradients and the properties of the atomization surface [Rojas *et al.* (1997)]. Other errors can arise from diffusion through the sample injection hole, assuming only one kinetic order and of course only one atomization mechanism, and small sistematic errors [Bass and Holcombe (1988b)].

Other related models have been also derived [Yan-Zhong *et al.* (1995); Bozdogân (1999); Xiu-Ping *et al.* (1990)]. Xiu-Ping *et al.* (1990) improved Smet's method for the calculation of the activation energy, but obviating first order kinetics. Other authors have developed kinetic models to determine the activation energy, establishing different relationships between parameters. Thus, Rojas and Olivares (1992) used a time-dependent parameter $W_m(t)$, for the atomization under isothermal conditions

$$W_m(t) = \frac{\left[k_2(t)A(t) + \left(\frac{dA}{dt}\right)\right]}{\left[\int_t^\infty k_2(\tau)A(\tau)d\tau - A(t)\right]^n} = \beta^{(1-n)}k_1(t), \qquad (7.33)$$

where $k_1(t)$ and $k_2(t)$ are respectively the rate constants for the supply and dissipation of the atomic vapor; β is a proportionality factor; n is the kinetics order; and $A(t)$ is the absorbance signal at time t. Assuming Arrhenius-type temperature dependence for k_1, one follows

$$\text{Ln } W_m(t) = \text{Ln}(v\beta^{(1-n)}) - \left[\frac{E_a}{RT(t)}\right], \qquad (7.34)$$

from which the activation energy of the atomization process, E_a, can be derived from the straight line obtained for the proper order, n, of an Arrhenius-type plot of Ln $W_m(t)$ *versus* $1/T$.

Other authors [Xiu-Ping *et al.* (1993a,b); Quan Zhe *et al.* (1994)] use only one absorbance peak profile to calculate the activation energy

$$\frac{\Delta \text{Ln}\left(\frac{dA}{dt} + k_2 A\right)}{\Delta \text{Ln}\left(\int_t^\infty k_2 A dt - A\right)} = n - \frac{\left(\frac{E_a}{R}\right)\Delta\left(\frac{1}{T}\right)}{\Delta \text{Ln}\left(\int_t^\infty k_2 A dt - A\right)}. \qquad (7.35)$$

If the analyte atomization from the atomizer surface is regulated by a simple kinetic model, the plot of $\Delta \text{Ln}(dA/dt + k_2 A)/\Delta \text{Ln}(\int_t^\infty k_2 A dt - A)$ *versus* $\Delta(1/T)/\Delta \text{Ln}(\int_t^\infty k_2 A dt - A)$ would provide a straight line from whose slope the activation energy, E_a, can be derived, while the kinetic order, n, results from the intersect.

Aller (2001), using one absorbance-time profile, has developed the following equation

$$\ln\left[\frac{\left(A_i^e + \int_0^t k_{2i}A_i^e\,dt\right)\sqrt{T}}{\left((A_p^e - A_i^e) + \int_1^{t_p} k_{2i}A_i^e\,dt\right)}\right] = \ln\left[2e^{-m}\sqrt{\frac{E_0}{\pi k}}\right] - \frac{E_0}{kT},$$

(7.36)

where A_i^e represents the experimental absorbance values at any time (or temperature) of the rising edge of the peak profile, k_{2i}, the kinetic constant of the dissipation process, k, the Boltzmann constant, m, a characteristic constant for each analyte, derived from intersect, and E_0, the activation energy at absolute zero, A_p^e, the absorbance value at the peak maximum. Equation (7.36) can be used to determine the activation energy at the absolute zero. The activation energy at any temperature, T, can be followed using the equation

$$E_a = E_0 + mkT.$$

(7.37)

The activation energy considered here needs to be interpreted as a temperature-dependent effective activation energy, as shown by Vyazovkin (2003). The constant values of the typical activation energy appears to exist only within undergraduate courses of physical chemistry that almost exclusively and, by no means, comprehensively treat gas phase reactions. It is worth to note that even for the gas phase reactions, the activation energy should show a temperature dependence because of the temperature dependence of the heat capacity of activation [Hulett (1964)]. As soon as the theory moves from vacuum to the condensed phase [Glasstone *et al.* (1941); Marcus (1964, 1993)], the free energy of activation becomes a function of temperature-dependent properties of the reaction medium [Vyazovkin (2000)], and the condensed phase reactions rate cannot in principle be separated from the properties of the reaction medium.

A comparison between several methods in the plot used for calculation of the activation energy is shown in Fig. 7.3.

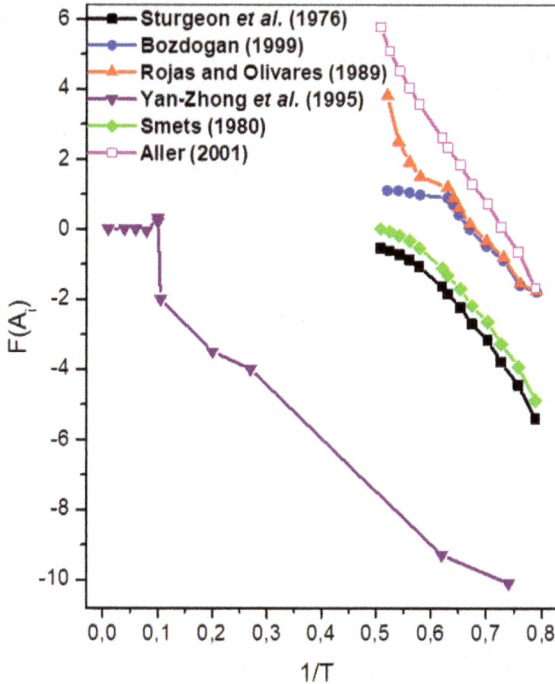

Fig. 7.3. Comparison between several theoretical methods for the calculation of the activation energy [Aller (2001)].

7.2.2.3. *Kinetic order*

Assuming that the activation energy, E_a, is not dependent upon the surface covered (σ), the temperature, T_p, for which the desorption rate, $\partial\sigma/\partial t$, is maximum can be deduced from the derivative of Eq. (7.31) with respect to time, t, and making it equal to zero. Increasing the amount of the injected sample (and hence σ), it's possible to follow [McNally and Holcombe (1987)] that:

- T_p shifts to higher temperatures for fraction values of the kinetic order ($0 \leq n < 1$),
- T_p is constant for kinetic orders equal to unity ($n = 1$), and
- T_p shifts to lower temperatures for kinetic orders higher than unity ($n > 1$).

Hence, the theoretical desorption kinetic orders can be derived from the shift suffered by T_p as a function of the amount of the analyte injected in the graphite tube, assuming only one desorption mechanism (i.e. E_a and v constants in Eq. (7.31)) [Wang *et al.* (1989)].

Besides the approaches from Rojas and Olivares (1992), Xiu-Ping *et al.* (1993a,b), Xiu-Ping and Zhe-Ming (1993) and Quan Zhe *et al.* (1994) (see the section before), there are other theoretical models allowing the derivation of the kinetic order of the atomization process from the absorbance peak profiles. Thus, Yan-Zhong *et al.* (1995) used the following equation for an isothermal atomisation process

$$\text{Ln}\left(\frac{dA}{dt} + k_2 A\right) = \text{Ln}\left(k_1 \varepsilon^{n-1}\right) + n\,\text{Ln}\left(\int_t^\infty k_2 A dt - A\right) \quad (7.38)$$

where ε is the proportionality constant (sensitivity) between the number of the analyte atoms in the optical pathway and the measured absorbance, and the plot of $\text{Ln}(\frac{dA}{dt} + k_2 A)$ versus $\text{Ln}(\int_t^\infty k_2 A dt - A)$ will provide a straight line with a slope equal to the kinetic order, n.

Cathum *et al.* (1991) proposed two methods to determine the kinetic order of the atomization process from the absorbance peak profile using the maximum absorbance. In the first method, they arrived at the following equation

$$\log(A_m T_m^2) = \log \int_{t_m}^\infty A dt + \log\left[\frac{\alpha E_a}{R}\right], \quad (7.39)$$

where the subscript m for each parameter refers to the maximum absorbance and α is the heating rate. The linearity of the plot of $\log(A_m T_m^2)$ versus $\log \int_{t_m}^\infty A\, dt$ verifies the first order assumption. The same authors [Cathum *et al.* (1991)] developed a second method to determine the kinetic order n_m of the atomization process for the maximum absorbance

$$n_m = \left[\frac{\alpha E_a}{RT_m^2}\right]\left[\frac{\int_{t_m}^\infty A dt}{A_m}\right]. \quad (7.40)$$

Histen and Holcombe (1998b), assuming a constant activation energy, E_a, for different amounts, c_0, of the analyte injected (which is not always true), used the following equation for the determination of the kinetic order

$$\text{Ln } A_\tau = n \text{ Ln } c_0 + \text{Ln}\left[v'k\left(\int_0^\tau e^{\left(-\frac{E_a}{RT}\right)}dt\right)\right],\qquad(7.41)$$

where, A_τ is the integrated absorbance between 0 and time τ, v' is a constant containing the pre-exponential and other proportionality and spectroscopic constants, k is a proportionality constant and c_0 is the initial sample concentration or mass.

Equation (7.41) needs to be applied in the first moments of the absorbance peak profile, because it is during this initial period of time that the atom generation process prevails over the dissipation process, and consequently the graphite surface covered by sample does not significantly change.

Bozdoğan (1999) determined the kinetic order using only one absorbance peak profile

$$\log\left(\frac{dA}{dt}\right) = \log k_1' + n\log(A_0 - A),\qquad(7.42)$$

where A_0 is the maximum absorbance, while k_1' is a proportionality constant including the rate constant of atoms formation, the surface area and the proportionality factor between the number of analyte atoms in the gas phase and the measured absorbance.

Aller (2001) proposed Eq. (7.43) to calculate the order of release, n, using only one peak profile, which needs to be non-temperature-dependent

$$\text{Ln}\left[\frac{\left(\frac{dA_i^e}{dt} + k_{2i}A_i^e\right)\left((A_p^e - A_i^e) + \int_t^{t_p} k_{2i}A_i^e\,dt\right)}{T^{m+0.5}}\right]$$

$$= \text{Ln}\left[p^{1-n}b\left(2\sqrt{\frac{E_0}{\pi k}}\right)^n e^{-mn}\right]$$

$$+ (1+n)\left(\text{Ln}\left[\frac{(A_p^e - A_i^e) + \int_t^{t_p} k_{2i}A_i^e\,dt}{\sqrt{T}}\right] - \frac{E_0}{kT}\right).\qquad(7.43)$$

where A_p^e and A_i^e refer to the absorbance at the peak and at any time, respectively, k_{2i} is the dissipation rate constant, t_p and t are the time at the peak and at any other moment of the absorbance-time profile, p is the proportionality factor between the measured absorbance and the number of the analyte atoms in the gas phase, b is a constant, E_0 and m and other terms have the same meaning as above (see page 302).

7.2.2.4. *Rate constants*

(a) Isothermal atomization

From the desorption rate and for a constant temperature, it is possible to derive

$$\text{Ln } A_t - \text{Ln } A_0 = -k''t \quad (n = 1) \qquad (7.44)$$

and

$$\frac{1}{(n-1)(A_t^{n-1})} - \frac{1}{(n-1)(A_o^{n-1})} = k''t \quad (n \neq 1), \qquad (7.45)$$

where n is the kinetics order, k'' the rate constant, while A_t and A_0 represent, respectively, the integrated absorbance during the time t and for all the analyte injected without losses. Equations (7.44) and (7.45) have been used to determine k_1 and k_2 (see Eq. (7.14)) using the rising and the falling edges of the absorbance-time profile, respectively [Zheng *et al.* (1984); Rayson and Johnson (1991); Slaveykova and Tsalev (1991, 1992)]. Other models [Chung (1984)] determine k_1 using absorbance at a particular temperature and the maximum absorbance, while other authors [Mandjukov *et al.* (1992)], studied the analyte losses prior to the atomization stage in the presence of chemical modifiers. In the last case, the model assumes the formation of a solid solution analyte-modifier and stabilization of analyte through the droping of the partial pressure of the analyte vapor at equilibrium. Hadgu *et al.* (1995) used similar expressions to study the analyte losses by diffusion through the end of the tube and the injection hole, but using a k_2 value dependent on the diffusion coefficient and the dimensions of the graphite tube.

(b) Non-isothermsal atomization: Increasing temperature

Several models have been developed to study the non-isothermal atomization processes, where the effect of temperature on the diffusion mechanism was undertaken [Akman *et al.* (1991); Rojas and Olivares (1989)], obviating the need for the first-order kinetics [Xiu-Ping *et al.* (1990); Cathum *et al.* (1991); Chakrabarti and Cathum (1991); Rojas and Olivares (1992); Rojas (1992)], and including dissipation of the analyte atoms by other processes different to diffusion, such as expansion, adsorption–desorption, and convection [Xiu-Ping *et al.* (1993a)].

Monte Carlo simulation is of the greatest interest to perform theoretical studies of the atomization process at growing temperatures [Güell and Holcombe (1991, 1992)]. By these methods, the effect of parameters, such as axial and longitudinal thermal gradients, partial covering of the atomizer, time-dependent analyte radial distribution, convection losses, gas phase reactions, analyte losses through the injection hole, variations in the original sample localization, analyte-graphite interactions, etc., is possible to be determined [Black *et al.* (1986)]. However, as uncertainity increases with the complex developed methods due to the high number of assumptions involved, small differences can be derived from application of several methods [Fonseca *et al.* (1994a)].

The relationship between the analytical signals and the atomization rate has been studied by Gilmutdinov and Shlyakhtina (1991). Transport of the atoms up to the gas phase and the time constant of the recorder devices, generate broadening, shift and decreased peak heights. Other models have been proposed [Akman *et al.* (1988, 1991)] for the first order kinetics. Mathematical treatments based on the first and second derivatives (Fig. 7.4) have been used to determine k_1 and k_2 [Rojas and Olivares (1989)].

The absorbance change rate shows the maximum and minimum values for the times t_x and t_y, respectively (Fig. 7.4) (where t_m and t_p represent the time values where absorbance and temperature reach the maximum and stable values). The same authors [Rojas and Olivares

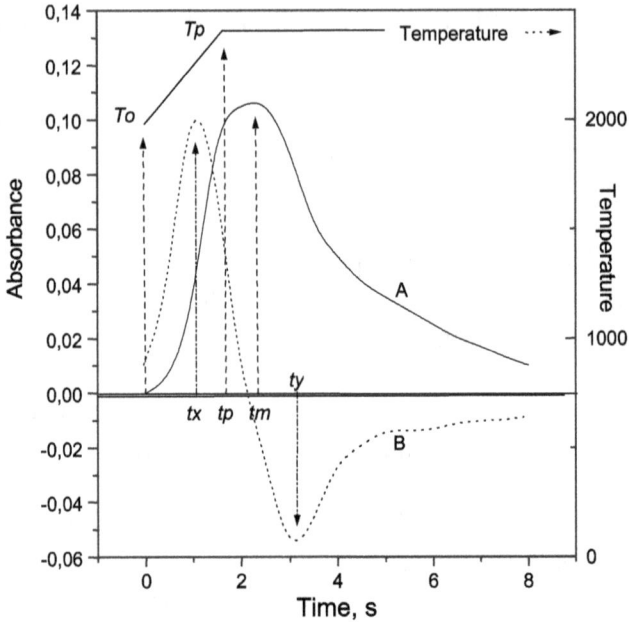

Fig. 7.4. Experimental absorbance pulse (A) and derivative of absorbance with time (B), showing the characteristic time values: t_p (time at the maximum temperature), t_m (time for the maximum peak absorbance), t_x (time for a maximum absorbance rate), and t_y (time for a minimum absorbance rate) [Rojas and Olivares (1989)].

(1995)] and others [Buhay *et al.* (2005)] have also developed kinetic mechanistic models based on two precursors.

The vapor atomic loss mechanism varies with the physico-chemical properties of the analyte species and the atomizer surface, the heating rate and the atomizer temeparture. Losses through the sample injection hole and through the graphite walls represent about 20% of the total [Sturgeon and Chakrabarti (1977)]. The main process contributing to the vapor atomic loss is diffusion towards the colder parts of the atomizer, where a condensation process can occur. However, in general, the dissipation process is the result of the combined action of at least three mechanisms: diffusion, thermal expansion, and convection. Each one is characterized by a kinetic constant, k_d, k_e, and k_c, respectively. So, k_2 represents the total kinetic constant for the dissipation process

$$k_2 = k_d + k_e + k_c. \tag{7.46}$$

Hence, the temperature-dependent kinetic constant k_2 can be calculated through the contribution of the three mechanisms remarked above from the following equation

$$k_2 T(t) = k_d^0 \left[\frac{T(t)}{T_0} \right]^y + \frac{CFT}{300 \, V} + \frac{\alpha}{T(t)} \tag{7.47}$$

where C is a proportionality factor, V, the atomizer volume, F, the gas flow, T, the temperature, α, the heating rate, y, a combination factor for the gas (diffusion factor) which varies usually between 1.5 and 2 (Table 7.2), while k_d^0 is the diffusion rate constant at the temperature $T_0 = 273.15$ K.

The value of k_d^0 can be calculated from the decaying zone of the absorbance-time profile

$$k_d^0 = \frac{8D}{l^2} = \frac{1}{\tau_2}, \tag{7.48}$$

where D is the diffusion coefficient at 273.15 K and 1 atm, and l is the length of the graphite tube. Values of the diffusion coefficients for several analytes have been recently investigated using experimental and theoretical approaches [Sadagov and Dědina (2002)]. It was stated that the diffusion rate of the atomic vapor grows with the amount of modifier (Pd), as a result of a gradual reduction in residence time, τ_2 and peak area [Sadagov and Katskov (2001)].

Table 7.2. Diffusion Factor, y, and Diffusion Rate Constant or Coefficient k_d^0 at 273.15 K [Rojas and Olivares (1992)]

Element	y	k_d^0
Cu	1.90 ± 0.04	0.020 ± 0.002
Li	1.80 ± 0.04	0.010 ± 0.002
Ni	1.65 ± 0.04	0.024 ± 0.002
Ru	1.70 ± 0.05	0.009 ± 0.002

7.3. Electrothermal Atomization Mechanisms: *Practical Considerations*

The vaporization Langmuir's theory, developed for ETAAS at atmospheric pressure, assumes an equilibrium vapor pressure above the limiting layer vaporization surface with a width of several times the mean free distance of the vaporizing particles. This is the result of multiple collisions between the vaporizing particles and the molecules existing in the graphite tube atmosphere. Vapor is put out of this layer by a diffusion mechanism based on a concentration gradient. Outside the limiting layer, transport of the particles can also be carried out by convection (thermal expansion, depending on the vaporizer design). According to this concept, the vaporization and atomization mechanism would be determined by the thermodynamic equilibrium existing in the limiting layer. Equilibrium is produced as a result of both homogeneous and heterogeneous reactions in which the sample components, gases of the atmosphere above the sample and the atomizer material can be implicated. The vapor partial pressure for each component in the limiting layer changes with the vaporizer temperature, the initial amount of the analyte, the presence of the active-chemically matrix components and the concentration of reactive impurities in the protective gas and/or the atomizer surface. The vaporization rate is strongly determined by the total vapor pressure of all the gaseous components above the sample surface containing analyte atoms [Katskov *et al.* (1994)]. Of course, several temperature-dependent atomization mechanisms can, simultaneously or not, occur. It is clear that the atomization mechanism for a particular element will depend on the experimental conditions used, among others, heating rate, initial analyte mass, and activity of the atomizer surface [Krakovská (1997)].

7.3.1. Basic Atomization Mechanisms

The general atomization mechanisms (Table 7.3) [Aller (2002)] are related to:

- reactivity of the solids formed on the atomizer surface,
- morphology of the atomizer surface (usually graphite),

Table 7.3. General Atomization Mechanisms [Aller (2002)]

Atomization Mechanisms	Example
Oxide thermal dissociation	$MO \xrightarrow{\Delta} M + 1/2O_2$
Oxide reduction	$MO + C^* \xrightarrow{\Delta} M + CO$
Sublimation	$M_{(s)} \rightarrow M_{(g)}$
Dimer dissociation	$M_{2(g)} \rightarrow 2M_{(g)}$

*Carbon (for graphite atomisers); Hydrogen, and/or Metal (for metal atomizers).

- adsorption processes, and
- homogeneous gas phase reactions.

From a general perspective, it is convenient to distinguish between those processes occurring in solid phase and those in gas phase.

7.3.1.1. *Solid phase reactions*

Solid phase reactions to yield atomic or molecular vapors have usually been classified into the following five categories [Majidi *et al.* (2000b)]:

- decomposition reactions,
- solid–solid reactions,
- solid–liquid reactions,
- solid–gas reactions, and
- catalytic reactions.

Each reaction usually implies at least one of the following mechanisms: adsorption/desorption processes, solid–solid interactions (homogeneous or heterogeneous), solid-phase nucleation (inside or outside the surface), and condensed phase diffusion (surface or bulk). By this way, adsorption of the analyte atoms on the tube walls, diffusion transport and heterogeneous distribution of the solid sample residue can produce an anisotropic distribution of the analyte atoms in the absorbent layer.

The global reaction rate is controlled by the rate of each implicated particular mechanism, which also depends on the chemical energy, the lattice structural parameters, and the structural defects (point, impurities, dislocations, electrons, and holes). The complexity of the electrothermal atomization processes is due to the strong relationship existing between all these mechanisms and the temperature, lattice strength, and lattice defect concentration. Thus, the heterogeneous exothermal reactions in the graphite atomizer affect the gas and substrate temperature, as well as its distribution in the analytical zone [Katskov *et al.* (1999)].

The graphite microstructure of the atomizer surface largely affects the reduction mechanism of oxides, through which the metal atoms are produced on before the generation of the gas phase analyte atoms. With the pyrolytic graphite coated graphite tubes, atoms are strongly fixed on the surface, while with the uncoated graphite tubes, atoms are intercalled, which is the cause of their thermal stability [Brennfleck *et al.* (1996)]. However, uncoated graphite atomizers favor reduction processes [Castro *et al.* (2005)].

Surface diffusion of the sorbed analyte species can control the nucleation and sintering rates of the *clusters* formed from the different analyte species that exist on the surface. Surface diffusion can also be associated with the incorporation rate of the sorbed species on the active centers. Generation of the active centers on the atomization surface is the result of some of the following processes:

- reactions between the graphite surface and oxygen, followed by de-sorption of CO and CO_2, or
- changes in the surface morphology, produced as a result of the lattice thermal stress.

Consequently, when the active centers are created during the heating of the atomizer, some competition between the nucleation rate (*cluster* growth) and the filling rate of the active centers exists. When the formation rate of the active centers is larger than the inclusion rate of the analyte adsorbates inside the *clusters*, filling of the active centers can predominate against the *cluster* growing. Consequently, the analyte desorption from the atomizer surface can become an important stage. On the other hand, chemi-sorption mechanisms are regulated by the geometry of the active center and the activation energy needed to dissociate a physic-sorbed molecule.

Chemi-sorption of oxygen on the active centers is usually behind the shifts in the appearance temperature of the gas phase analyte atoms (i.e. of the analytical signal). This is due to changes in the reduction rate of the analyte oxide on the graphite surface. In this situation, metal vaporization occurs as a result of one of the two following mechanisms:

- reduction of the metal oxides on the active centers (on the graphite surface), or
- thermal dissociation of the metal oxide.

Hence, blocking these active centers by the chemi-sorbed oxygen affects the reduction rate, which in turn governs the appearance temperature of the gas phase analyte atoms. The appearance temperature increase until the thermal dissociation of the analyte oxide become favorable, or alternatively the active centers, where reduction takes place, becomes accessible. The last condition is produced as a result of the oxygen desorption as CO or CO_2 from the graphite surface. Those analytes showing a shift in the appearance temperature are favorably vaporized in that temperature range most adequate for the chemi-sorption of oxygen (around 850 K, while total desorption is around 1200 K) [Salmon and Holcombe (1982)].

7.3.1.2. *Vapor phase reactions*

The multiple collisions that the gas phase species suffer against the atomizer surface provide many opportunities for interaction with the active centers. However, the macroscopic interactions between the atmosphere gases and the graphite surface can also affect the migration of the trapped gas phase species outside the graphite surface and inside the analysis volume. Interactions can produce a dense gaseous layer (Langmuir's layer) over the atomizer surface on which an equilibrium vapor pressure can exist. Formation of that layer results from the balance between the gas pressure and viscosity, while diffusion of the analyte species from this layer is forced according to the concentration gradient.

7.3.2. Atomization Mechanisms in the Absence of Chemical Modifiers

The atomization mechanisms have been experimentally studied using different analytical techniques [McNally and Holcombe (1991); Galbács *et al.* (1997); Styris and Redfield (1993); Eloi *et al.* (1995a,b); Masera *et al.* (1996); Majidi *et al.* (1997, 2000a,b)]. Mass spectrometry has proved to be very useful in providing information on the electrothermal atomization mechanisms. Real-time mass spectrometry has allowed the determination of the radial distribution of the analyte atoms and molecules in the graphite tube. On the other hand, many other techniques, such as X-ray diffraction, laser fluorescence, molecular spectroscopy, electron microscopy, electron beam microanalysis, radioactive isotopes tracers, Rutherford back-scattering spectrometry, Raman spectrometry, Fourier transform infrared spectroscopy (FT-IR) and shadow spectral filming techniques, have also provided very important contributions about the physical and chemical mechanisms occurring during the whole ET atomization process. Shadow spectral filming is a technique used to study the structure of the atom and molecule absorbent layers in a non-steady state in the ET atomizers [Gilmutdinov *et al.* (1991, 1992, 1993, 2005a,b); Voloshin *et al.* (2004)]. Obviously, ETAAS has also been largely used to study thermodynamic and physico-chemical properties of the atomized substances. Nonetheless, more information would be derived using ETAAS if a continuum emission source, a multi-channel detection device, and a controlled pressure atomizer could be used. On the other hand, using the method of the absolute rate analysis, important and unusual parameters, and relationships related to the atomization process have been derived [L'vov (1997a,b)]. In spite of the fact that the atomization mechanisms are usually dependent on several parameters (analyte bond type, experimental parameters and the existence of concomitant elements), using the above experimental techniques, some general conclusions have been established (Table 7.4) [Aller (2002)].

Table 7.4. Atomization Mechanisms for Some Analytes in the Absence of Chemical Modifiers [Aller (2002)]

Analyte	Atomization Mechanism
Alkaline	Thermal dissociation of metal hydroxides
Alkaline-earth	Thermal dissociation of oxides (in gas/condensed phase)
Fe, Co, Ni, Al, Ga, In, Bi, Ge, Tl, Pb, Se, As, Sb	Thermal dissociation of oxides, and/or Reduction of oxides
Cu	Sublimation of atoms and/or Dissociation of dimer
Mn, V, Y, Zn	Thermal dissociation of gas phase oxides
Cd, Au, Ag, Pt, Ir, Ru, Rh, Pd	Sublimation
U	Thermal dissociation of oxides, and/or Thermal dissociation of carbides
Sn	Reduction of oxide/chloride by CO Sublimation

7.3.2.1. *The alkaline metals*

Excepting lithium, the atomization mechanism for the other alkaline metals is thermal dissociation of hydroxide, but not of oxide [Styris (1986)].

7.3.2.2. *The alkaline-earth metals*

In the production of free atoms for the alkaline earth metals (Be, Mg, Ca, Sr, Ba), the following processes have been suggested:

- reduction of the analyte oxide by carbon, followed by sublimation of the metal,
- thermal dissociation of oxide in the solid phase (on the graphite surface) or in gaseous phase
- thermal dissociation of the metal chloride.

However, the predominant process seems to be thermal dissociation of metal oxides. Alkaline earth metal oxides are common intermediates, which are by heating (before vaporization) the corresponding chloride or nitrate [Ratliff (1996)]. Gas phase species related to the atomization of beryllium nitrate (5-20 ng Be) show the existence of oxide (BeO), dimer, tetramer and several carbides (Be_xC_y, where $x = 2$, $y = 1, 2, 4$, or $x = 1$, $y = 2$). As well, oxides as carbides appear in the temperature range where beryllium metal is also present. The atomization mechanisms for beryllium are similar to those of the other alkaline earth elements, with the particularity that beryllium forms polimeric oxides $(BeO)_n$ and the formation of the free atoms takes place according to the following reactions

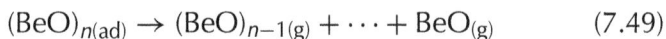

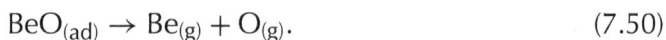

$$(BeO)_{n(ad)} \rightarrow (BeO)_{n-1(g)} + \cdots + BeO_{(g)} \qquad (7.49)$$

$$BeO_{(ad)} \rightarrow Be_{(g)} + O_{(g)}. \qquad (7.50)$$

So, the most important mechanism for the production of beryllium free atoms at atmospheric pressure is thermal dissociation of adsorbed oxide [Styris and Redfield (1987b)]. Beryllium oxide, which shows the lowest dissociation energy among all known oxides in the gas phase, also shows the lowest energy in the adsorbed phase. Formation of carbides takes place according to the following heterogeneous reaction

$$xBeO_{(g)} + (x + y)C_{(s)} \rightarrow Be_xC_{y(g)} + xCO_{(g)}. \qquad (7.51)$$

However, this reaction is different from the dissociative adsorption mechanism suggested for other alkaline earth elements. Atomization of Mg and Ca is produced by thermal dissociation in the gas phase of the corresponding oxide (MgO, CaO) after vaporization from the graphite surface [Chung (1984)].

A similar mechanism, together with some alternatives, has been assumed for strontium. Thus, direct formation of the gas phase atoms from the solid oxide occurs, i.e. desorption of the Sr atoms formed by dissociative chemisorption of oxide. Among the precursors of Sr atoms, carbides, oxides and hydroxides, are included [Prell *et al.* (1990)]. It is believed that BaO exists predominantly in the gas phase, and Ba free atoms are formed by thermal reduction of oxide in gas phase. Nonetheless, the low sensitivity found for the ETAAS determination of Ba suggests the formation of graphite-intercalated unvolatile carbides.

7.3.2.3. *The transition metals*

Among all the transition elements, we show as an example of the most frequently studied, the following: Fe, Co, Ni, Cr, Cu, Zn, Cd, Mn, V, Y, actinides (U).

Cobalt, Iron, Nickel

Using mass spectrometry, free Co atoms have been the only species detected during the atomization step of $Co(NO_3)_2$. Contrarily, during atomization of $CoCl_2$, important losses of Co (as chloride) occurred in the pyrolysis step. The most probable atomization mechanism for $Co(NO_3)_2$ in graphite atomizers, pyrolyrically coated or not, is the following [Chakrabarti and Cathum (1990)]

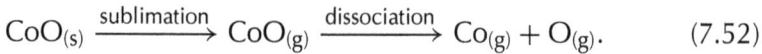

$$CoO_{(s)} \xrightarrow{\text{sublimation}} CoO_{(g)} \xrightarrow{\text{dissociation}} Co_{(g)} + O_{(g)}. \qquad (7.52)$$

Consequently, the cobalt atomic vapor is formed by sublimation or more exactly by dissociation of the gas phase oxide, $CoO_{(g)}$ [Chakrabarti and Cathum (1990)], whose activation energy, E_a, (380 KJ/mol) is closer to the dissociation energy of the gas phase oxide (361 KJ/mol) than to the sublimation heat (428.4 KJ/mol) of $Co_{(s)}$.

On the other hand, Alvarado (1996) proposes an alternative mechanism consisting in the reduction of cobalt oxide by carbon, where vaporization of the free metal is the limiting step of the global atomization process

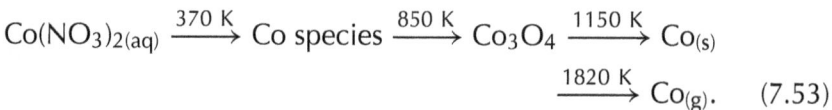

$$Co(NO_3)_{2(aq)} \xrightarrow{370 \text{ K}} Co \text{ species} \xrightarrow{850 \text{ K}} Co_3O_4 \xrightarrow{1150 \text{ K}} Co_{(s)}$$
$$\xrightarrow{1820 \text{ K}} Co_{(g)}. \qquad (7.53)$$

The atomization mechanisms for nickel and iron are similar to those proposed for cobalt [Alvarado (1996); Chung (1984)]. That is, reduction of the metal oxide by carbon to provide solid phase analyte atoms, which they are later vaporized.

Chromium

Evidences for the existence of two dependent-temperature mechanisms in the chromium atomization process have been provided [Fonseca *et al.* (1994b); Thomaidis and Piperaki (2000)].

- Chromium can form stable carbides by carbon reduction of chromium oxide in the temperature range 1000–1500 K [Wendl and Müller-Vogt (1984)]. Thus, thermal decomposition of the chromium carbide, is the first atomization mechanism

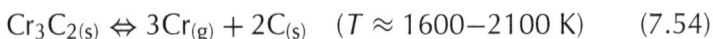

$$Cr_3C_{2(s)} \Leftrightarrow 3Cr_{(g)} + 2C_{(s)} \quad (T \approx 1600-2100 \text{ K}) \qquad (7.54)$$

and
- Thermal desorption of the adsorbed chromium atoms

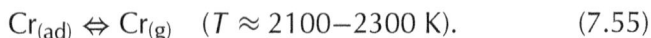

$$Cr_{(ad)} \Leftrightarrow Cr_{(g)} \quad (T \approx 2100-2300 \text{ K}). \qquad (7.55)$$

Copper

The proposed atomization mechanism for copper follows the reaction sequence [Wang *et al.* (1989)]

$$Cu(NO_3)_{2(s)} \rightarrow CuO_{(ad)} \rightarrow Cu_{(ad)} \rightarrow Cu_{(g)}. \qquad (7.56)$$

Nonetheless, other alternative mechanisms have also been suggested [Chung (1984)], at low pyrolysis temperatures

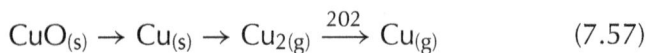

$$CuO_{(s)} \rightarrow Cu_{(s)} \rightarrow Cu_{2(g)} \xrightarrow{202} Cu_{(g)} \qquad (7.57)$$

and high pyrolysis temperatures

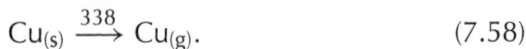

$$Cu_{(s)} \xrightarrow{338} Cu_{(g)}. \qquad (7.58)$$

For fast heatings, dimers formation is completed in a very short temperature range, due to the multiple collisions existing in the gas phase. For very low heating rates, different mechanisms can be observed. Thus, Chung (1984) and other authors, [Akman *et al.* (1991)], have suggested a dissociation process of Cu dimers (for high pyrolysis temperatures) and vaporization of the Cu atoms (for low pyrolysis temperatures). Readsorption of the Cu atoms on the atomizer graphite surface can occur, thus broadening the atomic absorption signal at atmospheric pressure.

Cadmium, Zinc

Cadmium atomization closely depends upon the physico-chemical properties of the atomization surface, because interactions between cadmium and the graphite surface on the active centers during the pyrolysis step produces Cd atoms [Alvarez *et al.* (1995)]. The predominant step in the atomization mechanism of Zn is a thermal dissociation process of the gas phase Zn oxides [Chung (1984)].

Manganese

With low heating rates and for high amounts of analyte, the following equilibrium,

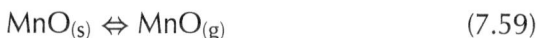

$$MnO_{(s)} \Leftrightarrow MnO_{(g)} \tag{7.59}$$

exists and under such a situation, the gas phase Mn atoms are formed via thermal dissociation of the gas phase oxide, $MnO_{(g)}$ [Akman *et al.* (1991)]. On the other hand, with ultra high heating rates and for small amounts of analyte, the equilibrium gas phase $\leftrightarrow$ solid phase is not attained. In this case, the dissociation energy of $MnO_{(g)}$ governs the atomization process. For different experimental conditions, different values of the atomization energy are obtained. So, differences found in the atomization mechanism can be assumed as due to the effect of the heating rate and the amount of analyte, because both affect the equilibrium Eq. (7.59).

Vanadium

The free atoms of vanadium formed in pyrolytic graphite coated graphite atomizers under an Ar atmosphere and at atmospheric pressure derived from the decomposition of $VO_{(g)}$, which in turn derived from $V_2O_{5(s)}$. Nonetheless, results derived from mass spectrometry suggest that the free atoms of $V_{(g)}$ originated from a sublimation process of $V_{(s)}$ [Styris and Kane (1982)], which can be explained by the fact that the appearance temperature (2200 K) for vanadium is similar to the melting point of this metal. On the contrary, it is also probable that gas vanadium atoms would be formed after dissociation of vanadium carbides in the temperature range 1000–1500 K [Wendl and Müller-Vogt (1984)].

Itrium

The atomization mechanism of Y from pyrolytic graphite surfaces or from metallic (tantalum, wolfram) surfaces follows the sequences [Prell and Styris (1991); Wahab and Chakrabarti (1981)]

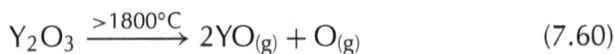

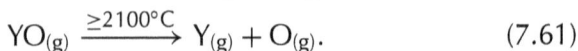

$$Y_2O_3 \xrightarrow{>1800°C} 2YO_{(g)} + O_{(g)} \tag{7.60}$$

$$YO_{(g)} \xrightarrow{\geq 2100°C} Y_{(g)} + O_{(g)}. \tag{7.61}$$

Under a thermal treatment of Y_2O_3 at atmospheric pressure, all the gas phase Y species observed (free Y atoms and YC_3 close to 2700 K; $YO_{(g)}$, $Y_2O_{2(g)}$ and $Y(OH)_3$ close to 2400 K) appear at higher temperatures than those noted for Eqs. (7.60) and (7.61). The compound YC_3 does not seem to participate in the atomization process. The formation of free Y atoms and $Y_2O_{2(g)}$ molecules implies that some gas phase interactions occur. Thus, the gas phase oxide $Y_2O_{2(g)}$ is a result of the interactions between the gas phase oxide molecules, $YO_{(g)}$, trapped in the Langmuir's film. Arising from multiple collisions against the graphite surface, the metal oxide molecules are dissociatively chemisorbed. The free Y atoms are formed via desorption at temperatures high enough to compensate for the related disociation energies.

Uranium

In the temperature range 2200–2600°C, uranium atoms are formed through a thermal dissociation from the oxide. However, at higher temperatures, uranium atoms are formed through a thermal disociation of the metal carbide [Goltz *et al.* (1995)].

7.3.2.4. *The precious metals: The platinum group*

The most studied metals from this group are: Au, Ag, Os, Ir, Ru, Rh, Pd and Pt.

Gold and silver

Elemental gold is present on the graphite surface before the vaporization process. So, this element is atomized through sublimation at the appearance temperature [Aller (1994)]. A first order has been observed for those conditions which favor the formation of adsorbed atoms (high dry temperatures, and/or very low Au concentrations) [Fonseca *et al.* (1993)]. However, vaporization of small amounts of gold (0.5– 5 ng) occurs with a fractionary order (0.80), decreasing even more (up to 0.5) with the amount of analyte (as much for wall atomization as for platform atomization) [Aller (1994)]. This suggests the existence of bidimensional sites on the graphite surface at the desorption temperature.

The shadow filming technique has been used to study the condensation process of the gold atomic vapor in ETAAS [Hughes *et al.* (1996a,b)]. The absence of uniformity noted in the distribution of the solid particles varies with the heating rate, temperature gradients in the tube, analyte mass, modifier mass, purge gas flow rate, and the presence and orientation of the sampling hole.

The activation energy is not dependent on the atomizer type and the analyte mass in a wide concentration range, although with high analyte masses, the desorption energy increases up to 66 Kcal/mol [Lynch *et al.* (1990)]. The differences between the activation energy for gold in the absence of modifier, usually ranging between 255.4 KJ/mol and 308.2 KJ/mol, regardless of the amount of gold present in the tube, have been provided [Aller (1994)]. Similar results have also been found for silver [Fonseca *et al.* (1993); Lynch *et al.* (1990); Chung (1984)]. So, the atomization mechanism suggested for silver included the direct vaporization of the metal atoms.

The atomization energies derived from the Arrhenius' plots suggest two kinetic processes with order unity. One of them is related to the dispersed particles desorption, and the other related to the atomization of small clusters, whose size and cohesive energy could increase with the initial mass of analyte (Ag). The desorption process could be governed by diffusion from the inside of the graphite to the surface [Rojas and Olivares (1995)].

The behavior of the metal drops on the graphite atomizer depends on their size, which affects the vapor pressure. Based on the Kelvin equation, Slaveykova *et al.* (1995) have proposed the following relationship between the vaporization process and the size of the evolved particles

$$\Delta H_r = \Delta H_v - \frac{2M\sigma}{\rho r}, \tag{7.62}$$

where ΔH_r is the vaporization enthalpy of the metal from the drop with radius r; ΔH_v is the vaporization enthalpy of the metal from a flat surface; M is the atomic mass; σ is the surface tension of melted metal; and ρ is the density of melted metal.

The platinum group

Platinum and iridium atomization mechanisms have been suggested to be similar to those assumed for silver and gold [Akman *et al.* (1991)]. Thus, some elements from this group (Ru, Rh, Pd, Ir and Pt) are reduced to the metallic state in the graphite atomizer, and this is followed by sublimation. On the contrary, for the osmium case, a volatile oxide is formed at low temperature. Later, some oxide is reduced to Os metal which is vaporized at temperatures above 2000°C [Byrne *et al.* (1997)]. Desorption of individual Pd atoms has been observed, which suggests strong interactions of Pd-grafite [Slaveykova *et al.* (1997)]. It was also shown that the atomization mechanism for low (1 ng) and high (8 ng) amounts of platinum changed progressively from an adsorption to an evaporation process [Eleni *et al.* (2005)].

7.3.2.5. *Metals from the groups III–V*

The most representative elements from these groups include Al, Ga, In, Tl, Ge, Sn, Pb, Bi, although Al is the most largely studied.

The spatial distribution of atoms and molecules in the atomizer suggests a non-uniform transversal and longitudinal distribution of some elements (Al, Ga, and In). On the other hand, free atoms (Al, In, Ga) distribution differs from that of their gas phase compounds. Inhomogeneity is explained assuming that analyte is vaporized as oxide, which is then dissociated or reduced to free metal atoms, later condensing on the atomizer walls. So, atom concentration is higher in the vicinity of the tube walls and is the minimum close to the injection hole. However, the analyte molecules are more concentrated along the tube axis and less close to the atomizer walls. Atom distribution close to the tube walls decreases in the wall atomization in comparison with the platform atomization. In the last one, the inverse atomization effect is noted, i.e. atoms are converted into the gaseous phase, not from the platform but from the atomizer wall. Atoms diffuse inside the tube and condense on the platform, which is at lower temperature, being then transferred to the gaseous phase after the platform is heated. Thus, the atomic vapor density increases in the transversal section of the tube after formation of the absorbent layers of the elements (Bi, Ge and Tl), but the molecular vapor density decreases.

Aluminium

Thermal dissociation of Al oxide has been proposed as the atomization mechanism for Al [Chung (1984); Styris and Redfield (1987a)]. However, other alternative atomization mechanisms are also plausible. The origin of the molecular absorption bands at 206.0, 214.0 and 255.0 nm has been suggested by L'vov as due to the presence of gas phase aluminium carbides, although it also seems to be related to some gas phase oxide, $Al_2O_{(g)}$ [Katskov *et al.* (1992)]. Splitting observed in the absorption profile for the atomization of high Al masses was initially explained as the evidence that Al oxides were reduced by carbon. Splitting is initiated as a result of rapid generation of atoms from an auto-catalytic reaction. In this process, a gas phase Al carbide is formed (Eq. (7.63)), which participates in the reduction of Al oxide (Eq. (7.64)). The subsequent decomposition of the gas phase carbide provides a graphite film on the oxide particle surface, stopping the reaction responsible for the splitting. An ulterior oxidation (Eq. (7.65)), removes that film, giving rise to a new splitting process as a result of the re-initialization of the auto-catalytic process [L'vov (1996)],

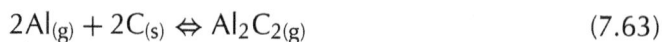

$$2Al_{(g)} + 2C_{(s)} \Leftrightarrow Al_2C_{2(g)} \qquad (7.63)$$

$$2Al_2O_{3(s)} + 3Al_2C_{2(g)} \Leftrightarrow 10Al_{(g)} + 6CO \qquad (7.64)$$

$$Al_2O_{3(s)} + 3C_{(s)} \Leftrightarrow 2Al_{(g)} + 3CO. \qquad (7.65)$$

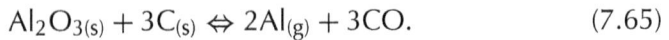

The splitting process is related to the re-distribution of the sample in the furnace, in such a way that splitting is more intensive the smaller the size of the sample microparticles [Bendicho and Loos-Vollebregt (1990)]. Spatial resolved determination of aluminium shows that the formation of the Al atoms is largely dominated by the oxygen partial pressure, pO_2, but not by the presence of gas phase species containing carbon. During the thermal treatment and before atomization, the dry residue from Al solution is transformed into Al_2O_3 in the temperature range between 500 °C and 1000°C. A very favorable mechanism for the formation of $Al_{(g)}$ is thermal dissociation of Al carbides, which are formed through the following global reaction

$$2Al_2O_{3(s)} + 9C_{(s)} \Leftrightarrow Al_4C_{3(s,l)} + 6CO_{(g)}. \qquad (7.66)$$

It is possible that part of Al_4C_3 would be vaporized during formation, then dissociated in the gas phase to form Al atoms

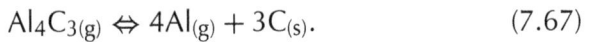

$$Al_4C_{3(g)} \Leftrightarrow 4Al_{(g)} + 3C_{(s)}. \qquad (7.67)$$

Al atoms react with Al_2O_3 to form $Al_2O(g)$ or $AlO(g)$

$$Al_2O_{3(s)} + 4Al_{(g)} \Leftrightarrow 3Al_2O_{(g)} \qquad (7.68)$$

$$Al_2O_{3(s)} + Al_{(g)} \Leftrightarrow 3AlO_{(g)}. \qquad (7.69)$$

After formation, Al_2O is reduced on the hot graphite surface to form Al atoms

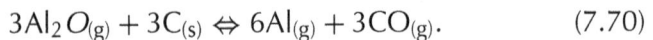

$$3Al_2O_{(g)} + 3C_{(s)} \Leftrightarrow 6Al_{(g)} + 3CO_{(g)}. \qquad (7.70)$$

So, for every four atoms of Al disappearing in reaction (7.68), six atoms are produced in reaction (7.70), deriving a 1.5 multiplicative factor. The above auto-catalytic mechanism requires that: (i) Al and Al_2O are simultaneously present in the gas phase; (ii) Al_2O is the main precursor for the formation of Al atoms; and (iii) the spatial distributions for Al and Al_2O should be complementary [Lamoureux *et al.* (1995); Hughes *et al.* (1996a,b); Castro and Aller (2003)]. However, alternative gas phase reactions are possible, where the gas phase aluminum atoms can also react with oxygen, according to the following thermodynamically favorable processes,

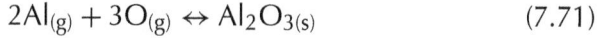

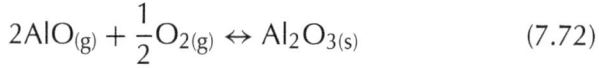

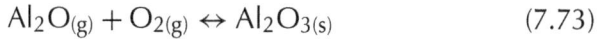

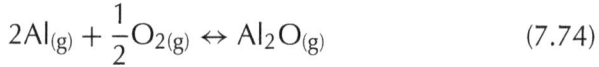

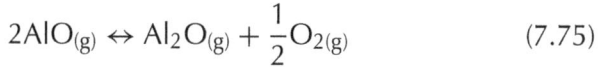

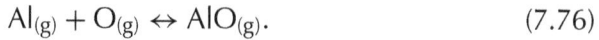

$$2Al_{(g)} + 3O_{(g)} \leftrightarrow Al_2O_{3(s)} \tag{7.71}$$

$$2AlO_{(g)} + \frac{1}{2}O_{2(g)} \leftrightarrow Al_2O_{3(s)} \tag{7.72}$$

$$Al_2O_{(g)} + O_{2(g)} \leftrightarrow Al_2O_{3(s)} \tag{7.73}$$

$$2Al_{(g)} + \frac{1}{2}O_{2(g)} \leftrightarrow Al_2O_{(g)} \tag{7.74}$$

$$2AlO_{(g)} \leftrightarrow Al_2O_{(g)} + \frac{1}{2}O_{2(g)} \tag{7.75}$$

$$Al_{(g)} + O_{(g)} \leftrightarrow AlO_{(g)}. \tag{7.76}$$

In conclusion, different aluminum oxides may be formed, which are regulated by both the oxygen content and temperature. The above mechanisms, proposed for the formation and dissipation of Al splitting, probably represents only a part of a more general and complex mechanism, which should include the participation of other species, such as AlH and Al_2. It is known that such species are present when microgram amounts of Al are atomized [Ohlsson (1992)], but they have never been considered in any of the mechanisms already proposed [Holcombe *et al.* (1991)]. The formation of $AlO_{(g)}$ and $AlO_{2(g)}$ decreases with increasing input amounts of hydrogen, but hydrogen cannot completely suppress the formation of potential interfering amounts of aluminium oxide. In any case, the presence of hydrogen and oxygen decreases sensitivity in the determination of aluminium [Persson *et al.* (1977a,b)].

Silicon

Silicon atoms are formed from $SiO_{2(s)}$ and $SiO_{(g)}$, and at extremely high temperatures from $SiC_{(s)}$ [Frech and Cedergren (1980)]. The main prerequisite for the formation of $SiC_{(s)}$ is an extremely low partial pressure of oxygen and high temperatures. It seems that $SiO_{(g)}$ is always formed between 1600 K and 2000 K, independently of the partial pressure of oxygen.

Tin

The precursors of Sn atoms have been established as $SnCl_{2(g)}$ and $SnO_{(g)}$. The atomization is produced in those atomizer regions where the temperature is lower. The proposed mechanisms involve the following reactions [Brown and Styris (1993)]

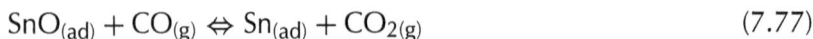

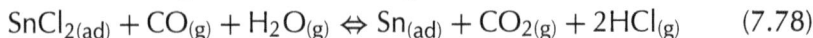

$$SnO_{(ad)} + CO_{(g)} \Leftrightarrow Sn_{(ad)} + CO_{2(g)} \tag{7.77}$$

$$SnCl_{2(ad)} + CO_{(g)} + H_2O_{(g)} \Leftrightarrow Sn_{(ad)} + CO_{2(g)} + 2HCl_{(g)} \tag{7.78}$$

$$Sn_{(ad)} \Leftrightarrow Sn_{(g)}. \tag{7.79}$$

In the presence of oxygen, similar atomization mechanisms have also been proposed [Müller-Vogt *et al.* (1996)]

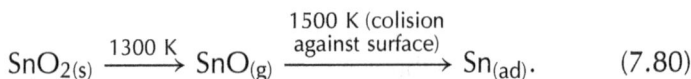

$$SnO_{2(s)} \xrightarrow{1300 \text{ K}} SnO_{(g)} \xrightarrow[\text{against surface}]{1500 \text{ K (colision}} Sn_{(ad)}. \tag{7.80}$$

Atomization of Sn in pyrolytic graphite coated pyrolytic graphite tubes seems to occur in two different manners, depending on the temperature. Thus, two kinetics with first and fractionary orders, and activation energies of 450 KJ/mol and 210 KJ/mol, for low and high temperatures, respectively, appear. The atomization rate-limiting stage at low temperatures is the metal oxide reduction by carbon

$$SnO_{2(l)} + C \Leftrightarrow Sn + CO_2. \tag{7.81}$$

At high temperatures, the atomization rate-limiting stage is controlled by the thermal dissociation of diatomic molecules

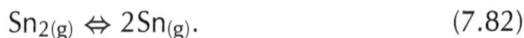

$$Sn_{2(g)} \Leftrightarrow 2Sn_{(g)}. \tag{7.82}$$

Lead

The existence of the gas phase oxide molecules, PbO, is an evident prove of the participation of that compound as an important precursor of Pb atoms. The two following reactions show two possible mechanisms [Gilchrist *et al.* (1992); Bass and Holcombe (1988a); Sturgeon *et al.* (1983)]

$$PbO_{(s)} \rightarrow Pb_{(g)} + \frac{1}{2}O_{2(g)} \quad (7.83)$$

$$PbO_{(s)} + C_{(s)} \rightarrow Pb_{(g)} + CO_{(g)}. \quad (7.84)$$

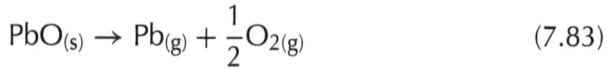

These mechanisms seem to exist for oxidative matrices (i.e. HNO_3). However, the most probable precursor of the Pb atoms is the gas phase lead chloride, $PbCl_{2(g)}$, for an hydrochloric matrix (i.e. HCl, NaCl) [Gilchrist *et al.* (1993)].

Indium

The atomization rate for In in non-coated graphite tubes, as compared to that for pyrolytic graphite coated pyrolytic graphite tubes, decreases from the first order kinetics at low temperatures to fractionary order kinetics at higher temperatures [Xiu-Ping *et al.* (1993b)]. This suggests that In atoms change from a dispersed state to an agglomerate state with temperature. However, using pyrolytic graphite coated graphite tubes atomization of In is regulated by a unique mechanism with a kinetics order close to 2/3. Liberalization of In atoms takes place via vaporization from the In melted drops.

The thermal decomposition of the metal oxide [Imai *et al.* (1998)], has also been proposed as an alternative way to generate atoms of In in a graphite tube

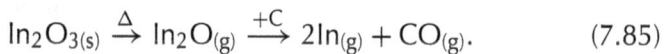

$$In_2O_{3(s)} \xrightarrow{\Delta} In_2O_{(g)} \xrightarrow{+C} 2In_{(g)} + CO_{(g)}. \quad (7.85)$$

If the graphite tube is coated with wolfram, the formation of the oxide, $In_2O(g)$, is avoided and the atomization process is as follows

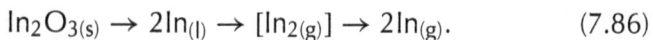

$$In_2O_{3(s)} \rightarrow 2In_{(l)} \rightarrow [In_{2(g)}] \rightarrow 2In_{(g)}. \quad (7.86)$$

7.3.2.6. *The non-metals (Se, As, Sb, Te)*

The atomization mechanisms for the non-metal elements are closely linked to either the thermal dissociation or reduction of the oxides.

Selenium

L'vov has suggested that the appearance temperature (400–500 K) of Se carbides agrees with the theoretical temperature for the thermal reduction of SeO_2. For this reason, atomization of Se in a graphite tube and at atmospheric pressure seems to be a consequence of the thermal dissociation of the Se oxide adsorbed on the graphite surface [Fischer and Rademeyer (1998a); Styris *et al.* (1991b)]. $Se_{(ad)}$, generated in the reaction of SeO_2 with carbon, can again form carbides. It has been noted that Se is lost as much Se molecules (dimers and oxides) as Se elemental before the atomization stage. The vaporization/sublimation of condensed phase Se produces gaseous Se polimers. This fact assumes Se dimer as the precursor of $Se_{(s,l)}$ atoms in the atomizer [Styris (1986)]. However, $SeO_{2(g)}$ is formed via sublimation from the condensed oxide and atomization at atmospheric pressure is the result of the thermal dissociation of these gas phase oxide [Droessler and Holcombe (1987)].

Arsenic

The atomization mechanisms for As are similar to those shown for Se [Styris (1986)], usually occurring through the vaporization of As_2O_3 as AsO and the dissociation of the last oxide [Akman *et al.* (1982)]. The nature of the graphite surface is very important in the As atomization [Korečkova *et al.* (1981)]. Thus, formation of As atoms seems to occur according to two atomization paths (low and high temperature) for non-pyrolytic graphite and pyrolytic graphite, but only to the high temperature mechanism for activated carbon [Imai *et al.* (2000)].

Antimonium

The atomization mechanisms suggested for antimonium include [Chung (1984)]: (i) thermal dissociation of Sb halide from HCl solutions; and (ii) gas phase dimer dissociation from HNO_3 solutions.

Tellurium

The main atomization reactions for the atomization of tellurium were suggested for both uncoated and Zr-treated tubes, and coated tubes [Müller-Vogt *et al.* (2000)].

(a) Uncoated and Zr-treated tubes,

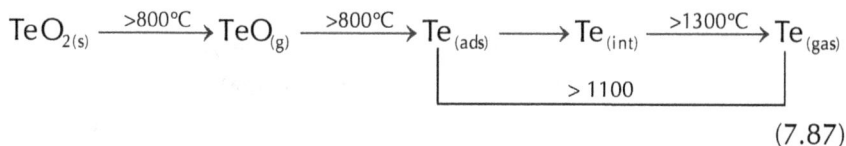

$$TeO_{2(s)} \xrightarrow{>800°C} TeO_{(g)} \xrightarrow{>800°C} Te_{(ads)} \longrightarrow Te_{(int)} \xrightarrow{>1300°C} Te_{(gas)}$$

$$\underset{> 1100}{\underline{\hspace{6cm}}}$$

$$(7.87)$$

and (b) coated tubes,

$$TeO_{2(s,l)} \xrightarrow{>500°C} TeO_{(g)} \xrightarrow{>700°C} Te_{(ads)} \xrightarrow{>900°C} Te_{(gas)}. \qquad (7.88)$$

7.3.3. Atomization Mechanisms in the Presence of Chemical Modifiers

The presence in the atomizer of any concomitant chemical species can strongly affect the atomization process. Consequently, the atomization mechanism of any analyte will also be altered by the presence of any chemical modifier. In these cases, some parameters characterizing the atomization process, such as activation energy, vary as a function of the type and amount of the chemical modifier used [Mazzucotelli and Grotti (1995); Aller (1997); LeBihan *et al.* (1998); Thomaidis and Piperaki (2000)], but in general, the analyte behavior is usually homogenized. Below, comments are given about the effects derived from the use of some of the most common chemical modifiers. The three following categories will be established: (a) *Non-metal inorganic chemical modifiers* {Gases (Oxygen), Phosphates (P)} , (b) *Metal inorganic chemical modifiers* {Ni, Mg, Pd salts} and (c) *Organic chemical modifiers.*

7.3.3.1. *Metal analytes*

(a) Non-metal inorganic chemical modifiers

One of the most widely used non-metal inorganic modifiers is oxygen. Gas phase inorganic modifiers were used, mixed together with a protective gas. Thus, using Ar as a purge gas containing 1% O_2, a shift towards higher values of the appearance temperature of the atomic absorption signal of Pb was observed [Sturgeon and Berman (1985)]. A similar effect, although in a lesser extent, was produced using oxygenated graphite surfaces. The amount of the adsorbed oxygen increases with the pretreatment stage temperature [Eloi *et al.* (1995a)]. To explain shifts, the existence of a heterogeneous reaction between O_2 and the condensed analyte was firstly assumed. Alternatively, a gas phase homogeneous reaction between the free analyte atoms and O_2 has also been considered. The last possibility is more probable because it is supported by the same shift in the appearance temperature for the vacuum atomization conditions. On the other hand, the content of CO and CO_2 grows with the amount of oxygen present in the atomizer. For this reason, the CO_2/CO ratio might be the most important factor controlling the shift in the appearance temperature. Consequently, if CO is formed from the reaction between CO_2 and graphite, when atomization is carried out using oxygen-free surfaces by incorporating 30% CO_2 in the purge gas, the same CO_2/CO ratio as that for oxygenated surfaces will be obtained. By modifying the CO_2/CO ratio, reduction of the analyte oxides is altered.

Phosphorous as a chemical modifier produces metal phosphates [Eloi *et al.* (1993)]. Thus, during the atomization of lead, Pb atoms appeared at 1150 K as a result of the thermal decomposition of the condensed lead phosphate. In a similar way, reduction of adsorbed PbO on the graphite active sites seems also to occur [Sturgeon *et al.* (1983)]. For the Cd case, solid $CdPO_2$ is formed, later changing to $CdPO_x$ at temperatures below 400 K. The existence of CdP_xO_y species has also been suggested, subsequently being transformed to P_3O_7 and gas phase Cd atoms at temperatures close to 1200 K [Hassell *et al.* (1991)].

(b) Metal inorganic chemical modifiers

• *Magnesium* as a chemical modifier

The use of magnesium as a chemical modifier shifts the Al peak profiles towards lower temperatures. This can be due to either an oxidation process of Al by $MgO_{(g)}$ forming non-stoichiometric Al oxides, or a rapid decomposition of Al_2O_3. The following reactions have been suggested [Styris and Redfield (1987a)]

$$MgO_{(g)} + 2Al_{(ad)} \rightarrow Al_2O_{(ad)} + Mg_{(g)} \qquad (7.89)$$

$$MgO_{(g)} + Al_2O_{(ad)} \rightarrow Al_2O_{2(ad)} + Mg_{(g)} \qquad (7.90)$$

$$MgO_{(g)} + Al_2O_{2(ad)} \rightarrow Al_2O_{3(ad)} + Mg_{(g)}. \qquad (7.91)$$

Stabilization of aluminium by magnesium during thermal treatment is achieved if Mg is injected in a concentration ratio of 1–500 Al/Mg wt/wt, owing to inhibition in the formation of gas phase aluminium hydroxide [Al(OH)$_3$], formed in the absence of Mg near to 1300 K. The existence of species such as MgOH$_{2(g)}$ and MgOH$_{(g)}$ near to 500 K suggests that MgO is hydrated at relatively low temperatures. Consequently, decreasing the amount of water in the atomizer, reduces the formation of Al(OH)$_{2(g)}$ and so inhibits losses of analyte.

Magnesium also produces a negative shift (400–600 K) in the appearance temperature of gas phase Be molecular species (oxides, carbides). As these species have been formed as a result of a desorption process, it is assumed that magnesium oxide protects those active sites on the graphite surface showing a high adsorption heat. In the presence of Mg, formation of Be(OH)$_2$ is inhibited and atomization of Be occurs via thermal decomposition of adsorbed monomeric oxide [Styris and Redfield (1987b)].

The behavior of magnesium as a chemical modifier was confirmed by Mofolo *et al.* (2001), who based their conclusion on the suppression of the molecular band at 205 nm (tentatively assigned to In$_2$) and the simultaneous appearance of In ion lines in the presence of MgO.

• *Palladium* as a chemical modifier

Palladium is the most widely used chemical modifier, and has many times been named the *universal* modifier. The effect of Pd on the electrothermal atomization of many analytes depends on both the amount and physical state of the chemical modifier [Alvarez *et al.* (1995)]. The compound PdCl$_2$ used as a chemical modifier is reduced on the graphite surface at 800°C, and the presence of metals in solution does not significantly alter that reduction process [Zhe-ming and De-qiang (1995)]. If the analyte and palladium form an intermetallic compound and its activity coefficient decreases, the vaporization or atomization of the analyte is delayed till higher temperatures are attained. At the melting point, the analyte atoms are strongly trapped on the intermetallic compound surface, resulting in vibration and vaporization [Yasuda *et al.* (1993, 1994)]. Atomization of some analytes can be carried out from specific alloying phases. Hence, by measuring the temperature of the atomic vapor from an alloy, it is possible to verify that the atomization starts from certain specific phases of the intermetallic compounds. Thus, during the atomization of Sn, several alloying phases, $PdSn \sim Pd_3Sn_2 \sim Pd_2Sn \sim Pd_3Sn$, have been observed for the Pd–Sn intermetallic compound. In a similar way, it was proved that Pb is atomized from the phases $Pd_3Pb_2 \sim Pd_3Pb$, and In from $PdIn_3 \sim Pd_2In_3 \sim PdIn \sim Pd_2In \sim Pd_3In$. The lattice of the intermetallic compound Pd–In vibrates several times, and after that the In atoms are converted into vapor. The intermetallic compound Pd–In existing after the In vaporization is pulled toward other phases, in which the atomic percentage of Pd increases, and eventually attaining 100% Pd [Yasuda *et al.* (1995)]. On the contrary, those systems, such as Pb–Fe, do not present a solid solution at any concentration range, and for this reason, Fe does not affect the atomization of Pb [Oishi *et al.* (1991)].

• *Other chemical modifiers* (**Vanadium**)

In the presence of vanadium (at 0.1%) as a chemical modifier, the order of the desorption process for gold atoms was unity, as much for wall atomization as for platform atomization. The activation energy of the atomization process in the presence of the modifier increases considerably (up to 741.6 KJ/mol), which suggests the formation of a very stable Au–V compound [Aller (1994)]. The atomization mechanism seems to be related to a first order desorption process, in which strong interactions take place.

• *Coated tubes*

The atomization mechanisms are dependent on the *nature of the atomizer surface*, i.e. on the material used to build the atomizer. Materials other than graphite have been used to make or to coat atomizers. Thus, some metals (W, Mo, Pd, Zr) and metal carbides showing high melting points have been used for these purposes, where the coating acts as a chemical modifier. Free atoms of analytes are usually produced by direct reduction of the nitrate salts by tungsten or hydrogen [Krakovská and Remeteiová (2000)]. The coated surfaces provide high thermal stabilization for many analytes [Volynsky (1998)], probably due to the blocking of the active centers, so avoiding analyte-graphite interactions, but favoring strong analyte-metal interactions [Quan Zhe *et al.* (1994)]. Atomization from a Zr- or Pd-coated pyrolytic graphite surface is produced from the analyte-modifier solid solutions. Atomization under these conditions shows first-order kinetics, as well at low as at high temperatures, suggesting the presence of only one atomization mechanism [Xiu-Ping *et al.* (1993a,b)]. The contrary is true for the non-coated tubes where a few releasing stages usually occur.

On the other hand, if the atomization surface is coated with any metal carbide (WC, MoC, etc.) or if these metals (W, Mo) are used as chemical modifiers, the analyte atoms are usually widely dispersed on the surface. However, the presence of Pd favors the formation of aggregates [Alvarez *et al.* (1996)]. The Zr carbide favors the formation of the gas phase Sn atoms, providing the shift of the atomic absorption signal toward lower appearance times [Xiu-Ping *et al.* (1993a,b); Xiu-Ping and Zhe-Ming (1993); Quan Zhe *et al.* (1994)].

(c) Organic chemical modifiers

Pyrolysis of an organic matrix provides two types of carbon: activated carbon (formed above 600 K and released at between 950 K and 1100 K), and amorphous carbon (thermally stable above 1100 K and formed during the releasing of the activated carbon) [Imai *et al.* (1996)]. The activated carbon (in condensed phase) acts as a reducer against metal oxides. The thermally stable amorphous carbon forms a thin layer and reduces directly and totally the metal oxides up to the elemental state. On the active centers of the amorphous carbon, very small drops are formed, which became vapor easier than they grow. By contrast, using pyrolytic graphite surfaces, which have been previously treated with an organic solution at temperatures above 1100 K, very large drops can be obtained due to collisions with other drops. Hence, the amorphous graphite surfaces play a key role in the movement, collision, and coalescence of the sample drops, thus reducing diffusion of the largest drops into the graphite surface [Imai *et al.* (1996)].

One of the most widely used organic modifiers is ascorbic acid. The atomization rate of an analyte grows with the presence of ascorbic acid, since its pyrolysis increases the number of active centers. By contrast, the atomization rate is low in the presence of the inorganic metal chemical modifiers, probably due to saturation of the active centers. The analyte atomization is simultaneously produced from the dispersed particles and from the small *clusters*. Atomization from the clusters would occur for a low number of the active centers and for the high amounts of the analyte injected. In this case, the *adatom–adatom* interactions are favored. Atomization from the dispersed particles prevails for the high amounts of the active centers and the heating rate of the atomizer, assuming the *adatom*-surface interaction [Rojas (1995)]. For the thermal pretreatment temperature below 970 K, the presence of ascorbic acid has little effect on the atomization of metals such as Mn. However, the chloride interference is overcome using high pretreatment temperatures. The presence of ascorbic acid stops the usual fast hydrolysis of a matrix of $MgCl_2(s)$. The residual amount of chloride grows with the presence of ascorbic acid, probably due to the possibility of ascorbic acid forming a complex with the Mn ion in the solid phase, so facilitating losses of the hydrated water during the pretreatment stage.

7.3.3.2. *Non-metal analytes*

Arsenic

Arsenic can be stabilized by graphite provided that an auxiliary oxidizing agent is present. Thus, nitric acid, hydrogen peroxide, potassium permanganate or oxygen facilitates separation of the graphite layers and creates more available active sites [Korečkova *et al.* (1981)].

Through the use of Cu or Co as a chemical modifier, As is atomized earlier than forming intermetallic compounds. However, in the presence of Pd or Ni, atomization of As starts in the vicinity of the melting point of intermetallic compounds [Hirano *et al.* (1994)]. The activity coefficient of arsenic is smaller than unity if intermetallic compounds are formed. However, the effective temperature of the atomic vapor of any analyte can be increased by the addition of metal chemical modifiers for a rapid atomization. On the other hand, the unexpected increase in the effective temperature of the atomic vapor can also be due to some atomizer factors. It is true that Mn and other metals obtain energy from As or the As-Pd alloy during the atomization process, so reducing the As atom density in the upper energy level.

The appearance temperature of the As free atoms are not dependent on the Pd oxidation state (Pd metal, PdO), being about 150 K above the temperature (1450 K) found in the absence of palladium. Nonetheless, PdO prevents the freeing of the As(s,l) atoms and formation of $As_2(g)$ but not decomposition of As_2O_3. In conclusion, PdO is more strongly associated with As(s,l) than with $As_2O_3(s)$. Formation of a solid solution or any stoichiometric compound containing oxygen is largely behind the above results, probably according to the following reactions [Styris *et al.* (1991a)],

$$x PdO_{(s)} + y As_{(s,l)} \rightarrow [Pd_x As_y O_z] + (x - z)O_{(g)} \qquad (7.92)$$
$$x PdO_{(s)} + y As_{(g)} \rightarrow [Pd_x As_y O_z] + (x + y - z)O, \qquad (7.93)$$

where brackets represent the compound or the solid solution formed. Free As atoms appear at high temperatures, very close to 1600 K after transformation of the compound $[Pd_x As_y O_z]$. The same compound or solid solution can also be formed if the reduced chemical modifier is used (Pd^0). This modifier prevents the formation of As oxides and dimers, $As_2(g)$, while a direct interaction with the condensed phase oxide can exist, following the reaction,

$$2x Pd_{(s)} + y As_2 O_{3(s)} \rightarrow 2[Pd_x As_y O_z] + (3y - 2z)O. \qquad (7.94)$$

The formation of a solid solution is responsible, at least partly, of an increase in the appearance temperature of about 210–240 K as well as in the integrated absorbances, when large amounts of palladium are used.

Selenium

Using Pd as a chemical modifier, it was observed that Se was atomized from some particular solid phases, $Se_x Pd_y O_z$ [Styris *et al.* (1991b);

Majidi and Robertson (1991)]. The presence of Pd^0 seems to be implicated in a solid heterogeneous reaction with Se. Nonetheless, at temperatures below 400 K, Pd^0 interacts with some selenium oxides, mainly SeO_2. It has been proposed [Styris *et al.* (1991b)] that stabilization of SeO_2 by Pd^0 involves the following reactions.

(i) *Inhibition of the selenium oxide hydration:*

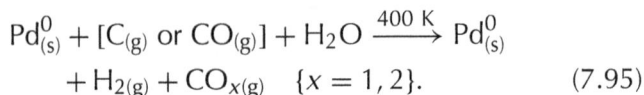

$$Pd^0_{(s)} + [C_{(g)} \text{ or } CO_{(g)}] + H_2O \xrightarrow{400 \text{ K}} Pd^0_{(s)}$$
$$+ H_{2(g)} + CO_{x(g)} \quad \{x = 1, 2\}. \tag{7.95}$$

(ii) *Formation of compounds [Pd, Se, O]:*

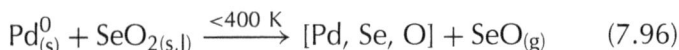

$$Pd^0_{(s)} + SeO_{2(s,l)} \xrightarrow{<400 \text{ K}} [Pd, Se, O] + SeO_{(g)} \tag{7.96}$$

and

$$Pd^0_{(s)} + SeO_{2(g)} \xrightarrow{>400 \text{ K}} [Pd, Se, O] + SeO_{(g)}. \tag{7.97}$$

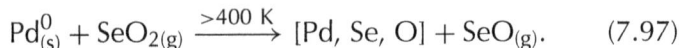

(iii) *Dissociation followed by trapping of Se and Pd (in the Langmuir layer):*

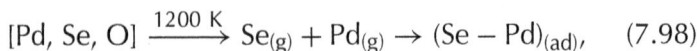

$$[Pd, Se, O] \xrightarrow{1200 \text{ K}} Se_{(g)} + Pd_{(g)} \rightarrow (Se - Pd)_{(ad)}, \tag{7.98}$$

where $(Se-Pd)_{(ad)}$ means that Se and Pd are re-adsorbed on the graphite surface.

(iv) *Decomposition and desorption of free species:*

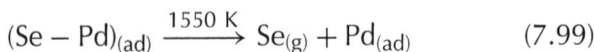

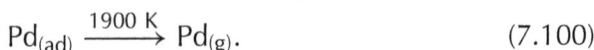

$$(Se - Pd)_{(ad)} \xrightarrow{1550 \text{ K}} Se_{(g)} + Pd_{(ad)} \tag{7.99}$$

$$Pd_{(ad)} \xrightarrow{1900 \text{ K}} Pd_{(g)}. \tag{7.100}$$

Using PdO as a chemical modifier, the main Se species observed were SeO and SeO_2 and similar equations were also proposed for the stabilization process. In conclusion, the formation of a solid solution or a compound between Se and the modifier is thought to result in the stabilization of this analyte. Nonetheless, other authors [Fischer and Rademeyer (1998b)] have suggested that stabilization from Pd (as nitrate), is only due to some physical processes, but not to the formation of any compound. The presence of other chemical modifiers, such as Ni (as nitrate), inhibits the formation of the usual precursors (Se_2, SeO_2, SeO) in the atomization of Se, which are currently found in the absence of a modifier. Nickel selenide is formed and atomization of Se is produced at high temperatures by decomposition of that selenide [Styris (1986)]. However, if Ni is added as $NiCl_2$, inhibition of precursors is not produced.

References

Aller, A.J. (1994). Vaporization mechanisms for gold and for gold in the presence of vanadium in a graphite atomizer. *Anal. Chim. Acta,* **292**: 317–324.

Aller, A.J. (1997). Atomization characteristics of selenium from a graphite tube in the presence of calcium and chemical modifiers. *Anal. Sci.,* **13**: 183–187.

Aller, A.J. (2001). A model for the determination of the activation energy and the order of release of the atom formation in electrothermal atomization atomic absorption spectroscopy. *Spectrochim. Acta, Part B,* **56**: 1441–1457.

Aller, A.J. (2002). Electrothermal atomization mechanisms: theoretical and practical considerations. *Trends in Appl. Spectrosc.,* **4**: 101–112.

Alvarez, M.A. and Carrión, N. and Gutierrez, H. (1995). Effects of atomization surfaces and modifiers on the electrothermal atomization of cadmium. *Spectrochim. Acta, Part B,* **50**: 1581–1594

Alvarez, M.A.; Carrión, N. and Gutierrez, H. (1996). Effects of atomization surfaces and modifiers on the kinetics of copper atomization in electrothermal atomic absorption spectrometry. *Spectrochim. Acta, Part B,* **51**: 1121–1132.

Alvarado, J. (1996). The mechanism of atomisation of iron, cobalt and nickel during electrothermal atomisation atomic absorption spectrometry. *Química Analítica,* **15**: 173–177.

Akmam, S.; Genç, Ö.; Özdural, A.R. and Balkis, T. (1980). Theoretical analysis of atom formation-time curves for HGA-74. *Spectrochim. Acta, Part B,* **35**: 373–378.

Akmam, S.; Genç, Ö. and Balkis, T. (1982). Atom formation mechanisms of As with different techniques in atomic absorption spectroscopy. *Spectrochim. Acta, Part B,* **37**: 903–912.

Akman, S.; Bektas, S. and Genç, Ö. (1988). A novel approach to the interpretation of graphite furnace atomic absorption signals. *Spectrochim. Acta, Part B,* **43**: 763–772.

Akmam, S.; Genç, Ö. and Bektas, S. (1991). Investigation of the atomization mechanism of copper, platinum, iridium and manganese in graphite furnace atomic absorption spectrometry. *Spectrochim. Acta, Part B,* **46**: 1829–1839.

Bass, D.A. and Holcombe, J.A. (1988a). Mechanism of lead vaporization from an oxygenated graphite surface using mass spectrometry and atomic absorption. *Anal. Chem.,* **60**: 578–582.

Bass, D.A. and Holcombe, J.A. (1988b). Anomalous curving of Arrhenius plots for data obtained from electrothermal vaporization at atmospheric pressure. *Spectrochim. Acta, Part B*, **43**: 1473–1483.

Bendicho, C. and Loos-Vollebregt, M.T.C. de (1990). Investigations of the carbothermal reduction mechanism of aluminium oxide in graphite furnace-atomic absorption spectrometry. *Spectrochim. Acta, Part B*, **45**: 547–559.

Black, S.; Riddle, M.R. and Holcombe, J.A. (1986). A Monte Carlo simulation for graphite furnace atomization of copper. *Appl. Spectrosc.*, **40**: 925–933.

Bozdoğân, A.E. (1999). A method for the determination of the kinetic parameters relevant to atom formation processes in electrothermal atomic absorption spectrometry. *Spectrochim. Acta, Part B*, **54**: 557–569.

Brennfleck, U.; Müller-Vogt, G. and Wendl, W. (1996). Stabilization of the oxide forming elements, lead, thallium and tin, in graphite tubes for graphite furnace atomic absorption spectrometry by intercalation. *Spectrochim. Acta, Part B*, **51**: 1139–1145.

Broek, W.M.G.T. van den and Galán, L. de (1977). Supply and removal of sample vapor in graphite thermal atomizers. *Anal. Chem.*, **49**: 2176–2186.

Brown, G.N. and Styris, D.L. (1993). Elucidation of mechanisms that control the electrothermal atomization of tin chloride. *J. Anal. At. Spectrom.*, **8**: 211–216.

Buhay, O.M.; Rogulsky, Yu. V.; Kulik, A.N.; Kalinkevich, A.N. and Sukhodub, L.F. (2005). Simulation of atomic absorption signals: A kinetic model with two independent sources. *Spectrochim. Acta, Part B*, **60**: 491–503.

Byrne, J.P.; Grégoire, D.C.; Benyounes, M.E. and Chakrabarti, C.L. (1997). Vaporization and atomization of the platinum group elements in the graphite furnace investigated by electrothermal vaporization — inductively coupled plasma — mass spectrometry. *Spectrochim. Acta, Part B*, **52**: 1575–1586.

Campbell, W.C. and Ottaway, J.M. (1974). Atom-formation processes in carbon-furnace atomizers used in atomic-absorption spectrometry. *Talanta*, **21**: 837–844.

Castro, M.A. and Aller, A.J. (2003). Mechanistic study of the aluminum interference in the determination of arsenic by electrothermal atomic absorption spectrometry. *Spectrochim. Acta, Part B*, **58**: 901–918.

Castro, M.A.; Aller, A.J.; McCabe, A.; Smith, W.E. and Littlejohn, D. (2005). Spectrometric and morphological characterization of condensed phase zirconium species produced during electrothermal heating on a graphite platform. *J. Anal. At. Spectrom.*, **20**: 385–394.

Cathum, S.J.; Chakrabarti, C.L. and Hutton, J.C. (1991). Investigation of the order of copper atomization at the absorbance maximum in graphite furmace atomic absoption spectrometry. *Spectrochim. Acta, Part B*, **46**: 35–44.

Chakrabarti, C.L. and Cathum, S.J. (1990). Mechanism of cobalt atomization from different atomizer surfaces in graphite furnace atomic absorption spectrometry. *Talanta*, **37**(12): 1111–1117.

Chakrabarti, C.L. and Cathum, S.J. (1991). Arrhenius plots for activation energy of atomization in graphite furnace atomic absorption spectrometry. *Talanta*, **38**: 157–166.

Chakrabarti, C.L.; Chang, S.B.; Lawson, S.R. and Wong, S.M. (1983). Computer modeling of atomization processes in graphite furnace atomic absorption spectrometry. *Spectrochim. Acta, Part B*, **38**: 1287–1300.

Chakrabarti, C.L.; Wan, C.C.; Teskey, R.J.; Chang, S.B.; Hamed, H.A. and Bertels, P.C. (1981). Mechanism of atomization at constant temperature in capacitive discharge graphite furnace atomic absoption spectrometry. *Spectrochim. Acta, Part B*, **36**: 427–438.

Chung, C.-H. (1984). Atomization mechanism with Arrhenius plots taking the dissipation function into account in graphite furnace atomic absorption spectrometry. *Anal. Chem.*, **56**: 2714–2720.

Droessler, M.S. and Holcombe, J.A. (1987). Mass spectral and atomic absorption studies of selenium vaporization from a graphite surface. *Spectrochim. Acta, Part B, **42**: 981–994.

Eleni, P.S.; Thomaidis, N.S. and Piperaki, E.A. (2005). Investigation of the mechanism of the electrothermal atomization of platinum in a graphite furnace from aqueous solutions and serum samples. *J. Anal. At. Spectrom.*, **20**: 111–117.

Eloi, C.C.; Robertson, J.D. and Majidi, V. (1993). Investigation of high temperature reactions on graphite with Rutherford backscattering spectrometry: interaction of cadmium, lead and silver with a phosphate modifier. *J. Anal. At. Spectrom.*, **8**: 217–222.

Eloi, C.C.; Robertson, J.D. and Majidi, V. (1995a). An RBS investigation of the effects of oxygen and hydrogen pre-treatment of pyrolytically coated graphite on Pb atomization. *Anal. Chem.*, **67**: 335–340.

Eloi, C.C.; Robertson, J.D. and Majidi, V. (1995b). RBS investigation of high temperature reactions on graphite substrates. *Nucl. Instrum. Meth. Phys. Res. B*, **99**: 468–471.

Falk, H. and Schnürer, C. (1989). On loss in graphite furnaces: model calculations and atomic absorption experiments. *Spectrochim. Acta, Part B*, **44**: 759–770.

Fischer, J.L. and Rademeyer, C.J. (1998a). Kinetics of selenium atomization in electrothermal atomization atomic absorption spectrometry (ETA-AAS). Part 1: selenium without modifiers. *Spectrochim. Acta, Part B*, **53**: 537–548.

Fischer, J.L. and Rademeyer, C.J. (1998b). Kinetics of selenium atomization in electrothermal atomization atomic absorption spectrometry (ETA-AAS). Part 2: selenium with palladium modifiers. *Spectrochim. Acta, Part B*, **53**: 549–567.

Fonseca, R.W.; McNally, J. and Holcombe, J.A. (1993). Mechanisms of vaporization for silver and gold using electrothermal atomization. *Spectrochim. Acta, Part B*, **48**: 79–89.

Fonseca, R.W.; Pfefferkorn, L.L. and Holcombe, J.A. (1994a). Comparisons of selected methods for the determination of kinetic parameters from electrothermal atomic absorption data. *Spectrochim. Acta, Part B*, **49**: 1595–1608.

Fonseca, R.W.; Wolfe, K.I. and Holcombe, J.A. (1994b). Mechanisms of vaporization for Cr using electrothermal atomization. *Spectrochim. Acta, Part B*, **49**: 399–408.

Frech, W.; Li, K.; Berglund, M. and Baxter, D. (1992a). Effect of modifier mass and temperature gradients on analyte sensitivity in electrothermal atomic absorption spectrometry. *J. Anal. At. Spectrom.*, **7**: 141–145.

Frech, W. and Baxter, D. (1996). Spatial distribution of non-atomic species in graphite furnaces. *Spectrochim. Acta, Part B*, **51**: 961–972.

Frech, W. and Cedergren, A. (1980). Investigations of reractions involved in flameless atomic absorption procedures. Part 7. A theoretical and experimental study of factors influencing the determination of silicon. *Anal. Chim. Acta*, **113**: 227–235.

Frech, W.; L'vov, B.V. and Romanova, N.P. (1992b). Condensation of matrix vapours in the gaseous phase in graphite furnace atomic absorption spectrometry. *Spectrochim. Acta, Part B*, **47**: 1461–1469.

Frech, W.; Zhou, N.G. and Lundberg, E. (1982). A critical study of some methods used to investigate atom formation processes in GFAAS. *Spectrochim. Acta, Part B*, **37**: 691–702.

Fuller, C.W. (1974). A kinetic theory of atomisation for non-flame atomic-absorption spectrometry with a graphite furnace. The kinetics and mechanism of atomisation for copper. *Analyst*, **99**: 739–744.

Fuller, C.W. (1977). Electrothermal atomization for atomic absorption spectrometry. The Chemical Society. Londres.

Galbács, G., Sneider, J., Oszkó, A., Vanhaecke, F. and Moens, L. (1997). X-ray photoelectron spectroscopic and atomic force microscopic studies of pyrolytically coated graphite and highly oriented pyrolytic graphite used for electrothermal vaporization. *J. Anal. At. Spectrom.*, **12**: 951–955.

García-Olalla, C. and Aller, A.J. (1992). Peak profile characteristics and atomization mechanisms for selenium in the presence of mercury for graphite furnace atomic absorption spectrometry. *Fresenius' J. Anal. Chem.*, **342**: 70–75.

Gilmutdinov, A.Kh. and Shlyakhtina, O.M. (1991). Correlation between analytical signal and rate of sample atomization in electrothermal atomic-absorption spectrosmetry. *Spectrochim. Acta, Part B*, **46**: 1121–1141.

Gilmutdinov, A.Kh.; Zakharov, Yu.A.; Ivanov, V.P. and Voloshin, A.V. (1991). Shadow spectral filming: A method of investigating electrothermal atomization. Part 1. Dynamics of formation and structure of the absorption layer of thallium, indium, gallium and aluminium atoms. *J. Anal. At. Spectrom.*, **6**: 505–519.

Gilmutdinov, A.Kh.; Zakharov, Yu.A.; Ivanov, V.P. and Voloshin, A.V. (1992). Shadow spectral filming: A method of investigating electrothermal atomization. Part 2. Dynamics of formation and structure of the absorption layer of aluminium, indium, gallium molecules. *J. Anal. At. Spectrom.*, **7**: 675–683.

Gilmutdinov, A.Kh.; Zakharov, Yu.A.; Ivanov, V.P. and Voloshin, A.V. (1993). Shadow spectral filming: A method of investigating electrothermal atomization. Part 3. Dynamics of longitudinal propagation of an analyte within graphite furnaces. *J. Anal. At. Spectrom.*, **8**: 387–395.

Gilmutdinov, A.Kh.; Mrasov, R.M.; Somov, A.R.; Chakrabarti, C.L. and Hutton, J.C. (1995). Three-dimensional modeling of the analyte dynamics in electrothermal atomizers for analytical spectrometry: influence of physical factors. *Spectrochim. Acta Part B*, **50**: 1637–1654.

Gilmutdinov, A.Kh.; Radziuk, B.; Sperling, M. and Welz, B. (1996). Spatially and temporally resolved detection of analytical signals in graphite furnace atomic absorption spectrometry. *Spectrochim. Acta Part B*, **51**: 1023–1044.

Gilmutdinov, A.Kh.; Voloshin, A.V. and Zakharov, Yu.A. (2005a). Shadow spectral imaging of absorbing layers in a transversely heated graphite atomizer. Part 1. Analyte atoms. *Spectrochim. Acta, Part B,* **60**: 511–518.

Gilmutdinov, A.Kh.; Voloshin, A.V. and Zakharov, Yu.A. (2005b). Shadow spectral imaging of absorbing layers in a transversely heated graphite atomizer. Part 2. Molecules and condensed-phase species. *Spectrochim. Acta, Part B,* **60**: 1423–1431.

Gilchrist, G.F.R., Chakrabarti, C.L., Ashley, J.T.F. and Hughes, D.M. (1992). Vaporization and atomization of lead and tin from a pyrolytic graphite probe in graphite furnace atomic absorption spectrometry. *Anal. Chem.,* **64**: 1144–1153.

Gilchrist, G.F.R.; Chakrabarti, C.L.; Cheng, J. and Hughes, D.M. (1993). Mechanisms of atomization of lead from nitric acid, hydrochloric acid and sodium chloride matrices in atomic absorption spectrometry using a graphite probe furnace. *J. Anal. At. Spectrom.,* **8**: 623–631.

Glasstone, S.; Laidler, K.J. and Eyring, H. (1941). *The Theory of Rate Processes.* McGraw-Hill, New York.

Goltz, D.M.; Grégoire, D.C.; Byrne, J.P. and Chakrabarti, C.L. (1995). Vaporization and atomization of uranium in a graphite tube electrothermal vaporizer: a mechanistic study using electrothermal vaporization inductively coupled plasma mass spectrometry and graphite furnace atomic absorption spectrometry. *Spectrochim. Acta, Part B,* **50**: 803–814.

Güell, O.A. and Holcombe, J.A. (1991). Monte Carlo study of the relationship between atom density and absorbance in electrothermal atomizers. *Appl. Spectrosc.,* **45**: 1171–1176.

Güell, O.A. and Holcombe, J.A. (1992). Monte Carlo study of analyte desorption, adsorption and spatial distribution in electrothermal atomizers. *J. Anal. At. Spectrom.,* **7**: 135–140.

Güell, O.A.; Holcombe, J.A. and Rademeyer, C.J. (1993). Effect on electrothermal atomization signals of contoured tube shapes and isothermality. *Anal. Chem.,* **65**: 748–751.

Hadgu, N.; Ohlsson, K.E.A. and Frech, W. (1995). Diffusion vapour transfer modelling for an end-capped atomizer. Part 1. — Atomizer with closed injection port. *Spectrochim. Acta, Part B,* **50**: 1077–1093.

Hadgu, N.; Ohlsson, K.E.A. and Frech, W. (1996). Diffusion vapour transfer modelling for end-capped atomizers. Part 2. Atomizer with open injection port. *Spectrochim. Acta, Part B,* **51**: 1081–1093.

Hadgu, N.; Gustafsson, J.; Frech, W. and Axner, O. (1998). Rubidium atom distribution and non-spectral interference effects in transversely heated graphite atomizers evaluated by wavelength modulated diode laser absorption spectrometry. *Spectrochim. Acta, Part B,* **53**: 923–943.

Hassell, D.C.; Majidi, V. and Holcombe, J.A. (1991). Temperature programmed static secondary ion mass spectrometric study of phosphate chemical modifiers in electrothermal atomizers. *J. Anal. At. Spectrom.,* **6**, 105–108.

Hassell, D.C.; Rettberg, T.M.; Fort, F.A. and Holcombe, J.A. (1988). Low-pressure vaporization for graphite furnace atomic absorption spectrometry. *Anal. Chem.,* **60**: 2680–2683.

Hirano, Y.; Yasuda, K. and Hirokawa, K. (1994). Lessening unexpected increases of atomic vapor temperature of arsenic in graphite furnace atomic absorption spectrometry. *Anal. Sci.,* **10**: 481–484.

Histen, T.E. and Holcombe, J.A. (1998a). Impact of non-linear heating on electrothermal atomic absorption spectrometry. *J. Anal. At. Spectrom.,* **13**: 769–775.

Histen, T.E. and Holcombe, J.A. (1998b). Simple approach for the determination of the order of release in electrothermal atomic absorption spectrometry. *Spectrochim. Acta, Part B,* **53**: 911–921.

Holcombe, J.A.; Rayson, G.D. and Akerlind, Jr., N. (1982). Time and spatial absorbance profiles within a graphite furnace atomizer. *Spectrochim. Acta, Part B,* **37**: 319–330.

Holcombe, J.A. and Rayson, G.D. (1983). Analyte distribution and relations within a graphite furnace atomizer. *Prog. Anal. At. Spectrosc.,* **6**: 225–251.

Holcombe, J.A.; Styris, D.L. and Harris, J.D. (1991). Mass spectrometric investigations of aluminium oxide reduction by gaseous aluminium carbides in electrothermal atomization. *Spectrochim. Acta, Part B,* **46**: 629–639.

Holcombe, J.A. and Histen, T.E. (1996). Impact of source and free atom density spatial non-uniformities in electrothermal atomic absorption spectrometry. *Spectrochim. Acta, Part B,* **51**: 1045–1053.

Hsiech, C. and Pardue, H.L. (1993). Improved data-processing method for atomic absorption spectroscopy with electrothermal atomization. *Anal. Chem.,* **65**: 1809–1813.

Hughes, D.M., Chakrabarti, C.L., Goltz, D.M., Sturgeon, R.E. and Grégoire, C. (1996a). Investigation of vapor condensation in graphite furnace atomic absorption spectrometry by the shadow spectral digital imaging technique. *Appl. Spectrosc.,* **50**: 715–731.

Hughes, D.M.; Chakrabarti, C.L.; Lamoureux, M.M.; Hutton, J.C.; Goltz, D.M.; Sturgeon, R.E.; Grégoire, C. and Gilmutdinov, A.Kh. (1996b). Digital imaging of formation and dissipation processes for atoms and moleculaes and condensed-phase species in graphite furnace atomic absorption spectrometry: a review. *Spectrochim. Acta, Part B,* **51**: 973–997.

Huie, C.W. and Curran, Jr., C.J. (1988). Spatial mapping of analyte distribution within a graphite furnace atomizer. *Appl. Spectrosc.,* **42**: 1307–1311.

Huie, C.W. and Curran, Jr., C.J. (1990). The effect of internal gas flow on the spatial distribution of sodium atom within a graphite furnace atomizer. *Appl. Spectrosc.,* **44**: 1329–1336.

Hulett, J.R. (1964). Deviations from the Arrhenius equation. *Q. Rev. Chem. Soc.,* **18**: 227–242.

Imai, S.; Haegawa, N.; Nishiyama, Y.; Hayashi, Y. and Saito, K. (1996). Effect of ascorbic acid and sucrose on electrothermal atomic absorption signals of indium. *J. Anal. At. Spectrom.,* **11**: 601–606.

Imai, S.; Harada, M.; Nishiyama, Y. and Hayashi, Y. (1998). Effect of the surface treatment of a graphite furnace with a refractory element (hafnium, titanium, tungsten and zirconium) by a one-drop coating method on the atomization mechanism of indium in electrothermal atomic absorption spectrometry. *Anal. Sci.,* **14**: 769–778.

Imai, S.; Ito, Y.; Tani, M.; Yonetani, A.; Nishiyama, Y. and Hayashi, Y. (2000). Desorption mechsnism of arsenic from non-pyrolytic graphite, pyrolytic graphite and pyrolyzed ascorbic acid in electrothermal atomic absorption spectrometry. *Anal. Sci.,* **16**: 1189–1194.

Jackson, J.G.; Fonseca, R.W. and Holcombe, J.A. (1995a). Mass spectral studies of thermal decomposition of metal nitrates. *Spectrochim. Acta, Part B,* **50**: 1449–1457.

Jackson, J.G.; Novichikhin, A.; Fonseca, R.W. and Holcombe, J.A. (1995b). Mass spectral studies of thermal decomposition of metal nitrates: an introduction to the discussion of two mechanisms. *Spectrochim. Acta, Part B,* **50**: 1423–1426.

Jackson, J.G.; Fonseca, R.W. and Holcombe, J.A. (1995c). Migration of Ag, Cd, and Cu into highly oriented pyrolytic graphite and pyrolytic coated graphite. *Spectrochim. Acta, Part B,* **50**: 1837–1846.

Jóźwiak, W.K. and Maniecki, T.P. (2005). Influence of atmosphere kind on temperature programmed decomposition of noble metal chlorides. *Termochim. Acta,* **435**: 144–154.

Katskov, D.A.; Darangwa, N. and Grotti, M. (2006). Chemically assisted release of transition metals in graphite vaporizers for atomic spectrometry. *Spectrochim. Acta, Part B,* **61**: 554–564.

Katskov, D.A.; Mofolo, R. and Tittarelli, P. (1999). Effect of beryllium nitrate vaporization on surface temperature in the pyrocoated graphite furnace. *Spectrochim. Acta, Part B,* **54**: 1801–1811.

Katskov, D.A.; Shtepan, A.M.; Grinshtein, I.L. and Pupyshev, A.A. (1992). Atomization of aluminium oxide in electrothermal atomic absorption analysis. *Spectrochim. Acta, Part B,* **47**: 1023–1041.

Katskov, D.A.; Shtepan, A.M.; McCrindle, R.I. and Marais, P.J.J.G. (1994). Applications of a two-step atomizer and related techniques for investigating the processes of sample evaporation and atomization in electrothermal atomic absorption spectrometry. *J. Anal. At. Spectrom.,* **9**: 321–331.

Korečkova, J.; Frech, W.; Lundberg, E.; Persson, J.-A. and Cedergren, A. (1981). Investigations of reractions involved in electrothermal atomic absorption procedures. Part 10. Factors influencing the determination of arsenic. *Anal. Chim. Acta,* **130**: 267–280.

Krakovská, E. (1997). Tungsten atomizer — the theory of atomization reactions. *Spectrochim. Acta, Part B,* **52**: 1327–1332.

Krakovská, E. and Remeteiová, D. (2000). Tungsten atomizer — theory of atomization mechanism of some volatile analytes. *Fresenius' J. Anal. Chem.,* **366**: 127–131.

Lamoureux, M.M.; Chakrabarti, C.L.; Hutton, J.C.; Gilmutdinov, A.Kh.; Zakharov, Y.A. and Grégoire, D.C. (1995). Mechanism of aluminium spike formation and disipation in electrothermal atomic absorption spectrometry. *Spectrochim. Acta, Part B,* **50**: 1847–1867.

Le Bihan, A.; Le Garrec, H.; Cabon, J.Y. and Guern, Y. (1998). Activation energies of metal atomization and nitrate and sulfate decomposition in concentrated matrices (10^{-1} M). *Spectrochim. Acta, Part B,* **53**: 1347–1353.

Lonardo, R.F.; Yuzefovsky, A.I.; Irwin, R.L. and Michel, R.G. (1996). Laser-excited atomic fluorescence spectrometry in a pressure-controlled electrothermal atomizer. *Anal. Chem.,* **68**: 514–521.

L'vov, B.V. (1970). *Atomic Absorption Spectrochemical Analysis,* 2nd edition. Adam Hilger, London.

L'vov, B.V. (1978). Electrothermal atomization — the way toward absolute methods of atomic absorption analysis. *Spectrochim. Acta, Part B,* **33**: 153–193.

L'vov, B.V.; Bayunov, P.A. and Ryabchuk, G.N. (1981). A macrokinetic theory of sample vaporization in electrothermal atomic absorption spectrometry. *Spectrochim. Acta, Part B,* **36**: 397–425.

L'vov, B.V. and Ryabchuk, G.N. (1982). A new approach to the problem of atomization in electrothermal atomic absorption spectrometry. *Spectrochim. Acta, Part B,* **37**: 673–684.

L'vov, B.V.; Nikolaev, V.G.; Novichikhin, A.V. and Polzik, L.K. (1988). Effect of platform material on sample vaporization rate in graphite furnace atomic absorption spectrometry. *Spectrochim. Acta, Part B,* **43**: 1141–1146.

L'vov, B.V. (1991). Mechanism of the thermal decomposition of nitrate from graphite furnace mass spectrometry studies. *Mikrochim. Acta* **104**: 299–308.

L'vov, B.V. and Novichikhin, A.V. (1995a). Mechanism of thermal decomposition of hydrated copper nitrate in vacuo. *Spectrochim. Acta, Part B,* **50**: 1459–1468.

L'vov, B.V. and Novichikhin, A.V. (1995b). Mechanism of thermal decomposition of anhydrous metal nitrates. *Spectrochim. Acta,* **50**: 1427–1448.

L'vov, B.V. (1996). Gaseous carbide theory. Has it been buried prematurely?. *Spectrochim. Acta, Part B,* **51**: 533–541.

L'vov, B.V. (1997a). Interpretation of atomization mechanisms in electrothermal atomic absorption spectrometry by analysis of the absolute rates of the processes. *Spectrochim. Acta, Part B,* **52**: 1–23.

L'vov, B.V. (1997b). Forty years of electrothermal atomic absorption spectrometry. Advances and problems in theory. *Spectrochim. Acta, Part B,* **52**: 1239–1245.

L'vov, B.V. (1998). Advances and problems in the investigation of the mechanisms of solid-state reactions by analysis of absolute reaction rates. *Spectrochim. Acta, Part B,* **53**: 809–820.

L'vov, B.V. and Ugolkov, V.L. (2003). The self-cooling effect in the process of dehydration of $Li_2SO_4 \cdot H_2O$, $CaSO_4 \cdot 2H_2O$ and $CuSO_4 \cdot 5H_2O$ in vacuum. *J. Therm. Anal. Calor.,* **74**: 697–708.

L'vov, B.V. and Ugolkov, V.L. (2004). Kinetics and mechanism of free-surface decomposition of group IIA and IIB hydroxides analyzed thermogravimetrically by the third-law method. *Thermochim. Acta,* **413**: 7–15.

L'vov, B.V.; Ugolkov, V.L. and Grekov, F.F. (2004). Kinetics and mechanism of free-surface vaporization of zinc, cadmium and mercury oxides analyzed by the third-law method. *Thermochim. Acta,* **411**: 187–193.

Lynch, S.; Sturgeon, R.E.; Luong, V.T. and Littlejohn, D. (1990). Comparison of the energetics of desorption of solution and vapour phase deposited analytes in graphite furnace atomic absorption spectrometry. *J. Anal. At. Spectrom.,* **5**: 311–319.

Majidi, V.; Smith, R.G.; Bossio, R.E.; Pogue, R.T. and McMahon, M.W. (1996). Observation of pre-atomization events on electrothermal atomizer surfaces. *Spectrochim. Acta, Part B*, **51**: 941–959.

Majidi, V.; Holcombe, J.A.; Vandervoort, K.G.; Butcher, D.J. and Robertson, J.D. (1997). Electrothermal vaporization and characterization of the graphite surface at elevated temperatures. *Appl. Spectrosc.*, **51**: 408A–423A.

Majidi, V. and Robertson, J.D. (1991). Investigation of high temperature reactions on solid substrates with Rutherford backscattering spectrometry: interaction of palladium with selenium on heated graphite surfaces. *Spectrochim. Acta, Part B*, **46**: 1723–1733.

Majidi, V.; Xu, N. and Smith, R.G. (2000a). Electrothermal vaporization, part 1: gas phase chemistry. *Spectrochim. Acta, Part B*, **55**: 3–35.

Majidi, V.; Smith, R.G.; Xu, N.; McMahon, M.W. and Bossio, R. (2000b). Electrothermal vaporization, part 2: surface chemistry. *Spectrochim. Acta, Part B*, **55**: 1787–1822.

Mandjukov, P.B.; Vassileva, E.T. and Simeonov, V.D. (1992). Regular solution theory in model interpretation of the analyte losses during preatomization sample treatment in the presence of chemical modifiers in electrothermal atomization atomic spectrometry. *Anal. Chem.*, **64**: 2596–2603.

Marcus, R.A. (1964). Chemical and Electrochemical Electron-Transfer Theory. *Annu. Rev. Phys. Chem.* **15**: 155–196.

Marcus, R.A. (1993). Electron transfer reactions in chemistry. Theory and experiment. *Rev. Mod. Phys.* **65**: 599–610.

Masera, E.; Mauchien, P. and Lerat, Y. (1996). Imaging of analyte distribution in a graphite tube atomizer by laser induced fluorescence. *Spectrochim. Acta, Part B*, **51**: 1007–1022.

Mazzucotelli, A. and Grotti, M. (1995). Effects of interfering elements and chemical modifier on the activation energy of electrothermal atomization of selenium. *Spectrochim. Acta, Part B*, **50**: 1897–1904.

McAllister, T. (1994). Equilibrium and mass spectrometry of nitrate decomposition in electrothermal atomic absorption spectrometry. *J. Anal. At. Spectrom.*, **9**: 427–430.

McNally, J. and Holcombe, J.A. (1987). Existence of microdroplets and dispersed atoms on the graphite surface in electrothermal atomizers. *Anal. Chem.*, **59**: 1105–1112.

McNally, J. and Holcombe, J.A. (1991). Topology and vaporization characteristics of palladium, cobalt, manganese, indium, and aluminium on a graphite surface using electrothermal atomic absorption. *Anal. Chem.*, **63**: 1918–1926.

Mofolo, R.; Katskov, D.A.; Tittarelli, P. and Grotti, M. (2001). Vaporization of indium nitrate in the graphite atomizer in the presence of chemical modifiers. *Spectrochim. Acta, Part B*, **56**: 375–391.

Müller-Vogt, G.; Kübler, M.; Lussac, C.; Wendl, W. and Würfel, P. (2000). Chemical reactions of tellurium in graphite tubes of atomic absorption spectrometry. *Spectrochim. Acta, Part B*, **55**: 501–508.

Müller-Vogt, G.; Weigend, F. and Wendl, W. (1996). Role of oxygen in the determination of oxide forming elements by electrothermal atomic absorption spectrometry. Part 3. Effect of oxygen on the reactions of tin in uncoated, pyrolytically coated and zirconium carbide coated graphite tube atomizers. *Spectrochim. Acta, Part B*, **51**: 1133–1137.

Musil, J. and Rubeska, I. (1982). Mathematical model of electrothermal atomisation signals based on free atom redeposition. *Analyst*, **107**: 588–590.

Ohlsson, K.E.A. (1992). Aluminium atom formation in electrothermal graphite atomizer atomic absorption spectrometry studied by *in situ* spectroscopic measurements of aluminium and aluminium hydride. *J. Anal. At. Spectrom.*, **7**: 357–363.

Oishi, K.; Yasuda, K. and Hirokawa, K. (1991). Beginning of atomization in presence of matrix modifier in graphite furnace atomic absorption spectrometry. *Anal. Sci.*, **7**: 883–887.

Paveri-Fontana, S.L.; Tessari, G. and Torsi, G. (1974). Time-resolved distribution of atoms in flameless spectrometry. A theoretical calculation. *Anal. Chem.*, **46**: 1032–1038.

Persson, J.-A.; Frech, W. and Cedergren, A. (1977a). Investigations of reractions involved in flameless atomic absorption procedures. Part IV. A theoretical study of factors influencing the determination of aluminium. *Anal. Chim. Acta*, **92**: 85–93.

Persson, J.-A.; Frech, W. and Cedergren, A. (1977b). Investigations of reractions involved in flameless atomic absorption procedures. Part V. An experimental study of factors influencing the determination of aluminium. *Anal. Chim. Acta*, **92**: 95–104.

Prell, L.J.; Styris, D.L. and Redfield, D.A. (1990). Mechanisms controlling atomisation of strontium and associated interferences by calcium in electrothermal atomic absorption spectrometry. *J. Anal. At. Spectrom.*, **5**: 231–238.

Prell, L.J. and Styris, D.L. (1991). Mechanism of electrothermal atomization for yttrium by mass spectrometry. *Spectrochim. Acta, Part B*, **46**: 45–49.

Quan Zhe; Ni Zhe-ming and Yan Xiu-Ping (1994). Influence of atomizer surface on the kinetics of tin atomization in electrothermal atomic absorption spectrometry. *Canadian J. Appl. Spectrosc.*, **34**: 54–59.

Ratliff, J. (1996). Investigation of the molecular species produced by the vaporization of the group IIA metal nitrates from a graphite surface. *Anal. Chim. Acta*, **333**: 285–293.

Rayson, G.D. and Holcombe, J.A. (1982). Tin atom formation in a graphite furnace atomizer. *Anal. Chim. Acta*, **136**: 249–260.

Rayson, G.D. and Johnson, C. (1991). A new approach to the investigation of preatomization analyte loss mechanisms occurring within a grapfite furmace atomizer. *Appl. Spectrosc.*, **45**: 1305–1309.

Rojas, D. (1992). A study of nickel electrothermal atomization. *Spectrochim. Acta, Part B*, **47**: 1423–1433.

Rojas, D. (1995). Electrothermal atomization of silver in graphite furnaces. Part 2. Effects of copper, ascorbic acid and Triton X-100. *Spectrochim. Acta, Part B*, **50**: 1031–1044.

Rojas, D. and Olivares, W. (1989). Simple mathematical treatment for non-isothermal atomisation. *J. Anal. At., Spectrom.*, **4**: 613–617.

Rojas, D. and Olivares, W. (1992). A method for the determination of the kinetic order and energy of the atom formation process in electrothermal atomization atomic absorption spectrometry (ETA-AAS). *Spectrochim. Acta, Part B*, **47**: 387–397.

Rojas, D. and Olivares, W. (1995). Electrothermal atomization of silver in graphite furnaces. Part 1. — A two-precursor mechanism. *Spectrochim. Acta, Part B*, **50**: 1011–1030.

Rojas, D.; Sánchez, M.A. and Olivares, W. (1997). Influence of vapor redeposition and modifiers on the Arrhenius plots of copper in graphite furnace atomic absorption spectrometry. *Spectrochim. Acta, Part B*, **52**: 1269–1281.

Sadagoff, Y. M. and Dedina, J. (2002). Atom diffusion in furnaces — models and measurements. *Spectrochim. Acta, Part B*, **57**: 535–549.

Sadagoff, Y. M. and Katskov, D. A. (2001). Effect of palladium modifier on the analyte vapor transport in a graphite furnace atomizer. *Spectrochim. Acta, Part B*, **56**: 1397–1405.

Salmon, S.G. and Holcombe, J.A. (1982). Alteration of metal release mechanisms in graphite furnace atomizers by chemisorbed oxygen. *Anal. Chem.*, **54**: 630–634.

Slaveykova, V.I. and Tsalev, D.L. (1991). Kinetic approach to the interpretation of analyte losses during the preatomization thermal treatment in electrothermal atomization atomic absorption spectrometry. *Spectrosc. Lett.*, **24**(1): 139–159.

Slaveykava, V.I. and Tsalev, D.L. (1992). Simplified kinetic model describing the analyte losses during pre-atomization thermal treatment in electrothermal atomic absorption spectrometry. *J. Anal. At. Spectrom.*, **7**: 365–370.

Slaveykova, V.I.; Manev, S.G. and Lazarov, D.L. (1995). Application of the Kelvin equation to vaporization of silver and gold in electrothermal atomic absorption spectrometry. *Spectrochim. Acta, Part B*, **50**: 1725–1732.

Slaveykova, V.I.; Manev, S.G. and Lazarov, D.L. (1997). Palladium release in electrothermal atomic absorption spectrometry. *Spectrosc. Lett.*, **30**: 297–307.

Smets, B. (1980). Atom formation and dissipation in electrothermal atomization. *Spectrochim. Acta, Part B*, **35**: 33–42.

Sturgeon, R.E.; Chakrabarti, C.L. and Landford, C.H. (1976). Studies on the mechanism of atom formation in graphite furnace atomic absorption spectrometry. *Anal. Chem.*, **48**: 1792–1807.

Sturgeon, R.E. and Chakrabarti, C.L. (1977). Mechanism of atom loss in graphite furnace atomic absorption spectrometry. *Anal. Chem.*, **49**: 1100–1106.

Sturgeon, R.E. and Berman, S.S. (1985). Absorption pulse shifting in graphite furnace atomic absorption spectrometry. *Anal. Chem.*, **57**: 1268–1275.

Sturgeon, R.E.; Mitchell, D.F. and Berman, S.S. (1983). Atomization of lead in graphite furnace atomic absorption spectrometry. *Anal. Chem.*, **55**: 1059–1064.

Sturgeon, R.E.; Siu, K.W.M. and Berman, S.S. (1984). Oxygen in the high-temperature graphite furnace. *Spectrochim. Acta Part B*, **39**: 213–224.

Styris, D.L. (1986). Elucidating atomization mechanisms by simultaneous mass spectrometry and atomic absorption spectrometry. *Fresenius' J. Anal. Chem.*, **323**: 710–715.

Styris, D.L. and Kaye, J.H. (1982). Mechanisms of vaporization of vanadium pentoxide from vitreous carbon and tantalum furnaces by combined atomic absorption/mass spectrometry. *Anal. Chem.*, **54**: 864–869.

Styris, D.L. and Redfield, D.A. (1987a). Mechanisms of graphite furnace atomization of aluminium by molecular beam sampling mass spectrometry. *Anal. Chem.*, **59**: 2891–2897.

Styris, D.L. and Redfield, D.A. (1987b). Mechanisms controlling graphite furnace atomization and stabilizatiuon of beryllium. *Anal. Chem.*, **59**: 2897–2903.

Styris, D.L.; Prell, L.J. and Redfield, D.A. (1991a). Mechanisms of palladium-induced stabilization of arsenic in electrothermal atomic absorption spectrometry. *Anal. Chem.*, **63**: 503–507.

Styris, D.L.; Prell, L.J.; Redfield, D.A.; Holcombe, J.A.; Bass, D.A. and Majidi, V. (1991b). Mechanisms of selenium vaporization with palladium modifiers using electrothermal atomization and mass spectrometric detection. *Anal. Chem.*, **63**: 508–516.

Styris, D.L. and Redfield, D.A. (1993). Perspectives on mechanisms of electrothermal atomization. *Spectrochim. Acta Rev.*, **15**: 71–123.

Tessari, G. and Torsi, G. (1975). Time-resolved distribution of atoms in flameless spectrometry: Experimental. *Anal. Chem.*, **47**: 842–849.

Thomaidis, N.S. and Piperaki, E.A. (2000). Effect of chemical modifiers on the kinetic parameters characterizing the electrothermal atomization of chromium. *Spectrochim. Acta, Part B*, **55**: 611–627.

Torsi, G. and Tessari, G. (1973). Influence of heating rate on analytical response in flameless atomic absorption spectrometry. *Anal. Chem.*, **45**: 1812–1816.

Torsi, G. and Tessari, G. (1975). Time-resolved distribution of atoms in flameless spectrometry: recovery of the source parameters from the response function. *Anal. Chem.*, **47**: 839–842.

Torsi, G.; Valcher, S.; Reschiglian, P.; Cludi, L. and Patauner, L. (1995). High current power supply for electrothermal atomic absorption spectrometry. *Spectrochim. Acta, Part B*, **50**: 1679–1685.

Torsi, G.; Rossi, F.N.; Melucci, D.; Reschiglian, P.; Locatelli, C. and Di Cintio, D. (2000). Absorbance *vs.* time curves at high heating rates in electrothermal atomic absorption spectrometry. *Spectrochim. Acta, Part B*, **55**: 65–73.

Torsi, G.; Zattoni, A.; Locatelli, C. and Valcher, S. (2005). A new heating strategy in electrothermal atomic absorption spectrometry for better absorbance-time curves at high atomization rate. *Spectrochim. Acta, Part B*, **60**: 285–289.

Voloshin, A.V.; Gilmutdinov, A.Kh.; Zakharov, Yu.A. and Sevast'yanov, A.A. (2004). Effect of the Pd-Mg modifier, magnetic field, and gas flow on the dynamics of matrix vapors in a transversely heated graphite furnace atomizer. *J. Anal. Chem.*, **59**: 234–242.

Volynsky, A.B. (1998). Graphite atomizers modified with high-melting carbides for electrothermal atomic absorption spectrometry. III. Practical aspects. *Spectrochim. Acta, Part B*, **53**: 1607–1645.

Volynsky, A.B.; Sedykh, E.M.; Spivakov, B.Ya. and Havezov, I. (1986). Factors influencing the free oxygen content in an electrothermal atomizer. *Anal. Chim. Acta*, **174**: 173–182.

Vyazovkin, S. (2000). On the phenomenon of variable activation energy for condensed phase reactions. *New J. Chem.* **24**: 913–917.

Vyazovkin, S. (2003). Reply to "What is meant by the term 'variable activation energy' when applied in the kinetics analyses of solid state decompositions (crystolysis reactions)?" *Thermochimica Acta,* **397**: 269–271.

Wahab, H.S. and Chakrabarti, C.L. (1981). Mechanism of yttrium atom formation in electrothermal atomization from metallic and metal-carbide surfaces of a heated graphite atomizer in atomic absorption spectrometry. *Spectrochim. Acta, Part B,* **36**: 475–481.

Wang, P.; Majidi, V. and Holcombe, J.A. (1989). Copper atomization mechanisms in graphite furnace atomizers. *Anal. Chem.,* **61**: 2652–2658.

Welz, B.; Radziuk, B. and Schlemmer, G. (1988). Evaluation of a mathematical model for peak interpretation in graphite furnace atomic absorption spectrometry based on free analyte atom redeposition on carbon surfaces. *Spectrochim. Acta, Part B,* **43**: 749–762.

Wendl, W. and Müller-Vogt, G. (1984). Chemical reactions in the graphite tube for some carbide and oxide forming elements. *Spectrochim. Acta, Part B,* **39**: 237–242.

Xiu-Ping, Y.; Tiezheng, L. and Zhijun, L. (1990). Improvement of the Smets method in electrothermal atomic-absorption spectrometry. *Talanta,* **37**: 167–171.

Xiu-Ping, Y.; Zhe-Ming, N.; Xiao-Tao, Y. and Guo-Qiang, H. (1993a). Approach to the determination of the kinetic parameters for atom formation in electrothermal atomic absorption spectrometry. *Spectrochim. Acta, Part B,* **48**: 605–624.

Xiu-Ping, Y.; Zhe-Ming, N.; Xiao-Tao, Y. and Guo-Qiang, H. (1993b). Kinetics of indium atomization from different atomizer surfaces in electrothermal atomic absorption spectrometry (ETAAS). *Talanta,* **40**: 1839–1846.

Xiu-Ping, Y. and Zhe-Ming, N. (1993). Electrothermal atomization of lead from different atomizer surfaces. *Spectrochim. Acta, Part B,* **48**: 1315–1323.

Yan-Zhong, L., Zhe-Ming, N. and Xiu-Ping, Y. (1995). Determination of kinetic parameters for atom formation at constant temperature in graphite furnace atomic absorption spectroscopy. *Spectrochim. Acta, Part B,* **50**: 725–737.

Yasuda, K.; Hirano, Y.; Hirokawa, K.; Kamino, T. and Yaguchi, T. (1993). Observation on vaporization of atoms in tin-palladium alloy in graphite furnace atomic absorption spectrometry by means of electron microscopy. *Anal. Sci.,* **9**: 529–532.

Yasuda, K.; Hirano, Y.; Kamino, T. and Hirokawa, K. (1994). Relationship between the formation of intermetallic compounds by matrix modifiers and atomization in graphite furnace-atomic absorption spectrometry, and an observation of the vaporization of intermetallic compounds by means of electron microscopy. *Anal. Sci.,* **10**: 623–631.

Yasuda, K.; Hirano, Y.; Kamino, T.; Yaguchi, T. and Hirokawa, K. (1995). Observation of vaporization in palladium-indium intermetallic compounds by graphite furnace atomic absorption spectrometry using transmission electron microscopy. *Anal. Sci.,* **11**: 437–440.

Zheng, Y.; Woodriff, R. and Nichols, J.A. (1984). Mechanism of atomic vapor loss for aluminium and potassium in a constant-temperature carbon-tube furnace. *Anal. Chem.,* **56**: 1388–1391.

Zhe-ming, N. and De-qiang, Z. (1995). Influence of sample deposition and coating with Zr and Pd on the atomization kinetics of germanium in graphite furnace atomic absorption spectrometry. *Spectrochim. Acta, Part B*, **50**: 1779–1786.

Zhou, N.G.; Frech, W. and de Galan, L. (1984). On the relationship between heating rate and peak height in electrothermal atomic absorption spectroscopy. *Spectrochimica Acta Part B: Atomic Spectroscopy,* **39**: 225–235.

Chapter 8

Analytical Characteristics

8.1. Introduction: *Validation of Analytical Methods*

The first goal of an analytical method is to provide an accurate analysis. So, all developed analytical methods need to be validated early. Application of a complete validation process does not exclude all the potential problems, but it should be able to address the most common ones. By applying a validation procedure, it is expected to prove if any analytical method is suitable or not for the proposed aim. The validation procedure may be very tedious, but by using a validated analytical method, the best analysis data are generated. The first stage in the development and validation of any analytical method is to establish the minimum requirements to be accepted [Penninckx *et al.* (1996)]. Once the validation study is completed, it is necessary to check in the same laboratory if the developed method is capable to provide a good quantitation procedure. To obviate the need for repetitive studies and to find validated data, which should be generated under equivalent conditions to those of the final procedure, studies covering *selectivity, trueness (accuracy, precision), linearity, working range, detection limit, quantitation limit, stability, traceability and robustness* should be carried out [Green (1996); Aller (2002)].

• *Selectivity*

Selectivity represents the capability of an analytical method to measure accurately the analyte response in the presence of all sample potential components.

• *Trueness*

Trueness of a measurement can be described using parameters referring to *accuracy* (usually expressed in terms of bias), and *precision* (usually expressed in terms of imprecision (uncertainty) and computed as a standard deviation).

Accuracy of an analytical method is defined as the concordance between the average value obtained from a large serie of data and a reference value accepted as the true value. Accuracy is usually determined through some of the following four ways.

(i) Comparing the experimental result obtained by analyzing a certified reference material with an average value previously accepted as the true value.

(ii) By comparing the results obtained after applying the new analytical method with the data derived using an alternative method whose accuracy is known.

(iii) The most widely used method is adding known amounts of analyte from a standard to a blank, which is used as a reference to compare the experimental result and to deduce the analyte concentration in the sample.

(iv) The standard addition method can also be used to determine the recovery of the analyte spiked. This methodology is used whenever it would be imposible to prepare a blank containing the same matrix composition as the sample.

The accuracy criterion for an analytical method considers that the average recovery should be about $100 \pm 2\%$. For an analytical method for impurities, the average recovery should be between 0.1% absolute of the theoretical concentration or of 10% relative, which is the large, for impurities in the range of 0.1–2.5% (in weight).

Precision is defined as the concordance existing between the data from a replicate measurement obtained under selected conditions. The precision of an analytical method is a measure of the dispersion in the results obtained from the multiple analysis derived from a homogeneous sample. In order to be of significance, the precision study should be carried out using the same type of sample and the same preparation procedures of standards, which will be used in the final method. Precision can usually be expressed in terms of repeatability and reproducibility.

Repeatability (*intra-laboratory precision*) is used to express the variation in replicate procedures performed within a few days (same analytical run) and with the same operational conditions (analyst, instrument, reagents).

Reproducibility refers to *inter-laboratory precision*, being performed at two or more laboratories, and consequently within a long time period with different instrumentation, analysts, etc. It is usually focalized as a part of cross-interlaboratory studies in measuring the bias of results than in determining differences in precision. Reproducibility and repeatability are really of limited utility, because the former is not founded on a definite and quantitative basis and the latter is void of general significance.

• Linearity

The linearity study verifies that sample solutions show an analyte concentration range for which the analytical response is directly proportional to the analyte concentration. Acceptation of the linearity data is usually judged by examining the determination coefficient (r^2) and the interseption (y) of the regression line for the plot of the analytical response versus the analyte concentration. A determination coefficient of >0.999 is considered to be the evidence of an accepted fit of the data to the regression line.

• Working range

The working range of an analytical method represents the analyte concentration range for which accuracy, linearity, and precision are assumed to be acceptable. In practice, the working range is determined using the data from the linearity and accuracy studies. Assuming that linearity is acceptable and accuracy has been obtained as noted before, the only factor that needs to be evaluated is precision. The precision data should be obtained from the analysis by triplicate of samples *spiked* in the accuracy study. However, it should be noted that precision can change with the analyte concentration. In conclusion, the working range acceptable for an analytical method is defined as the concentration range for which linearity and accuracy, obtained according to the crietria above mentioned, provide a precision $<3\%$ RSD (relative standard deviation). For an impurities method, the working range acceptable should provide a precision $<10\%$ RSD.

• Detection limit

The limit of detection of an analytical method represents the lowest analyte concentration which provides a detectable signal different from the average background of the system. Usually, the detection limit is considered to take a value equal to the averaged background signal plus three times the standard deviation from all background signals.

• *Quantitation limit*

The quantitation limit represents the lowest analyte concentration which can be quantitatively determined with an accuracy and precision previously fixed. The quantitation limit is usually determined by lowering the analyte concentration till a value for which the precision of the method is unacceptable. It can also be possible to be defined as the analyte concentration which provides an analytical signal equal to the averaged background signal plus ten times the standard deviation from all background signals. Sometimes, the quantitation limit can also be determined as the analyte concentration which provides an accepted signal-to-background ratio.

• *Stability*

This criterion is related to the conservation of solutions, as much for samples as for standards. Stability should be at least for 48 hours under storage conditions clearly established.

• *Traceability*

This criterion is related to the possibility of making a characterization (or control) of any (or all) stages from any analytical procedure. In other words, the whole analytical procedure should be checked at any time of the analysis. So, *traceability* can be related to state references or international units through an unbroken chain of comparisons all having known uncertainties.

Traceability is not an end in itself but serves the purpose of the assessment of general features of the result, by defining *a range within which the true value of a measurement can be expected to lie*, at a stated confidence level. Traceability, expressed in a quantitative form, assumes a crucial role in characterizing a measure, a method, a technique or a laboratory. Traceability is strictly connected to accuracy. If the defined range for traceability is smaller than a certain value assumed as a limit of acceptability, it can be said that the measure is certainly accurate, while, in the case of a range greater than this limit, nothing testifies to the accuracy of the result. Traceability and accuracy are not the same thing because a measure may be traceable but not accurate; however, the range expressing traceability allows an assessment of accuracy. Traceability also gives a measure of the dispersion range (more specifically called "uncertainty range"). Such a range allows an effective evaluation of accuracy and uncertainty (precision), because it takes into account all the factors affecting the result, what is even more important from the point of view of uncertainty.

• *Robustness (ruggedness)*

Robustness of an analytical method represents its ability to be unaffected by changes in those operational parameters (*organic content %, pH, buffer concentration, temperature, sample volume, etc.*) that usually affect analytical methods. The effect from any of these parameters on the analytical method can be evaluated separately (each to each) or simultaneously (altogether) as a part of a factorial experiment, but introducing *quantitatively defined* variations in all the operational parameters. A rigorous way to express robustness is in terms of the corresponding variation in the measured amount of analyte.

Ruggedness is the reproducibility of test results obtained by the analysis of samples in the face of *unintentional* variations caused by different laboratories, analysts, etc. *Ruggedness* should then be considered less rigorous than *robustness*.

In the sections that follow, some of the most important analytical characteristics, such as calibration, sensitivity, atomisation efficiency and precision, are covered.

8.2. Calibration

Calibration is the way to quantify the existing relationship between the analytical signal and the amount of analyte (usually expressed as concentration units). Two calibration methods can be distinguished: absolute and relative methods. However, in atomic absorption spectrometry, only the relative methods are used (at least until now).

Absolute calibration methods do not establish a comparison between the measurements from the samples and standards. On the contrary, measurements are related to the theoretical principles and fundamental physical and chemical constants. Absolute calibration methods are easily to carry out using techniques analyzing samples in solution, even though they are not always the most adequate due to the virtually existence of the matrix effect. Contrarily, it is difficult to be used with those techniques analyzing direct solid samples.

Relative calibration methods make use of comparison between measurements. Thus, analytical signals from the samples are compared with those from the standards. The material used for calibration purposes should be a standard prepared in the laboratory from analytical grade chemicals. Relative calibration methods are applied to all the atomic spectroscopic techniques. Sometimes, however, limitations can be present, fundamentally if some errors in the reference material composition used to prepare lab-samples exist.

8.2.1. Relative Calibration Methods

Quantitative determinations of any analyte using physico-chemical techniques are carried out by comparing the analytical signals derived from the sample with those from the standards. Hence, to make this comparison a relationship between the analytical signal (y) and the analyte concentration (x) in the standard needs to be established. This relationship can be mathematically expressed in a generalized manner as follows

$$y = f(x). \tag{8.1}$$

In the simplest case, Eq. (8.1) shows a linear relationship

$$y = a + bx, \tag{8.2}$$

where the slope, b, is the proportionality factor (hence, related to the sensitivity), and the intersect a represents the analytical signal provided by a blank, which in the most favorable case takes the value zero. By plotting values of the parameter y_i obtained from several standards (generally in solution) containing growing concentrations of analyte (x_i) versus the respective concentrations of analyte (x_i), a calibration line is obtained for a proper relationship.

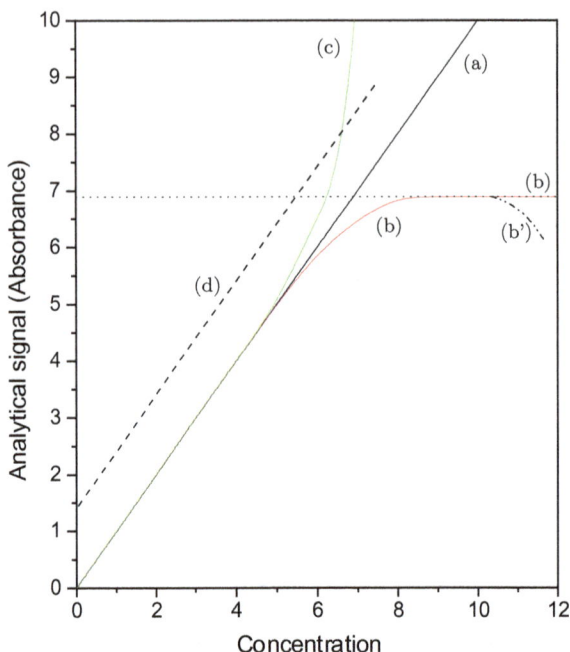

Fig. 8.1. Types of calibration lines: (a) ideal; (b) normal; and (c) complex. Calibration line (b′) showing the *roll-over* effect. Calibration lines derived from uncontaminated (a, b, c) and contaminated (d) standards.

The most widely used calibration methods in atomic absorption spectrometry are the following.

(i) **The calibration line method (external method).** In this method several standard solutions are used to construct a calibration line. The calibration lines can be constructed with solutions prepared from pure chemicals alone or together with other compounds simulating the sample matrix. The plot of the analytical signal versus the analyte concentration in the standard (Fig. 8.1) provides an experimental point series, which can be fitted to a straight line by applying the minimum-square method. This straight line constitutes the calibration line, whose slope is related to the technique sensitivity. By comparison of the analytical signals with the calibration line, it is possible to derive the amount of analyte present in it. For preparation of the calibration line, it would be

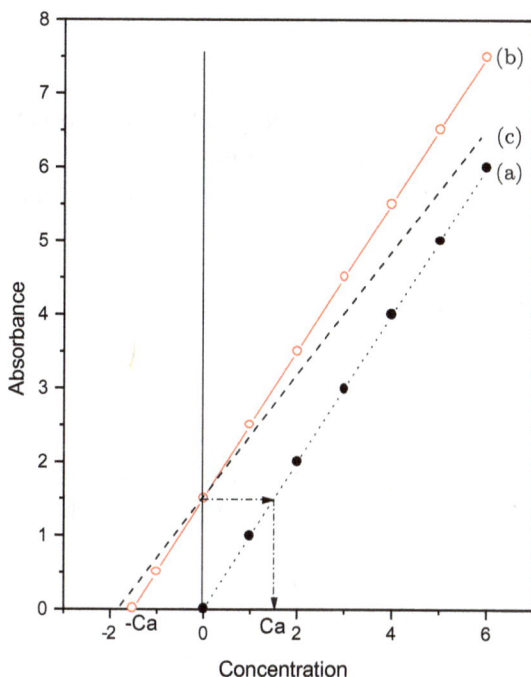

Fig. 8.2. Calibration line derived from aqueous standards (a), and using the standard addition method in absence (b), and in presence (c) of interferences. Extrapolation of the line derived from the standard addition method to estimate the analyte concentration (c_a) of the sample.

appropriate to use at least seven concentration levels with three replicas per level and three determinations by reply.

(ii) **The standard addition method.** It consists in analyzing directly the sample solution and various aliquots from such solution spiked with known and growing amounts of analyte. The analytical signal obtained in each case is plotted against the standard analyte concentration added, and then extrapolating till the abscise axis from which the analyte concentration in the sample is deduced (Fig. 8.2). At least, a data series of seven points needs to be plotted. This calibration method is used with techniques needing samples in solution to work. Nonetheless, it could also be used with solid samples if a good mixing procedure is available, but this is usually very difficult to achieve.

- **The advantages** provided by this calibration method are as following:

  — good fit of the matrix composition, which is very adequate in analyzing complex samples;
  — it can really be considered as an *absolute* calibration method;
  — the standard addition method has also been used to carry out multi-element analysis, which represents a powerfull alternative in many cases [Jochum *et al.* (1981)].

- **The main disadvantage** of the standard addition method relates to the construction of the calibration line, which can be used for only one sample and new calibration lines have to be prepared to analyze additional samples. Nonetheless, other problems can also be outlined. Thus, to derive the analyte concentration of the sample, it is necessary to make a linear extrapolation, which is not always correct, whether we assume that the calibration line always curve for low analyte concentrations. Other additional problems are due to the background signals. The standard addition method assumes that the analytical signals are only due to the analyte. If this is not the case, a second plot from the standard addition method would need to be prepared, by adding the analyte to the aliquots from a solution without the analyte, but containing all the other matrix components. This second plot would be extrapolated till $y = 0$ and the value (a/b) obtained is a measure of the background signals (interferences). By substrating this value from that obtained from the first plot, a correct value for the analyte concentration would be derived. An estimation of the uncertainty associated with the standard addition method can be derived when a matrix effect is observed [Bagur *et al.* (2005)]. However, interest in this procedure is many times short-lived because it is usually impossible to obtain an analyte-free matrix. On the other hand, analyte present in the sample can show different chemical forms. The analyte species added from the standard solution should be the same as from that of the sample, or at least it should show a similar behavior in the atomizer, which it is not always true.

(iii) **The internal standard method.** It is currently used in atomic emission spectrometry, but it could also be used in atomic absorption spectrometry if, but only if, a simultaneous multi-element detection system is incorporated. The procedure consists in adding one or more elements as well to samples as to standards, in order to compensate for variations in the analyte response from the matrix components, and instrumental effect. This method will be so much adequate as the behavior of analyte and internal standard does in respect to the atomizer conditions and the interferent effects. The internal standard can also be used to compensate for the effects from the differences in the amount of solvent from a sample to the other. Of course, matching samples and standards can also be used to reduce all these effects, but this procedure does not consider specific variations occurring during the introduction of each sample.

In multi-element analysis techniques, the internal standard method is very attractive, because several channels from the spectrometer can be used for the reference measurements. However, some difficulties exist in selection the most adequate internal standard for the simultaneous determination of several analytes in a wide in selecting concentration range. In the multi-element sequential techniques, there is also the need to use the internal standard method. However, the normal working mode excludes the simultaneous control of the signals derived from the analyte and the internal standard.

8.2.2. Quality of the Calibration Line

The general quality of the calibration line can be tested using as reference a quality coefficient (QC), which should take a threshold value. Those calibration lines with a QC value above (in 5%) the threshold value would be considered as problematic lines, while those lines with a smaller QC value would be adequate. An adapted version of the QC [Penninckx *et al.* (1995); Vankeerberghen *et al.* (1996)] assumed a constant absolute standard deviation, where the QC can be calculated using the following equation

$$QC = \sqrt{\frac{\sum_{i=1}^{n} \left(\frac{y_i - \bar{\bar{y}}_i}{\bar{y}}\right)^2}{n-1}}, \qquad (8.3)$$

where n represents the number of data (included the blank), y_i is the absorbance measured for the standard y, $\bar{\bar{y}}_i$ is the absorbance for the standard y established by the model, while the parameter $\bar{y}$ is the average of the absorbance measured.

To detect non-linearity of the calibration line, the slope ranking method (SRM) is usually employed [Vankeerberghen *et al.* (1992)]. The procedure is based on measuring the relative value of the slopes, m_{0i}, related to the lines obtained by connecting each data to the origin. For linear calibration lines, the relative variation of the slopes m_{0i} follows a random sequence, due to the random error of the measurements. On the contrary, the relative value of the slopes m_{0i} of a non-linear calibration line follows a sistematic sequence. If the values of the slopes m_{0i} follow a decreasing sequence, the line is a curve in the full range covered or a blank problem exists. One the most important advantageous from the SRM is the independence from the model used to fit the points. The disadvantages of the SRM are related to:

- the effect of the variance model on the probability of the different values relative to the slopes,
- probability (usually large) existing from which it can be concluded that there is a blank problem when the calibration line is deviated from the linear model, and
- difficulty in evaluating the relatively large series of the calibration data.

Nonetheless, other currently used methods for small samples ($6 \leq n \leq 10$) are the ANOVA lack-of-fit procedure for a first order model, and the test of significance of b_2 for the second order models [Massart *et al.* (1988)]. In the ANOVA lack-of-fit procedure, dispersion of the data around the regression line is compared with the experimental error. When dispersion of the data is significant larger, the model is not correct. However, when a lack of fit is detected, it is not possible to distinguish between the existence of an outlier and the lack of linearity of the calibration line.

When the least square (LS) method is applied, a problem usually presents, arising from the fact that the LS regression line obtained is affected by the presence of outliers. Some diagnostic procedures for outliers compare the LS residuals standardized for a reference value (2 or 3), while others apply robust regression methods: the simple mean method, the least median of squares (LMS) method for the detection of outliers, and the fuzzy calibration. However, the *fuzzy* calibration and the least median of squares method are the most effective for detecting outliers in the routine analysis.

In the second order models, most of the calibration curved lines in ETAAS could be fitted to a second degree polynomial

$$A = b_1 C + b_2 C^2, \tag{8.4}$$

where A is absorbance, C is the analyte concentration, b_1 represents sensitivity in the origin and b_2 represents sensitivity reduction for an absorbance unity. Nonetheless, an hyperbolic or exponential model can also be used. The quadratic term, b_2, of the second order model has been used to detect the non-linearity in the full concentration range [Vankeerberghen *et al.* (1996)]. If b_2 differs significantly from zero ($\alpha = 5\%$), we need to conclude that the calibration line is a calibration curve. Two modes of prediction of the β error in the test of significance of b_2 have been proposed [Kuttatharmmakul *et al.* (1998)]:

- prediction of the β error derived from the plots between the lack of linearity (expressed as an average relative error of prediction) and the measurement precision (expressed as % RSD for the concentration in the middle of the calibration range), and
- prediction using the significance test of b_2, which basically is the application of a t test.

For comparison of two analytical methods, a regression procedure that takes into account the errors in both methods has been proposed [Zwanziger and Sârbu (1998)]. When two analytical methods are compared, it does not matter which is taken as x or which as y, because both are affected from errors. That is, we can write, $y = f(x)$ or $x = f(y)$. The general linear function, $Ax + By + C = 0$, represents the straight line equation whose coefficients, A and B, are different from zero. Hence, $y = (A/B)x - (C/B) = mx + n$. The A/B ratio can be used to make a comparison (i.e., to show the lack of similarity) between the two analytical methods. Thus, for an ideal case, when the two methods provide practically similar results, A and B take the same absolute value. On the contrary, the larger the difference between A and B, the more different would be the two methods. However, how can we objectively prove the meaning of the difference between the absolute values of the two coefficients, A and B? This can be done by using the *informational analysis* of variance, which is based on the calculation of the *informational energy* (a measure of uncertainty or randomness of a probability system), $(E = \sum_{i=1}^{n} P_i^2)$. This is a valid procedure with free distribution and small assumptions, and is very robust and unaffected by the data range.

The null hypothesis is $P_A = P_B$, where the probabilities P_A and P_B are calculated as follows: $P_A = \frac{A}{A+B}$, and $P_B = \frac{B}{A+B}$. The null hypothesis would be accepted if the theoretical *informational energy* is minimal (or $E = \frac{1}{2}$) and equal to the empirical *informational energy*, $\epsilon = \frac{(A^2+B^2)}{(A+B)^2}$, with the two coefficients A and B taking the same absolute value. On the other hand, if $E \neq \epsilon$, the null hypothesis is rejected. Hence, the difference between the two analytical methods would be taken as significant [Zwanziger and Sârbu (1998)].

Matrix interferences that introduce relative systematic errors can be detected by comparison of the slopes of the calibration line and the standard addition line. Basically, this comparison is performed by means of a t-test. Before applying the t-test, variance of residuals is compared using the F-test. If the variances are not statically different, a globalize variance is determined and the classical t-test is applied [Vankeerberghen *et al.* (1996)]. If variances are significantly different, the t value would be calculated using the following equation

$$t = \frac{|b_1 - b_2|}{\sqrt{\sigma_{b_1}^2 + \sigma_{b_2}^2}}, \tag{8.5}$$

where b_1 and b_2 are the slopes of the calibration line and of the standard addition line, respectively, and $\sigma_{b_1}^2$, $\sigma_{b_2}^2$ are their corresponding variances. The number of the freedom degrees (*df*) used [Snedecor and Cochran (1980)] can be derived as follows

$$\frac{1}{df} = \frac{1}{(\sigma_{b_1}^2 + \sigma_{b_2}^2)^2} \left[\frac{(\sigma_{b_1}^2)^2}{n_1 - 2} + \frac{(\sigma_{b_2}^2)^2}{n_2 - 2} \right] \tag{8.6}$$

which is an average value of the freedom degrees of both variances, where, n_1 and n_2 represent the number of data of the calibration line and the standard addition line, respectively. As the difference between both slopes must be statistically significant, without being relevant for the considered problem, the difference in the percentage is also compared with a limiting value (10%) which is considered acceptable. A difference that is statistically significant below that limiting value (10%) points to the conclusion that matrix interferences are present.

8.2.3. Linearization of the Calibration Line: *Expanding the Working Concentration Range*

ETAAS exhibits a linear working concentration range of usually 1.5 orders of magnitude. This short working range represents one of the bigger problems found with ETAAS measurements.

The non-linearity of the calibration line represents a shortcoming in the use of the characteristic mass, m_0, as the only calibration parameter, especially for those samples providing relatively large integrated absorbances. The origin of the non-linearity of the calibration line in ETAAS for relatively high analyte concentrations derived from the presence of straight light, which produces a negative curvature [Loos-Vollebregt and Galan (1986)]. Among the most common sources of stray light we can include [Larkins (1988); L'vov et al. (1992a)].

- The use of non-monochromatic radiation.
- Scattered radiation from the imperfections of the grid or any other part of the instrument.
- Non-absorbent lines or lines with a very small absorption coefficient located very close to the analytical line.

- Lines from the inert gas and from the impurities present in the cathodic material of the HCL.
- Above certain absorbance values, electronic reading devices do not show a linear response.
- Due to the transient nature of the measurement, auto-absorption and auto-inversion processes can be derived in the roll-over effect.
- Hyperfine structure broadening from the absorption and/or emission lines.
- Radiation interacting with an inhomogeneous atom density distribution into the atomizer (analytical volume).
- Irregular distribution of the absorption coefficients through the radiation cross section.

Voigtman *et al.* (1994) carried out a theoretical study on the behavior of the experimental calibration lines using a Zeeman spectrometer in the presence of three different types of stray light. The stray light effect is directly related to the characteristic mass and the roll-over absorbance of the Zeeman effect, while the Zeeman sensitivity ratio affects directly the characteristic mass. The main source of stray light in the measurements of the Zeeman effect seems to be from the radiation polychromaticity source, even though the spectral bandwidth of the radiation source is relatively narrow [Voigtman *et al.* (1994); Loos-Vollebregt and Galan (1979, 1980a,b)].

Approximations to extend the linear working range of the calibration lines can be grouped into the three following categories: (i) *manipulation of the sample or the experimental parameters*; (ii) *theoretical algorithms*, and (iii) *non-linear calibration methods*.

(i) *Manipulation of the sample or the experimental parameters.* One of the most effective trends would probably include the use of the wavelength modulation methods [Harnly and O'Haver (1981)]. Absorbance measurements are carried out on different points of the same line profile. In order to determine low analyte concentrations, absorbance values are taken from the line center, while for high analyte concentrations, absorbance values are taken from the line profile wings. This procedure extends the dynamic range several orders of magnitude. On the contrary, continuum radiation sources are not very suitable for the determination of those elements absorbing at wavelenghts below 250 nm. Based on the fact that the sample optical thickness is a multi-valued function of the signal, a methodology has been developd for wavelength-modulated diode laser absorption spectrometry [Gustafsson *et al.* (2000)].

Other alternatives incorporate some dilution (mainly automatically) stages of the sample, or some modifications of the instrumental conditions, such as the working pressure in the atomizer (decreased [Wang and Holcombe (1992)] or increased [L'vov (1978)]); the gas flow regime during atomization [Carnrick *et al.* (1986); Irwin *et al.* (1990)]; and the use of less sensitive non-resonance lines [Botha and Fazakas (1984); Belarra *et al.* (1996a,b)].

The extending of the dynamic range has also been studied by applying the three field alternating current Zeeman effect (3-field *ac* Zeeman-effect) [Loos-Vollebregt *et al.* (1989, 1993); Loos-Vollebregt and Galan (1982, 1984)]. A dynamic range of 5–10 times longer can be obtained for the analytes studied. However, this method has a loss of sensitivity due to a great decrease in the slope of the calibration line. An additional disadvantage of the 3-field *ac* Zeeman method is the great modification of the conventional instrumentation, which is not necessary for other linearization procedures. This method offers no advantage over normal Zeeman for integrated absorbance [Harnly (1994)].

(ii) *Theoretical algorithms.* The linearization method of the calibration line in Zeeman ETAAS used by L'vov *et al.* (1992a,b,c, 1993) recomposes those theoretical values of the absorbance (A) satisfying Bouguer–Lambert–Beer's law from the experimental values ($A_0 = \log \frac{I_0}{I}$). Thus, for the simplest case, one obtains

$$A = \log \frac{I_0 + \alpha I_0}{I + \alpha I_0} = \log \frac{1 + \alpha}{10^{-A_0} + \alpha} \tag{8.7}$$

where $\alpha = \frac{1}{10^{A_{lim}} - 1}$ is the stray light fraction, which causes the curvature of the calibration lines. The L'vov algorithm used to linearize the analytical curves is particularly based on three parameters, namely, the *roll-over* absorbance (A_r), the limiting absorbance or maximum value of the calibration straight line (A_{lim}) and the Zeeman sensitivity ratio (R).

If auto-inversion exists, every point A_z from the inverted pulse (dip) or points from the normal pulse out of the dip, from the raising edge of the peak profile, are re-calculated as a corrected and normalized absorbance, $A_{c,n}$, using the *roll-over* absorbance, A_r

$$A_{c,n} = (1 - 10^{-A_r}) \log \frac{10^{A_r} - 1}{10^{A_r - A_z} - 1}. \tag{8.8}$$

In order to improve the *quasi*-theoretical approximation of the raising edge of the curve, the value A_r is replaced by the parameter $A'_r = A_r + \Delta$, The parameter Δ (=0.01) was introduced in order to eliminate uncertainty from the re-calculation of the absorbance, A_r values close to the *roll-over* dip.

Hence, for the raising edge of the pulse we will have

$$A_{c,n} = (1 - 10^{-A'_r}) \log \frac{10^{A'_r} - 1}{10^{A'_r - A_z} - 1}. \tag{8.9}$$

In the same way, the points A_z from the declining part of the dip, which correspond to the raising edge of the pulse are transformed into the values $A_{c,n}$ through the A'_r parameter and the Zeeman sensitivity ratio, R_d (ratio between the Zeeman and conventional AAs, i.e. $R = A_z/A = A_z(A_z + A_{BG})$). The subscript d means that the value of R in the dip differs from the value at low concentrations. The algorithm found for this case is as follows

$$A_{c,n} = (1 - 10^{-A'_r}) \log \frac{10^{A'_r} - 1}{10^{0.01} - 1}$$
$$+ \frac{R_d(1 - 10^{-A'_r})}{1 - R_d} \log \frac{10^{A'_r} - 1}{10^{A_z + 0.01} - 1}. \tag{8.10}$$

This procedure enlarges the linear working range of the calibration line in one order of magnitude.

L'vov *et al.* (1993) also developed an algorithm to correct automatically the absorption pulses, describing a simple method for the direct determination of the parameter R_d

$$R_d = \cfrac{1}{\left[1 + \cfrac{Q_3}{\left(\frac{0.0049m}{m_0} - Q_1 - Q_2\right)}\right]} \tag{8.11}$$

where, Q_1, Q_2 and Q_3, represent the integrated absorbance re-calculated, area of a rectangular absorbance pulse (intermediate between Q_1 and Q_3) and the dip absorbance, respectively; while m and m_0 are the analyte mass for the dip and the characteristic mass measured for the linearized pulse. If the algorithm is adequately selected, the integrated absorbance re-calculated in this manner will be proportional to the analyte mass. Other authors [Yuzefovsky *et al.* (1994); Su *et al.* (1994); Berglund (1996)] also used the computational procedure from L'vov for the linearization of the calibration curve and to provide an improvement in the calculation of the Zeeman sensitivity ratio, R. This linearization technique covers an analytical range of 3–4 magnitude orders above the detection limit.

The L'vov algorithm has been refined by Yuzefovsky *et al.* (1994, 1996, 1997), including the Newton method of successive approximations to approach a solution to the theoretical expression of L'vov. This alternative introduces, instead the parameter R, an additional fourth parameter, the sensitivity ratio at the *roll-over* point ($R' = A_r/A_{\text{lim}}$, where A_{lim} is the absorbance limit value with the field *off*) which was calculated using Newton's method and considered in the algorithm along with the *roll-over* absorbance

$$A_{c,n} = A_{c,n}^{\text{int}}$$

$$- \frac{A_{0,n}^{\text{int}}\left(\frac{1-R'}{R'}\right) + A_Z - \log(1+\alpha) + \log[10^{-(1+\alpha)A_{c,n}^{\text{int}}/R'} + \alpha]}{\frac{1-R'}{R'} - \left(\frac{1+\alpha}{R'}\right)([10^{-(1+\alpha)A_{c,n}^{\text{int}}/R'}])[10^{-(1+\alpha)A_{c,n}^{\text{int}}/R'} + \alpha]^{-1}} \tag{8.12}$$

The Newton approximation method allows to use A_Z for calculation of $A_{c,n}$ through the use of the running values of $A_{c,n}$ initially calculated from Eq. (8.8), {similarly from Eqs. (8.9) or (8.10)}, and denoted in

Eq. (8.12) as $A_{C,n}^{int}$. The number of iterative steps of the Newton approximation method depends on the procedure efficiency to obtain the corrected and normalized absorbance value. Yuzefovsky *et al.* (1996) found a satisfactory result using values for $R' < 1$.

L'vov *et al.* (1995a) have studied the effect of the slitwidth and the HCL current on the linearization process close to the *roll-over* point, making use of the equations

$$A_{off} = \log \frac{1 + \alpha}{10^{-(1+\alpha)A_0/R} + \alpha} \tag{8.13}$$

$$A_{on} = \log \frac{1 + \alpha}{10^{-(1+R)(1+\alpha)A_0/R} + \alpha} \tag{8.14}$$

$$A_z = A_{off} - A_{on}, \tag{8.15}$$

where $(1 + \alpha)/R$ is the normalization factor, and

$$\alpha = \frac{R(1 - R)^{(1-R)/R}}{10^{A_r/R} - 10^{(1-R)A_r/R}}. \tag{8.16}$$

Later, L'vov *et al.* (1996) replace the value of A_0 by A_0', which are interrelated

$$A_0' = \frac{A_0}{1 + \beta A_0}, \tag{8.17}$$

where β is a parameter by which it is possible to reduce or enlarge the superior zone of the calibration line along the A_0 axis. If there are stray light and auto-absorption, the fit is more suitable using the variable β parameter. This algorithm permits the linear working range to be enlarged in one order of magnitude, being of general applicability. In other words, applicability is for any element and for any measurement condition (lamp current and slitwidth), as well as for those cases where the curvature of the calibration line is due to chemical effects, which depend on the analyte mass and on the inhomogeneous distribution of atoms across the cross-section of the graphite tube.

(iii) *Non-linear calibration methods.* Miller-Ihli *et al.* (1984) make a comparison between the interpolation and fitting methods by non-pondered least squares for the nonlinear calibration. Among the interpolation methods, the linear interpolation by which the data points are interconnected by linear segments, the cubic fitting methods using cubic polynomials between the points based on the local first and second derivatives, and the Stineman interpolation are included [Lonardo et al. (1996)]. Harnly et al. (1996) fit the experimental data to a hyperbole

$$12.0x^2 - 36.2xy + 24.2y^2 + F = 0 \qquad (8.18)$$

to carry out calibration, even using a continuum radiation source and an array detector [Wichems et al. (1998)]. In Eq. (8.18), the term F can take different values.

8.3. Sensitivity

Sensitivity is directly related to the slope (b, in Eq. (8.2)) of the calibration line. A method is more sensitive than other if the same change in concentration, c, or quantity, q, causes a larger change in the absorbance measured, A; in other words, the derivative dA/dc or dA/dq is larger. Sensitivity, here represented by S, corresponds to the minimum concentration necessary to produce a minimum detectable change in the absorbance. It is possible to say that sensitivity, S, represents the minimum detectable change in the absorbance per concentration unit. By definition, c_S can be mathematically expressed as follows

$$c_S = \frac{3\sigma_D(Q_A)}{S}, \qquad (8.19)$$

where $\sigma_D(Q_A)$ is the standard deviation for the integrated absorbances, Q_A.

The method of the signal integration in ETAAS makes use of the characteristic mass, m_0, as an alternative parameter related to sensitivity. The characteristic mass represents the minimum amount of analyte required to produce a 1% transmittance variation (0.0044 absorbance). Taking into account the relationship between S and m_0, one obtains

$$S = \frac{0.0044}{m_0/V}, \qquad (8.20)$$

where V is the sample volume injected into the tube. In this case, c_S can be expressed as follows

$$c_S = \frac{3m_0\sigma_D(Q_A)}{0.0044V}.$$ (8.21)

Hence, c_S is regulated by the sample volume, sensitivity (characteristic mass) and the error (precision) derived from the integrated signal.

Analytical sensitivity is not dependent on the graphite tube length if the mean residence time of the analyte atoms in the tube is equal to, or higher than, the average atomization time. Nonetheless, an inverse relationship is shown with the transversal sectional area of the graphite tube. Sensitivity increases with the slitwidth and spectral linewidth (bandwidth). However, sensitivity decreases with the atomization temperature for volatile elements, but increases for the less volatile elements [Harnly and Radziuk (1995)].

8.3.1. Limit of Detection

The detection limit represents an essential characteristic of any analytical method when low levels of analyte are to be measured. The detection limit is defined as the lowest concentration (or alternatively the lowest mass) capable of being detected with a defined confidence level. The criteria based on concentration and mass provide the relative (c_L) and absolute (m_L) detection limits, respectively (Table 8.1) [Sturgeon (1990)]. The detection limit is used to show the capability of an analytical method for the detection of a particular analyte and fits two important proposals:

- the analyte level distinguishable from the background, and

Table 8.1. Approximate Relative (c_L in μg/l) and Absolute (m_L in pg) Detection Limits for Several Analytes [Sturgeon (1990)]

	Elements											
DL	Ag	Al	As	Au	Ca	Cd	Cu	Fe	Ni	Pb	Se	Zn
c_L	0.01–0.025	0.2	1	0.5	0.5	0.01–0.015	0.05	0.1	0.5–1	0.2–0.5	1	0.05
m_L	0.1–0.5	2–4	10	10	5	0.1–0.3	1	2	10	2–5	10–20	0.1–0.5

- the comparative detection capability of different analytical methods.

An analytical method shows a high (or *low*) sensitivity if it provides low (or *high*) values for the detection limit. ETAAS provides detection limits higher than those found for flame AAS (and AES), but in many cases are similar to those obtained for ICP-MS. However, the relatively long residence times of the analyte atomic vapor in the analysis volume, as well as the absence of mass transport losses, provide absolute detection limits (in pg) better than those derived for ICP-MS (Table 8.1).

By decreasing sufficiently, the analyte concentration in the sample, the analytical signal becomes so small that it cannot be differentiated from the signals (background) originating from a blank. The smaller signal (S_{LD}) being statistically different from that in the background can be defined as follows

$$S_{LD} = S_n + k\sigma_B \tag{8.22}$$

where S_n and σ_B represent the average background signal and the standard deviation of the background signals from a blank. The statistical factor k can be selected, arbitrarily, according to the confidence level desired, but obviously it would be ≥ 0. If $k \leq 1$, some signals would be incorrectly attributed to the positive deviations of the background, even though the analyte is not present (α error). If a large value for k ($k \geq 4$) is selected, the presence of the analyte could be occasionally ignored when it is really present (β error). In analytical chemistry (and also in spectrometry), it is usual to take the value $k = 3$ [IUPAC (1983)]. The value of 3 assumes that the random errors are normally distributed, and it provides a confidence level of 99.6% for the analyte detection [Epstein and Winefordner (1984)]. Nonetheless, in practice, other values (1:2:4:6:10) can also be used for k, which evolve the instrument detection limit ($k = 1$), the lower detection limit, c_L ($k = 2$), the method detection limit ($k = 4$), the purity guarantee limit, L_G ($k = 6$), and the quantification limit ($k = 10$), respectively. To show the detection capability of any analytical method, the value of the guarantee limit (L_G) is more informative than the lower detection limit c_L, because it is the level for which the analyte will be virtually always detected. However, the parameter c_L, which is more widely used than L_G, is adequate for making a comparison between the different analytical methods. The detection limit values of c_L or L_G can only be estimated through repetitive measurements of a blank and by making a calibration line. Hence, they cannot be exactly known and consequently they are always biased. Uncertainty of instrumental detection limits may depend on the α and β errors, the number of standard solutions and replications used for calibration, analyte concentration range and standard deviation of the adequated calibration line [Kuselman and Shenhar (1995)].

Considering the signal to background ratio, **S/N**, the detection limit can be defined as the concentration or amount of analyte which provides a signal to background ratio, **S/N**, equal to k, where k is the statistical significance level, being 2 or 3 in most of the cases [IUPAC (1976)]. However, the minium analyte concentration (or mass) detectable is, therefore, the concentration, c_L, or mass, m_L, theoretically expressed as

$$c_L = \frac{(S_{LD} - S_n)}{S} \tag{8.23}$$

or

$$m_L = \frac{(S_{LD} - S_n)}{S}, \tag{8.24}$$

where the corresponding signals S_{LD} and S_n refer to the analyte concentration (or mass) and to a blank, respectively, while S (sensitivity) is assumed to be constant for low values of c_L or m_L. Equations (8.23) and (8.24) correspond to Eq. (8.22) if we equate the sensitivity S to the statistical factor.

The limit of detection depends not only on the spectral selectivity and background noise but also on other analytical figures of merit, such as efficiency of detection and of measurement [Winefordner and Stevenson (1993); Winefordner *et al.* (1994)].

The detection limits have been grouped into two categories: intrinsic and extrinsic. The intrinsic c_L is regulated by the statistical nature of the measurement process, having only been defined for those methods using laser technology. The extrinsic c_L represents the limit value dependent on the experimental noises (shot and flash noises from the detector, and scattering and fluorescence from the emission source), the background emission, and other extrinsic noises relative to processes such as production, excitation, ionization and measurement of atoms (and ions) [Winefordner and Stevenson (1993)]. The detection limits in ETAAS are limited by the extrinsic noise of the atomizer (background shot noise) and/or shot and flicker noises from the source [Winefordner and Stevenson (1993)]. Consequently, the method of choice for the reduction of the detection limits depends on the noise type predominating in the analytical procedure.

The limit of detection in the absence of background depends on the three following experimental parameters [L'vov *et al.* (1995b)]: the baseline offset compensation (boc) time, t_{boc}, the integration time, t_{int}, and the energy parameter E, which can be evaluated *a priori* (by the criterion 3σ). Thus, the relative detection limit can be calculated using the following equation

$$c_L \equiv 3\Delta Q_A = \left(\frac{3a}{\sqrt{f}}\right) 10^{-\frac{E}{b}} t_{int} \sqrt{\frac{1}{t_{int}} + \frac{1}{t_{boc}}}. \qquad (8.25)$$

Nonetheless, if large background absorption exists, the detection limit can be evaluated by the following equation

$$c_L \equiv \left(\frac{3a}{\sqrt{f}}\right) 10^{-\frac{E}{b}} t_{int} \sqrt{\frac{10^{A_{bg}}}{t_{int}} + \frac{1}{t_{boc}}}, \qquad (8.26)$$

where a and b are typical empirical coefficients for the spectrometer used, A_{bg} is the background absorbance, f is the times the absorption signal is registered per second, and ΔQ_A is the total error of the integrated area due to the baseline offset compensation and the signal reading errors.

In hydride generation ETAAS, the detection limit decreases for large sample volumes. Tyson *et al.* (1998) have looked into the detection limit variation as a function of the sample volume, taking into account the blank contribution, which is proportional to the sample volume. The results found support the following equation

$$c_L = \left(\frac{3\sqrt{2}\sigma_{A,0}}{SV}\right) + 3\sqrt{2}KC_b, \tag{8.27}$$

where $\sigma_{A,0}$ is the standard deviation of the blank signal; S is the method sensitivity; V is the sample volume injected; K is a constant; and C_b is the analyte concentration added to the reactive volume necessary to prepare a sample volume equal to V.

In conventional solid sampling procedures, it is difficult to derive the limit of detection by applying the classical approach using the threefold standard deviation of blank measurements, because no blank sample is usually available. In these cases, it is possible to use the term "zero-mass response" proposed by Kurfürst (1998) or, alternatively, to estimate the critical measurement value, based on the tolerance interval of a calibration curve in the range of the presumed detection limit [Lücker *et al.* (2000)].

8.3.1.1. *Mass detection limit*

By taking into account the measurement error, L'vov *et al.* (1995b) have derived the following equation to optimize the measurement conditions for the mass detection limit, m_L

$$m_L = 0.076 \times 10^{-\frac{E}{29}} \left(\frac{m}{Q_A}\right) t_{int} \sqrt{\frac{10^{\tilde{A}_{bg}}}{t_{int}} + \frac{1}{t_{boc}}}, \tag{8.28}$$

where E is the energy parameter, and t_{boc} is the time of baseline offset compensation. Only one measurement (integrated absorbance, Q_A) is needed in order to calculate the inverse sensitivity, m/Q_A, which is then used for the optimization of the integration time, t_{int}, and the evaluation of $\tilde{A}_{bg}$ using the following equation

$$\tilde{A}_{bg} = [A_{bg(max)} + \hat{A}_{bg}]/2, \tag{8.29}$$

where $A_{bg(max)}$ is the peak height of the background absorption pulse and $\hat{A}_{bg}$ is the mean background absorption

$$\hat{A}_{bg} = Q_{bg}/t_{int}, \tag{8.30}$$

while Q_{bg} is the integrated area of the background absorption pulse.

Other authors [Le Bihan *et al.* (1995)] obtained for some elements (Mn, Al), a similar mass detection limit to that calculated from only one atomization pulse using the following equation

$$m_L = 3\sigma\sqrt{v_s t_{int}}(m/Q_A),\qquad(8.31)$$

where σ is the standard deviation of the instantaneous absorbances from a blank, v_s is the measurement rate, and the other parameters have the usual meaning.

The detection limits for ETAAS using continuum emission sources and a photodiode detector are similar to those obtained using line sources [Smith and Harnly (1994)]. The detection limits in single-element ETAAS measurements using diode lasers are improved by applying wavelength modulation techniques, because the influence of various sources of noise on the absorption signal is very small [Schnürer-Patschan *et al.* (1993)].

8.3.2. Characteristic Mass

The characteristic mass, m_0 ($= \frac{0.0044m}{Q_A}$) is defined as the mass (m) of analyte required to give a time integrated absorbance, Q_A, of 0.0044 s or, alternatively, an absorbance, Q, of 0.0044, which represents 1% transmittance. The experimental characteristic mass depends on the atomization temperature, but also on the effective stray light [Su *et al.* (1993)]. The gas phase temperature is really the most important parameter affecting the characteristic mass. Other factors, such as the tube dimensions, will also affect the tube length, l, being inversely proportional to the characteristic mass ($m_{0(calc)} \propto l^{-2}$) [Berglund and Baxter (1992)]. However, this relationship is true only in theory, because by extrapolating for very long tubes, the value of $m_{0(calc)}$ approximates asymtotically to a minimum value, due to losses through the sample injection hole.

The largest potentiality of the characteristic mass is related to the practical application of optimizing the atomization temperature. It is common practice to select as the optimum atomization temperature for a particular analyte, the temperature that provides the highest integrated absorbance and consequently the lowest value for the characteristic mass. When the integrated absorbance for growing temperatures achieves a constant value, the lowest temperature providing that constant absorbance value is selected, which is based on the pragmatic rule of preserving and optimizing the tube life. However, it is convenient to keep in mind that the temperature selected according to this criterion not necessarily needs to provide the highest atomization efficiency.

L'vov *et al.* [L'vov *et al.* (1986); L'vov (1990)] have provided an equation to calculate the theoretical characteristic mass, m_0, using either peak heights (absorbances)

$$m_{0(PH)} = 4.064 \left[\frac{A_r \Delta \nu_D}{H_{(a,\omega)} \gamma \delta f} \right] \left[\frac{Z_{(T)}}{g_1 \exp\left(\frac{-E_1}{kT}\right)} \right] r^2 \qquad (8.32)$$

or integrated absorbances

$$m_{0(PA)} = 5.08 \times 10^{-13} \left[\frac{M_a D \Delta \nu_D Z_{(T)}}{H_{(a,\omega)} \gamma \delta f g_1 \exp\left(\frac{-E_1}{kT}\right)} \right] \left(\frac{r^2}{l^2} \right), \qquad (8.33)$$

where M_a is the molar mass of analyte (g/mol); D is the diffusion coefficient (cm^2/s) of atoms at the absolute temperature, T(K), of the gas phase; $\Delta \nu_D$ is the Doppler width of the absorption line profile (cm^{-1}); $H_{(a,\omega)}$ is the Voigt integral for the contour of the absorption line at a distance from the line center of $\omega = 0.72a$ (where a is the dissipation constant of the Voigt profile) ($a \propto T^{-1.2}$); γ is the coefficient taking into account the hyperfine multiplicity of the analytical line and the Doppler width of the emission source; δ is the correction factor for adjacent lines in the light source spectrum; f is the oscillator strength; $Z_{(T)}$ is the partition function at the temperature T(K); g_1 and E_1 are the statistical weight and the energy (J) of the lower level for the analytical line, respectively; k is the Boltzmann constant (1.38 × 10^{-23} J/K); r is the radius of the graphite tube (cm); and l is the width of the absorption layer (cm) (it is assumed to be equal to the length of the graphite tube).

The characteristic mass shows better stability than 10–15%, over the temperature range shown for the following elements {Ag (1700–2700 K), Cd (1300–2100 K), Cr (2300–3100 K), Ga (1800–2800 K), Ge (2400–3200 K) and In (1700–2700 K)}. The characteristic mass related to many elements (Au, Bi, Cd, Co, Cu, Mn, Ni, Pb, Tl) increases with the lamp current and the slitwidth [L'vov *et al.* (1995c)]. The characteristic mass values obtained by different workers from different samples and on different days were also generally stable [Shuttler *et al.* (1991); Zheng *et al.* (1993)]. The differences found for the experimental characteristic masses of the same element can be attributed to differences in the auto-absorption of the emission lines from the HCL [L'vov *et al.* (1992d)]. The stability of the characteristic mass with time, for different sample types, is a very important characteristic to carry out absolute analysis. Nonetheless, it has been suggested that the use of the spectroscopic constant, K (see Eqs. (1.28), (1.30) and (1.37) in Chapter 1), instead of m_0, is more useful in comparing the data obtained with different atomizers [Torsi (1995)]. A combination of convective-diffusive vapor-transport models has been recently used for calculation of the characteristic mass of various analytes with theoretical results very close to the experimental ones [Bencs *et al.* (2015)].

8.4. Atomization Efficiency

The atomization efficiency, ε_a, can be defined as the ratio between the experimental, $m_{0(exp)}$, and the theoretical, $m_{0(theor)}$, characteristic masses

$$\varepsilon_a(\%) = 100\{m_{0(exp)}/m_{0(theor)}\}. \qquad (8.34)$$

The atomization efficiency depends on several experimental parameters: temperature, atomizer surface, atomization mode (wall, platform), and analytical signal (absorbance, integrated absorbance). The atomization efficiency needs to be calculated at a constant temperature, usually showing values between 35% (In) and (50–60%) (Ag, Cd, Mn, Zn) [Sturgeon and Berman (1983)]. However, the atomization efficiency increases (about 20–50%) for Cr, Ga, Ge, In and Sn with the atomization temperature, achieving values even two times higher

(Au) for the highest atomization temperature [Wei-min and Zhe-ming (1996)]. The use of a chemical modifier (Pd, Ni) also increases the atomization efficiency [Zheng *et al.* (1993)], but this was not true for Ag and Cd [Zheng *et al.* (1993)]. For other elements (Bi, Pb, Mn, Ga and Cr), it had been found that the proportionality factor, P, establishing a relationship between the number of the absorbing atoms with the registered absorbance, is nearly constant in a wide temperature range [Wei-min and Zhe-ming (1995)], being related to the atomization efficiency, ε_a, through the following equation

$$\varepsilon_a = \frac{P k_2 Q_A}{N_0},\qquad (8.35)$$

where Q_A is the integrated absorbance, k_2 is the rate constant of the atom dissipation process, and N_0 is the total number of the analyte atoms injected into the graphite tube.

The atomization efficiency found for the platform atomization and using integrated absorbances is higher (about 20%) than that found for the wall atomization and using absorbances (peak heights) (Pd, Ni) [Zheng *et al.* (1993)], but not for other elements, such as Sn, whose result is the opposite [Wei-min and Zhe-ming (1997)]. The importance of optimizing the temperature program with respect to the atomization efficiency, ε_a, instead of $m_{0(exp)}$ is only apparent for routine analysis, because the potential matrix interferences decrease with temperature [Berglund and Baxter (1992)]. The procedure suggested to optimize the atomization temperature using the atomization efficiency as a reference criteria is the following:

- the integrated absorbance is measured (of course, after a proper background correction), Q_A, at an atomization temperature (T_1) and the correspoonding absolute characteristic mass determined, $m_u(T_1)$:

$$m_u(T_1) = Q_A m_0 \text{ (teor)}/0.0044 \qquad (8.36)$$

where, m_0(teor) is calculated for the same temperature;
- the first stage is repeated after increasing the atomization temperature up to T_2. If $m_u(T_2)$ is higher than $m_u(T_1)$, the higher temperature has obviously provided a better atomization efficiency;

- then the atomization temperature is raised until a value (T_i) for which $m_u(T_i)$ does not increase any more; in this case, the value of ε_a is optimum.

This procedure is simular to that of comparing the integrated absorbances for standard solutions with those for samples at each temperature. The parameter m_u is not dependent on the relative differences on the atomization efficiencies for the standard solutions and the samples. This way of operation is proper [Berglund and Baxter (1992)], whether important discrepancies between m_0(teor) and m_0(exp) exist or when absolute security is essential.

8.4.1. Absolute Analysis

The stability of both the atomization efficiency and the characteristic mass with temperature, time and the sample type is very important to achieve absolute analysis. Absolute analysis can be carried out if all the processes providing the analytical signal are known and they can be quantitatively related. Absolute analysis represents the capability to show a relationship between the analytical signal and the analyte concentration through the experimental characteristic mass. Hence, the capability of ETAAS to perform absolute analysis depends on its ability to obtain predicted experimental values for the characteristic mass. In this case, an equation capable of deriving the analyte concentration of the sample from only one measurement should be known [Torsi *et al.* (1998)]. Vecchietti *et al.* (1989) have proposed a method to carry out absolute analysis from standard solutions, being then extended to the case of those samples also containing Pd-based chemical modifiers [Fagioli *et al.* (1991)]. The equation proposed for the complete atomization in ETAAS is the following

$$\int_0^\infty N_t dt = (1/S_c)\left(\left[\int_0^\infty S_t dt\right]\left[\int_0^\infty R_t dt\right]\right) = N_0\tau_2/S_c \tag{8.37}$$

$$\int_0^\infty A_t dt = Q_A = KN_0\tau_2/S_c, \tag{8.38}$$

where Q_A is the integrated absorbance; K (cm^2 atom^{-1}) is a characteristic spectroscopic constant for the analytical line, whose value can be derived from fundamental constants; N_0 is the total number of the analyte atoms injected into the tube; τ_2 (s) is the average time spent by the atoms in the optical path (dissipation factor); and S_c (cm^2) is the cross-section of the atomizer. If the experimental conditions are under strict control, τ_2 can be considered as a constant, resulting in a direct proportionality between A_i and N_0 for the same type of atomizer. For a flash atomization, it is possible to determine the experimental value of τ_2. For elements with high and mean volatility, the term A_i/τ_2 seems to be constant for a wide temperature range.

Gilmutdinov and Harnly (1998) considered the basic instrumental characteristics to carry out absolute analysis, including a continuum emission source, a high-resolution spectrometer, a bidimensional solid state detector and an optical device capable of transmitting collimated radiation from the source through the atomizer. These characteristics allow us to carry out new concepts for the spectral and spatial integration, maintaining an adequate temporal resolution. Multi-dimensional integrated absorbances, based on the resolved intensities, would provide the capability for the absolute detection needed to perform absolute analysis. However, L'vov proposed that the most simple and efficient way for absolute analysis consists in implementation of the boosted-output HCLs in commercial instruments [L'vov (1999)].

8.5. Precision

Precision is affected by several factors: instrumental design, optic quality and the prevailing noise. Nonetheless, light polarization, circular dichroism of analyte, preamplifier noise, and the effect of the extinction ratio of polarizer on the calibration line, among others factors, may also contribute in an important manner [Kale and Voigtman (1995a)].

The statistical nature of the noise affects greatly the performance of any signal processing device and it should not be forgotten, as it is usual. The noise limiting the precision of any analytical signal can be composed of other several noises,

- the signal random fluctuations (noise $1/f$) induced by the source noise and the electronic components noise, and
- variations in the atomization conditions change the peak profile characteristics, mainly due to variations in the atomizer characteristics, in the experimental parameters of atomization, and in the matrix composition.

The instrumental error arises from the determination stage of the analyte, and it mainly represents the contribution of all the errors made in each stage of the analysis procedure, such as the sample introduction error, the atomization error and the photometric error.

The **sample introduction error** is due to differences in either the sample volume injected into the tube, the analyte retained on the inner surface of the capillary used in the injection system, or changes in the analyte concentration during the dosification stage for successive determinations.

The **atomization error** is the result of variations in the atomizer temperature, the magnetic field when Zeeman effect-based spectrometers are used, the platform position in the tube as well as the dry residue on the platform surface, and the coating of the pyrolytic graphite for successive firings. The atomization temperature can change with the type of analyte, its concentration, and the matrix composition. The highest atomization temperature can induce some interference, due to the vaporization of the less volatile matrix components, and also degrading of the atomizer surface.

The **photometric error**, which represents the error derived in the measurement of the absorbance signal, is dominated by the shot noise. All these factors can alter, from one measurement to the following, the shape of the analytical signal, i.e., its area, height, and peak position. Kale and Voigtman (1995b) have carried out an evaluation of the effect of the noise on the precision of the integrated absorbance signal, confirming again that the use of the integrated absorbance provides a better precision than that found using the peak height.

Obviously, both the atomization error and the dosification error affect in a similar manner (proportionally) the relative error in the determination of the analyte mass, independently of its concentration. On the contrary, the photometric error can drastically change with the analyte concentration. It is thought that the largest error in the determination of the analyte mass for high signals, and consequently the highest limit of the analytical working range, is affected by the photometric error. However, the contribution of the photometric error to the total error in the analyte determination for moderate and large signals is not significant [L'vov *et al.* (1994)]. Nonetheless, for small integrated signals (and hence around the limit of detection), the standard deviation is fundamentally affected by the photometric error [L'vov *et al.* (1995b)]. The main contribution to the photometric error, when integrated absorbances closed to the limit of detection region are measured, derives from the error in the compensation of the baseline, ΔQ_{boc}, and in the reading stage of the signal, ΔQ_{read}.

In the solid sampling analysis, it is usually considered that uncertainty (precision) is derived from: (i) calibration, (ii) the measurement stage of the analyte, and (iii) the control process of the analytical quality [Kurfürst *et al.* (1996)].

Selection of the most adequate strategy in order to minimize those errors derived from the matrix depends on the sample type, as well as from the precision and accuracy fixed for the corresponding analysis. The matrix effect can be compensated for by matching the composition of samples and standards. Nonetheless, this assumes a previous knowledge of the matrix composition. Alternatively, the standard addition method, or even an iterative procedure, could also be applied. On the other hand, for some types of samples, the best solution is to isolate the analyte from the matrix components. However, this procedure can take a lot of time.

8.6. Other Analytical Considerations

Detection of possible sources of bias (constant and/or proportional) errors may be evidenced by changes in the shape of the absorbance profiles [Llobat-Estellés *et al.* (2006)]. The typical general analytical characteristics (advantages and disadvantages) for ETAAS are included in Table 8.2 [Banks *et al.* (1992)].

The advantages inherent to ETAAS are mainly derived from the electrothermal atomization process, which provides high atomization efficiencies for many elements. Nonetheless, the disadvantages mainly derived from the excitation process and the detection systems. So, developments in the radiation sources, which could include the use of laser radiations and even continuum sources, as well as the incorporation of new detector devices, such as solid state detectors, will necessarily improve the application possibilities of ETAAS. Elements usually determined by ETAAS are included in Fig. 8.3.

Table 8.2. General Advantages and Disadvantages of ETAAS [Banks *et al.* (1992)]

Advantages	Disadvantages
• High analytical sensitivity	• Chemical and physical interferences
• Low detection limits (<pg and ng)	• High background
• Very small sample volumes (μL)	• Low precision (RSD = 10%) for a few elements
• Relatively free from spectral interferences	• Limitations to multi-element analysis
• Relatively good analytical accuracy	• Solid samples usually need to be dissolved
• High precision for most elements	• Low performance in the number of analysis
• Many elements can be determined	• Difficulties to use the internal standard method
• Relatively easy sample preparation	• A very short linear working range
• Applicability to both solutions and solids	• Problems to determine some biological and environmental important elements (Br, Cl, F, I and S)
• Moderate prices of instruments	• Low sensitivity for B, P, Ti, U and W
• Easy manipulation and maintenance	
• Possibility of automatization	
• Good established methodologies	

Ia	IIa	IIIa	IVa	Va	VIa	VIIa	VIII			Ib	IIb	IIIb	IVb	Vb	VIb
Li 670.78 (0.753)	**Be** 234.86 (1.36)														
Na 598.00 (0.655)	**Mg** 285.21 (1.81)														
K 766.49 (0.682)	**Ca** 422.67 (1.75)		**Ti** 364.27 (0.14)	**V** 318.40 (0.7)	**Cr** 357.87 (0.3)	**Mn** 279.48 (0.57)	**Fe** 371.99 (0.035)	**Co** 304.40 (0.023)	**Ni** 341.48 (0.14)	**Cu** 324.75 (0.32)	**Zn** 213.86 (1.5)	**Ga** 287.42 (0.32)		**As** 193.7 (0.123)	**Se** 196.1 (0.12)
Rb 780.02 (0.675)	**Sr** 460.73 (1.55)				**Mo** 313.26 (0.2)					**Ag** 328.07 (0.46)	**Cd** 228.80 (1.30)		**Sn** 286.33 (0.19)	**Sb** 231.15 (0.04)	**Te** 225.90 (0.002)
Cs 852.11 (0.73)										**Au** 342.80 (0.29)	**Hg** 253.65 (0.025)	**Tl** 276.79 (0.30)	**Pb** 283.31 (0.20)	**Bi** 306.77 (0.12)	

Fig. 8.3. Main elements which are capable of being easily determined by ETAAS together with the most intense wavelength (in nm) and the absorption oscillator strength (in brackets).

References

Aller, A.J. (2002). Analytical characteristics of electrothermal atomic absorption spectroscopy. *Current Topics in Anal. Chem.*, **3**: 41–53.

Bagur, G.; Sánchez-Viñas, M.; Gázquez, D.; Ortega, M. and Romero, R. (2005). Estimation of the uncertainty associated with the standard addition methodology when a matrix effect is detected. *Talanta*, **66**: 1168–1174.

Banks, P.R.; Liang, D.C. and Blades, M.W. (1992). Graphite furnace-capacitively coupled plasma spectrometry: a new, fast furnace technique. *Spectroscopy*, **7**: 36, 38, 40.

Belarra, M.A.; Resano, M. and Castillo, J.R. (1996a). Linearization of calibration curve for tin 224.6 nm line in graphite furnace atomic absorption spectrometry. *Química Analítica*, **15**: 32–37.

Belarra, M.A.; Resano, M. and Castillo, J.R. (1996b). Expanding the working concentration range in graphite furnace atomic absorption spectrometry: study of non-resonance lines of tin. *Spectrochim. Acta, Part B*, **51**: 697–705.

Bencs, L.; Laczai, N. and Ajtony, Z. (2015). Model calculation of the characteristic mass for convective and diffusive vapor transport in graphite furnace atomic absorption spectrometry. *Spectrochim. Acta, Part B*, **109**: 52–59.

Berglund, M. (1996). The importance of the rollover absorbance and the Zeeman sensitivity ratio related coefficient for the accuracy in linearization of calibration curve in electrothermal atomic absorption spectrometry with Zeeman effect background correction. *Spectrochim. Acta, Part B*, **51**: 429–439.

Berglund, M. and Baxter, D.C. (1992). Computer program (CHMASS) for calculating theoretical characteristic mass values in electrothermal atomic abpsortion spectrometry, *J. Anal. At. Spectrom.*, **7**: 461–470.

Botha, P.V. and Fazakas, J. (1984). The use of non-resonance lines for the determination of lead, indium and thallium by electrothermal atomic abpsortion spectrometry, *Spectrochim. Acta, Part B*, **39**: 379–386.

Carnrick, G.R.; Lumas, B.K. and Barnett, W.B. (1986). Analyses of solid samples by graphite furnace atomic absorption spectrometry using Zeeman background correction. *J. Anal. At. Spectrom.*, **1**: 443–447.

Epstein, M.S. and Winefordner, J.D. (1984). Summary of the usefulness of signal-to-noise treatment in analytical spectrometry. *Progress in Anal. At. Spectrom.*, **7**: 67–137.

Fagioli, F.; Locatelli, C.; Vecchietti, R. and Torsi, G. (1991). Absolute analysis in electrothermal atomic absorption spectrometry: Pb with matrix modifiers. *Appl. Spectrosc.*, **45**: 983–985.

Gilmutdinov, A.Kh. and Harnly, J.M. (1998). Multidimensional integration of absorbances: An approach to absolute analyte detection. *Spectrochim. Acta, Part B*, **53**: 1003–1014.

Green, J.M. (1996). A practical guide to analytical method validation. *Anal. Chem.*, **68**: 305A–309A.

Gustafsson, J., Chekalin, N., Rojas, D. and Axner, O. (2000). Extension of the dynamic range of the wavelength-modulated diode laser absorption spectrometry technique. *Spectrochim. Acta, Part B*, **55**: 237–262.

Harnly, J.M. (1994). Evaluation of calibration methods for Zeeman graphite furnace atomic absorption spectrometry using computer modelling. *Appl. Spectrosc.,* **48:** 1156–1165.

Harnly, J.M. and O'Haver, T.O. (1981). Extension of analytical calibration curves in atomic absorption spectrometry. *Anal. Chem.,* **53:** 1291–1298.

Harnly, J.M. and Radziuk, B. (1995). Effect of furnace atomization temperatures on simultaneous multielement atomic absorption measurement using a transversely-heated graphite atomizer. *J. Anal. At. Spectrom,* **10:** 197–206.

Harnly, J.M.; Smith, C.M.M. and Radziuk, B. (1996). Extended calibration ranges for continuum source atomic absorption spectrometry with array detection. *Spectrochim. Acta, Part B,* **51:** 1055–1079.

Irwin, R.; Mikkelsen, A.; Michel, R.G.; Dougherty, J.P. and Preli, F.R. (1990). Direct solid sampling of nickel-based alloys by graphite furnace atomic absorption spectrometry with aqueous calibration. *Spectrochim. Acta, Part B,* **45:** 903–915.

IUPAC (1976), Nomenclature, symbols, units and their usage in spectrochemical analysis – II. Data interpretation. *Pure Appl. Chem.,* **45:** 99–103.

IUPAC (1983). Limit of detection, a closer look at the IUPAC definition, *Anal. Chem.,* **55:** 712A, 714A, 716A, 718A, 720A, 722A, 724A.

Jochum, C.; Jochum, P. and Kowalski, B.R. (1981). Error propagation and optimal performance in multicomponent analysis. *Anal. Chem.,* **53:** 85–92.

Kale, U. and Voigtman, E. (1995a). Signal and noise analysis of non-modulated polarimeters using Mueller calculus simulations. *Analyst,* **120:** 325–330.

Kale, U. and Voigtman, E. (1995b). Signal processing of transient atomic absorption signals. *Spectrochim. Acta, Part B,* **50:** 1531–1541.

Kurfürst, U. (1998). Calibration in solid sampling analysis. In: U. Kurfürst, Editor, *Solid Sample Analysis,* Springer, Berlin, Heidelberg, New York, pp. 35–60.

Kurfürst, U.; Rehnert, A. and Muntau, H. (1996). Uncertainty in analytical results from solid materials with electrothermal atomic absorption spectrometry: a comparison of methods. *Spectrochim. Acta, Part B,* **51:** 229–244.

Kuselman, I. and Shenhar, A. (1995). Design of experiments for the determination of the detection limit in chemical analysis. *Anal. Chim Acta,* **306:** 301–305.

Kuttatharmmakul, S.; Massart, D.L. and Smeyers-Verbeke, J. (1998). Influence of precission, sample size and design on the beta error of linearization tests. *J. Anal. At. Spectrom.,* **13:** 109–118.

Larkins, P.L. (1988). The effect of spectral line broadening on the shape of analytical curves obtained using pulsed hollow-cathode lamps for background correction. *Spectrochim. Acta, Part B,* **43:** 1175–1186.

Le Bihan, A.; Le Garrec, H.; Cabon, J.Y. and Guern, Y. (1995). Noise studies for detection limits for some electrothermal atomic absorption determinations and calculation of the optimal detection limit from one atomization. *J. Anal. At. Spectrom.,* **10:** 993–997.

Lonardo, R.F.; Yuzefovsky, A.I.; Zhou, J.X.; McCaffrey, J.T. and Mitchel, R.G. (1996). Extension of working range in Zeeman graphite furnace atomic absorption spectrometry by nonlinear calibration with prior correction for stray light. *Spectrochim. Acta, Part B,* **51:** 1309–1323.

Loos-Vollebregt, M.T.C. de and Galan, L. de (1979). The shape of analytical curves in Zeeman atomic absorption spectrometry. I. Normal concentration range. *Appl. Spectrosc.*, **33**: 616–626.

Loos-Vollebregt, M.T.C. de and Galan, L. de (1980a). The shape of analytical curves in Zeeman atomic absorption spectrometry. II. Theoretical analysis and experimental evidence for absorption maximum in the analytical curve. *Appl. Spectrosc.*, **34**: 464–472.

Loos-Vollebregt, M.T.C. de and Galan, L. de (1980b). Construction and performance of an a.c. modulated magnet for Zeeman atomic absorption spectroscopy. *Spectrochim. Acta, Part B*, **35**: 495–506.

Loos-Vollebregt, M.T.C. de and Galan, L. de (1982). Correction for background absorption and stray radiation in a.c. modulated Zeeman atomic absorption spectrometry. *Spectrochim. Acta, Part B*, **37**: 659–672.

Loos-Vollebregt, M.T.C. de and Galan, L. de (1984). The shape of analytical curves in Zeeman atomic absorption spectrometry. II. Extended dynamic range. *Appl. Spectrosc.*, **38**: 141–148.

Loos-Vollebregt, M.T.C. de and Galan, L. de (1986). Stray light in Zeeman and pulsed hollow cathode lamp atomic absorption spectrometry. *Spectrochim. Acta, Part B*, **41**: 597–610.

Loos-Vollebregt, M.T.C. de; Koot, J.P. and Padnos, J. (1989). Non-linearity and range of analytical curves in Zeeman-effect atomic absorption spectrometry. *J. Anal. At. Spectrom.*, **4**: 387–391.

Loos-Vollebregt, M.T.C. de; Oosten, P. van; Koning, M.J. de and Padnos, J. (1993). Extension of the dynamic range in a.c. Zeeman electrothermal atomic absorption spectrometry. *Spectrochim. Acta, Part B*, **48**: 1505–1515.

Lücker, E.; Failing, K. and Schmidt, T. (2000). Determination of analytical limits in solid sampling ETAAS: a new approach towards the characterization of analytical quality in rapid methods. *Fresenius' J. Anal. Chem.*, **366**: 137–141.

L'vov, B.V. (1978). Electrothermal atomization — the way toward absolute methods of atomic absorption analysis. *Spectrochim. Acta, Part B*, **33**: 153–193.

L'vov, B.V. (1990). Recent advances in absolute analysis by graphite furnace atomic absorption spectrometry. *Spectrochim. Acta, Part B*, **45**: 633–655.

L'vov, B.V. (1999). A continuum source vs. line source on the way toward absolute graphite furnace atomic absorption spectrometry. *Spectrochim. Acta, Part B*, **54**: 1637–1646.

L'vov, B.V.; Polzik, L.K. and Kocharova, N.V. (1992a). Theoretical analysis of calibration curves for graphite furnace atomic absorption spectrometry. *Spectrochim. Acta, Part B*, **47**: 889–895.

L'vov, B.V.; Polzik, L.K.; Kocharova, K., Nemets, Yu.A. and Novichikhin, A.V. (1992b). Linearization of calibration curves in Zeeman atomic absorption spectrometry. *Spectrochim. Acta, Part B*, **47**: 1187–1202.

L'vov, B.V.; Polzik, L.K.; Fedorov, P.N. and Slavin, W. (1992c). Extension of the dynamic range in Zeeman graphite furnace atomic absorption spectrometry. *Spectrochim. Acta, Part B*, **47**: 1411–1420.

L'vov, B.V.; Kocharova, N.V.; Polzik, L.K.; Romanova, N.P. and Yarmak, Yu.I. (1992d). Quality control of hollow cathode lamps in Zeeman graphite furnace atomic absorption spectrometry. *Spectrochim. Acta, Part B*, **47**: 843–854.

L'vov, B.V.; Polzik, L.K.; Novichikhin, A.N.; Fedorov, P.N. and Borodin, A.V. (1993). Automatic correction of absorption pulses in Zeeman graphite furnace atomic absorption spectrometry. *Spectrochim. Acta, Part B*, **48**: 1625–1632.

L'vov, B.V.; Polzik, L.K.; Borodin, A.V.; Fedorov, P.N. and Novichikhin, A.V. (1994). Precision and detection limits in Zeeman graphite furnace atomic absorption spectrometry. *Spectrochim. Acta, Part B*, **49**: 1609–1627.

L'vov, B.V.; Polzik, L.K.; Novichikhin, A.V.; Borodin, A.V. and Dyakov, A.O. (1995a). Effectiveness of linearization of calibration curves in Zeeman graphite furnace atomic absorption spectrometry. *Spectrochim. Acta, Part B*, **50**: 1757–1768.

L'vov, B.V., Polzik, L.K.; Borodin, A.V.; Dyakov, A.O. and Novichikhin, A.V. (1995b). Detection limits in Zeeman-effect electrothermal atomic absorption spectrometry. *J. Anal. At. Spectrom.*, **10**: 703–709.

L'vov, B.V.; Polzik, L.K.; Fedorov, P.N.; Novichikhin, A.V. and Borodin, A.V. (1995c). Correction of characteristic mass in Zeeman graphite furnace atomic absorption spectrometry. *Spectrochim. Acta, Part B*, **50**: 1621–1636.

L'vov, B.V.; Nikolaev, V.G.; Norman, E.A.; Polzik, L.K. and Mojica, M. (1986). Theoretical calculation of the characteristic mass in graphite furnace atomic absorption spectrometry. *Spectrochim. Acta, Part B*, **41**: 1043–1053.

L'vov, B.V.; Polzik, L.K.; Novichikhin, A.V.; Borodin, A.V. and Dyakov, A.O. (1996). Improved algorithm for linearization of calibration curves in Zeeman graphite furnace atomic absorption spectrometry. *Spectrochim. Acta, Part B*, **51**: 609–618.

Llobat-Estellés, M.; Mauri-Aucejo, A.R. and Marin-Saez, R. (2006). Detection of bias errors in ETAAS. Determination of copper in beer and wine samples. *Talanta*, **68**: 1640–1647.

Massart, D.L., Vandeginste, B.G.M.; Deming, S.N.; Michotte, Y. and Kaufman, L. (1988). Chemometrics: A Textbook. Elsevier, Amsterdam, p. 172.

Miller-Ihli, N.J.; O'Haver, T.C. and Harnly, J.M. (1984). Calibration and curve fitting for extended range AAS. *Spectrochim. Acta, Part B*, **39**: 1603–1614.

Penninckx, W.; Vankeerberghen, P.; Massart, D.L. and Smeyers-Verbeke, J. (1995). Knowledge-based computer system for the detection of matrix interferences in atomic absorption spectrometric methods. *J. Anal. At. Spectrom.*, **10**: 207–214.

Penninckx, W.; Hartmann, C.; Massart, D.L. and Smeyers-Verbeke, J. (1996). Validation of the calibration procedure in atomic absorption spectrometric methods. *J. Anal. At. Spectrom.*, **11**: 237–246.

Schnürer-Patschan, C.; Zybin, A.; Groll, H. and Niemax, K. (1993). Improvement in detection limits in graphite furnace diode laser atomic absorption spectrometry by wavelength modulation technique. *J. Anal. At. Spectrom.*, **8**: 1103–1107.

Shuttler, J.L.; Schlemmer, G.; Carnrick, G.R. and Slavin, W. (1991). The stability of graphite furnace characteristic mass data. *Spectrochim. Acta, Part B*, **46**: 583–602.

Smith, C.M.M. and Harnly, J.M. (1994). Sensitivities and detection limits for graphite furnace atomic absorption spectrometry using a continuum source and linear photodiode arrray detection. *Spectrochim. Acta, Part B*, **49**: 387–398.

Snedecor, G.W. and Cochran, W.G. (1980). *Statistical Methods*, 7th edn. The Iowa State University Press, Ames, Iowa, USA, p. 97.

Sturgeon, R.E. and Berman, S.S. (1983). Determination of the efficiency of the graphite furnace for atomic absorption spectrometry. *Anal. Chem.*, **55**: 190–200.

Sturgeon, R.E. (1990). Status and future prospects for use of the graphite furnace for elemental trace analysis of liquids and solids. *Fresenius' J. Anal. Chem.*, **337**: 538–545.

Su, E.G.; Yuzefovsky, A.I.; Mitchel, R.G.; McCaffrey, J.T. and Slavin, W. (1993). Effect of stray light on characteristic mass in Zeeman graphite furnace atomic absorption spectrometry. *Microchem. J.*, **48**: 278–302.

Su, E.G.; Yuzefovsky, A.I.; Mitchel, R.G.; McCaffrey, J.T. and Slavin, W. (1994). Linearization of calibration curves of manganese, copper, silver, thallium and chromium in Zeeman graphite furnace atomic absorption spectrometry. *Spectrochim. Acta, Part B.*, **49**: 367–386.

Torsi, G. (1995). Comparison of theoretical and experimental spectroscopic constants in atomic absorption spectroscopy with electrothermal atomization. *Spectrochim. Acta, Part B*, **50**: 707–712.

Torsi, G.; Reschiglian, P.; Locatelli, C.; Rossi, F.N. and Melucci, D. (1998). Standardless elemental analysis through electrothermal atomic absorption spectroscopy. Applications to aqueous samples and particulate matter in air. *Spectroscopy Europe*, **10**: 16, 18, 20, 21.

Tyson, J.F.; Ellis, R.I.; McIntosh, S.A. and Hanna, C.P. (1998). Effect of sample volume on the limit of detection in flow injection hydride generation electrothermal atomic absorption spectrometry. *J. Anal. At. Spectrom.*, **13**: 17–21.

Vankeerberghen, P.; Smeyers-Verbeke, J. and Massart, D.L. (1992). The slope ranking method, an algorithm for diagnosing linearity of calibration lines. *Analusis*, **20**: 103–109.

Vankeerberghen, P.; Smeyers-Verbeke, J. and Massart, D.L. (1996). Decision support system for run suitability checking and explorative method validation in electrothermal atomic absorption spectrometry. *J. Anal. At. Spectrom.*, **11**: 149–158.

Vecchietti, R.; Fagioli, F.; Locateli, C. and Torsi, G. (1989). Experimental measurement of absolute number of atoms vaporized in a graphite cuvette. *Talanta*, **36**: 743–748.

Voigtman, E.; Yuzefovsky, A.I. and Michel, R.G. (1994). Stray light effects in Zeeman atomic absorption spectrometry. *Spectrochim. Acta, Part B*, **49**: 1629–1641.

Wang, P.X. and Holcombe, J.A. (1992). Pressure-regulated electrothermal atomizer for atomic absorption spectrometry. *Spectrochim. Acta, Part B*, **47**: 1277–1286.

Wei-min, Y. and Zhe-ming, N. (1995). Atomization efficiencies of bismuth, lead, manganese, chromium and gallium under stabilized temperature platform furnace conditions. *J. Anal. At. Spectrom.*, **10**: 493–499.

Wei-min, Y. and Zhe-ming, N. (1996). The possibility of standardless analysis in graphite furnace atomic absorption spectrometry: determination of gold in geological samples. *Spectrochim. Acta, Part B*, **51**: 65–73.

Wei-min, Y. and Zhe-ming, N. (1997). Atomization efficiencies for indium and tin from different atomizer surfaces in graphite furnace atomic absorption spectrometry. *Spectrochim. Acta, Part B,* **52**: 241–254.

Wichems, D.N.; Fields, R.E. and Harnly, J.M. (1998). Characterization of hyperbolic calibration curves for continuum source atomic absorption spectrometry with array detection. *J. Anal. At. Spectrom.,* **13**: 1277–1284.

Winefordner, J.D. and Stevenson, C. (1993). Linking principles with absolute detection power in atomic spectrometry: how far can we go?. *Spectrochim. Acta, Part B,* **48**: 757–767.

Winefordner, J.D.; Petrucci, G.A.; Stevenson, C.L. and Smith, B.W. (1994). Theoretical and practical limits in atomic spectroscopy. *J. Anal. At. Spectrom.* **9**: 131–143.

Yuzefovsky, A.I.; Su, E.G.; Michel, R.G.; Slavin, W. and McCaffrey, J.T. (1994). Newton approximation method for linearization of calibration curves in Zeeman graphite furnace atomic absorption spectrometry. *Spectrochim. Acta, Part B,* **49**: 1643–1656.

Yuzefovsky, A.I.; Lonardo, R.F.; Zhou, J.X.; Mitchel, R.G. and Koltracht, I. (1996). Maintenance of the slope of linearized calibration curves in Zeeman graphite furnace atomic absorption spectrometry. *Spectrochim. Acta, Part B,* **51**: 713–729.

Yuzefovsky, A.I.; Lonardo, R.F.; Zhou, J.X. and Mitchel, R.G. (1997). Newton method of successive approximations for the linearization of the calibration curves of chromium, copper, lead, manganese, silver, and thallium in Zeeman graphite furnace atomic absorption spectrometry. *Appl. Spectrosc.,* **51**: 738–743.

Zheng, Y.; Su, X. and Quan, Z. (1993). Factors influencing characteristic mass in the graphite furnace. *Appl. Spectrosc.,* **47**: 1222–1226.

Zwanziger, H.W. and Sârbu, C. (1998). Validation of analytical methods using a regression procedure. *Anal. Chem.,* **70**: 1277–1280.

Index

www.ingramcontent.com/pod-product-compliance
Lightning Source LLC
Chambersburg PA
CBHW050535190326
41458CB00007B/1789